SIMULATED ANNEALING – ADVANCES, APPLICATIONS AND HYBRIDIZATIONS

Edited by **Marcos de Sales Guerra Tsuzuki**

Simulated Annealing – Advances, Applications and Hybridizations
http://dx.doi.org/10.5772/3326
Edited by Marcos de Sales Guerra Tsuzuki

Contributors
T.C. Martins, A.K. Sato, M.S.G. Tsuzuki, Felix Martinez-Rios, Juan Frausto-Solis, Masayuki Ohzeki, Dursun Üstündag, Mehmet Cevri, Cristhian A. Aguilera, Mario A. Ramos, Angel D. Sappa, Yann-Chang Huang, Huo-Ching Sun, Kun-Yuan Huang, Hisafumi Kokubugata, Yuji Shimazaki, Shuichi Matsumoto, Hironao Kawashima, Tatsuru Daimon, F. Charrua Santos, Francisco Brojo, Pedro M. Vilarinho, Edo D'Agaro, Sergio Ledesma, Jose Ruiz, Guadalupe Garcia, Tiago Oliveira Weber, Wilhelmus A. M. Van Noije, Xiaorong Xie, Makoto Yasuda

Published by InTech
Janeza Trdine 9, 51000 Rijeka, Croatia

Publishing Process Manager Sandra Bakic
Typesetting InTech DTP Team
Cover InTech Design Team

Additional hard copies can be obtained from orders@intechopen.com

Simulated Annealing – Advances, Applications and Hybridizations,
Edited by Marcos de Sales Guerra Tsuzuki
p. cm.
ISBN 978-953-51-0710-1

For EU product safety concerns:
IN TECH d.o.o., Prolaz Marije Krucifikse Kozulić 3, 51000 Rijeka, Croatia,
info@intechopen.com or visit our website at intechopen.com.

Meet the editor

Dr Marcos de Sales Guerra Tsuzuki was born in São Paulo, Brazil, in 1963. He received his electrical engineer degree in 1985 from the Engineering School from University of São Paulo, São Paulo, Brazil, and a Master's degree in engineering in 1989 from Yokohama National University, Yokohama, Japan. In 1995, he received his Ph.D. degree from the Engineering School of University of São Paulo. He is an Associate Professor at the Department of Mechatronics and Mechanical Systems Engineering, University of São Paulo since 2000, where he teaches courses on CAD/CAM, computer graphics, image processing and medical imaging. His current research areas include solid modeling, geometric modeling, optimization problems, simulated annealing, bioengineering and biomedical inverse problems.

Contents

Preface

Global optimization is computationally extremely challenging and, for large instances, exact methods reach their limitations quickly. One of the most well-known probabilistic meta-heuristics is Simulated Annealing (SA). Proposed initially to solve discrete optimization problems, it was later extended to continuous domain. The significant advantage of SA over other solution methods has made it a practical solution method for solving complex optimization problems.

In this book, some advances in SA are presented. More specifically: criteria for the number of evaluated solutions required to reach thermal equilibrium; crystallization heuristics that add a feedback mechanism to the next candidate selection; and estimation of the equilibrium quantities by the average quantity through the non-equilibrium behavior. Subsequent chapters of this book will focus on the applications of SA in signal processing, image processing, electric power systems, operational planning, vehicle routing and farm animal mating. The final chapters combine SA with other techniques to obtain the global optimum: artificial neural networks, genetic algorithms and fuzzy logic.

This book provides the reader with the knowledge of SA and several SA applications. We encourage readers to explore SA in their work, mainly because it is simple and because it can yield very good results.

Marcos de Sales Guerra Tsuzuki
Department of Mechatronics and Mechanical Systems Engineering,
University of São Paulo,
Brazil

Section 1

Advances in SA

Chapter 1

Adaptive Neighborhood Heuristics for Simulated Annealing over Continuous Variables

T.C. Martins, A.K.Sato and M.S.G. Tsuzuki

Additional information is available at the end of the chapter

http://dx.doi.org/10.5772/50302

1. Introduction

Simulated annealing has been applied to a wide range of problems: combinatorial and continuous optimizations. This work approaches a new class of problems in which the objective function is discrete but the parameters are continuous. This type of problem arises in rotational irregular packing problems. It is necessary to place multiple items inside a container such that there is no collision between the items, while minimizing the items occupied area. A feedback is proposed to control the next candidate probability distribution, in order to increase the number of accepted solutions. The probability distribution is controlled by the so called crystallization factor. The proposed algorithm modifies only one parameter at a time. If the new configuration is accepted then a positive feedback is executed to result in larger modifications. Different types of positive feedbacks are studied herein. If the new configuration is rejected, then a negative feedback is executed to result in smaller modifications. For each non-placed item, a limited depth binary search is performed to find a scale factor that, when applied to the item, allows it to be fitted in the layout. The proposed algorithm was used to solve two different rotational puzzles. A geometrical cooling schedule is used. Consequently, the proposed algorithm can be classified as simulated quenching.

This work is structured as follows. Section 2 presents some simulated annealing and simulated quenching key concepts. In section 3 the objective function with discrete values and continuous parameters is explained. Section 4 explains the proposed adaptive neighborhood based on the crystallization factor. Section 5 explains the computational experiments and section 6 presents the results. Finally, section 7 rounds up the work with the conclusions.

2. Background

Simulated annealing is a probabilistic meta-heuristic with a capacity of escape from local minima. It came from the Metropolis algorithm and it was originally proposed in the area of combinatorial optimization [9], that is, when the objective function is defined in a discrete domain. The simulated annealing was modified in order to apply to the optimization of multimodal functions defined on continuous domain [4]. The choices of the cooling schedule and of the next candidate distribution are the most important decisions in the definition of a simulated annealing algorithm [13]. The next candidate distribution for continuous variables is discussed herein.

In the discrete domain, such as the traveling salesman and computer circuit design problems, the parameters must have discrete values; the next point candidate $\mathbf{x}_{k+1}$ corresponds to a permutation in the list of cities to be visited, interchanges of circuit elements, or other discrete operation. In the continuous application of simulated annealing a new choice of the next point candidate must be executed. Bohachevsky et al. [1] proposed that the next candidate $\mathbf{x}_{k+1}$ can be obtained by first generating a random direction vector $\mathbf{u}$, with $|\mathbf{u}| = 1$, then multiplying it by a fixed step size Δr, and summing the resulting vector to the current candidate point $\mathbf{x}_k$.

Brooks & Verdini [2] showed that the selection of Δr is a critical choice. They observed that an appropriate choice of this parameter is strictly dependent on the objective function $F(\mathbf{x})$, and the appropriate value can be determined by presampling the objective function.

The directions in [1] are randomly sampled from the uniform distribution and the step size is the same in each direction. In this way, the feasible region is explored in an isotropic way and the objective function is assumed to behave in the same way in each direction. But this is not often the case. The step size to define the next candidate point $\mathbf{x}_{k+1}$ should not be equal for all the directions, but different directions should have different step sizes; i.e. the space should be searched in an anisotropic way. Corana et al. [4] explored the concept of anisotropic search; they proposed a self-tuning simulated annealing algorithm in which the step size is configured in order to maintain a number of accepted solutions. At each iteration k, a single variable of $\mathbf{x}_k$ is modified in order to obtain a new candidate point $\mathbf{x}_{k+1}$, and iterations are subdivided into cycles of n iterations during which each variable is modified. The new candidate point is obtained from $\mathbf{x}_k$ in the following form $\mathbf{x}_{k+1} = \mathbf{x}_k + v \cdot \Delta r_i \cdot \mathbf{e}_i$. Where v is a uniform random number in $[-1, 1]$, and Δr_i is the step size along direction $\mathbf{e}_i$ of the i-th axis. The anisotropy is obtained by choosing different values of Δr_i for all the directions. The step size is kept fixed for a certain number of cycles of variables, and the fraction of accepted moves in direction $\mathbf{e}_i$ is calculated. If the fraction of accepted moves generated in the same direction is below 0.4, then the step size Δr_i along $\mathbf{e}_i$ is decreased. It is assumed that the algorithm is using too large steps along $\mathbf{e}_i$ thus causing many moves to be rejected. If the fraction is between 0.4 and 0.6 the step size is left unchanged. If the fraction is above 0.6 then Δr_i is increased. It is assumed that the step size is too small thus causing many moves to be accepted.

This procedure may not be the best possible to process the different behavior of the objective function along different axes. Ingber [7] proposed that the random variable should follow a Cauchy distribution with different sensitivities at different temperatures. The maximum step size is kept constant during the algorithm and it allows escaping from local minima even at low temperatures. The parameter space can have completely different sensitivities for each dimension, therefore the use of different temperatures for each dimension is suggested. This method is often referred to as very fast simulated re-annealing (VFSR) or adaptive simulated annealing (ASA). The sensitivity of each parameter is given by the partial derivative of the function with relation to the i-th dimension [3].

3. Integer objective function with float parameters

Irregular packing problems arise in the industry whenever one must place multiple items inside a container such that there is no collision between the items, while minimizing the area occupied by the items. It can be shown that even restricted versions of this problem (for instance, limiting the polygon shape to rectangles only) are NP complete, which means that all algorithms currently known for optimal solutions require a number of computational steps that grow exponentially with the problem size rather than according to a polynomial function [5]. Usually probabilistic heuristics relax the original constraints of the problem, allowing the search to go through points outside the space of valid solutions and applying penalization to their cost. This technique is known as external penalization. The most adopted penalization heuristic for external solutions of packing problems is to apply a penalization based on the overlapping area of colliding items. While this heuristic leads to very computationally efficient iterations of the optimization process, the layout with objective function in minimum value may have overlapped items [6].

Fig. 1 shows an example in which the cost function is the non–occupied space inside the container. As this space can change only by adding or removing areas of items, the cost function can assume only a finite set of values, becoming discontinuous. This particularity of the primal problem makes it difficult to evaluate the sensibility of the cost function related to the optimization variables.

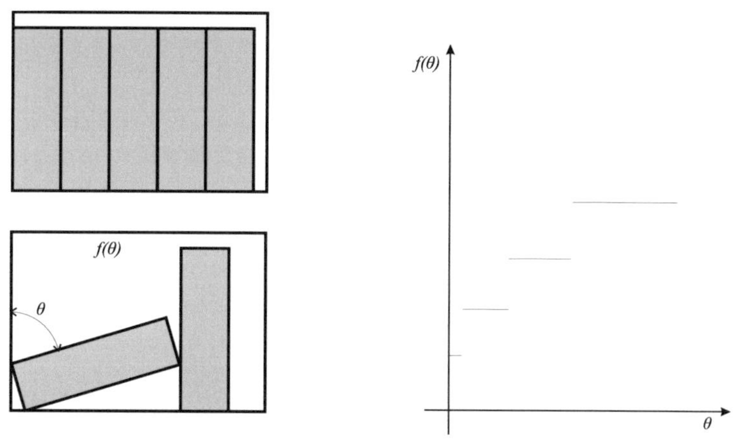

Figure 1. Objective function behavior.

Recently, researchers used the collision free region (CFR) concept to ensure feasible layouts; i.e. layouts in which the items do not overlap and fit inside the container [11]. This way, the solution has discrete and continuous components. The discrete part represents the order of placement (a permutation of the items indexes - this permutation dictates the order of placements) and the translation that is a vertex from the CFR perimeter. The continuous part represents the rotations (a sequence of angles of rotations to be applied to each item). The translation parameter is converted to a placement point at the perimeter of the CFR for its respective item. Fig. 2 shows the connection between the CFR and the translation parameter. Notice that the rotation parameter is cyclic in nature. All arithmetic operations concerning this parameter is performed in modulus 1 (so they always remain inside the interval $[0, 1[$).

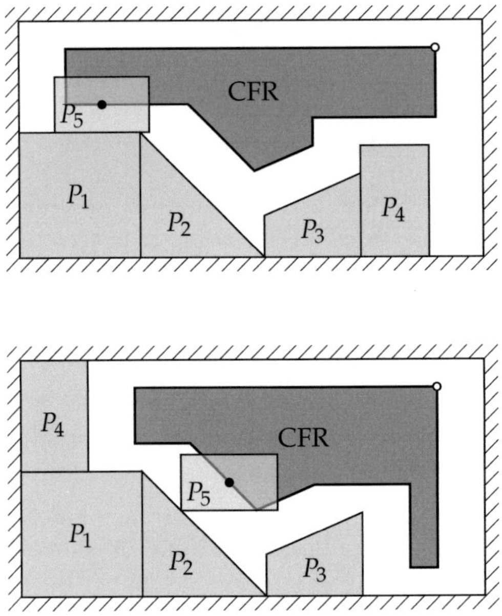

Figure 2. Consider that the container is rectangular and items P_1, P_2, P_3 and P_4 are already placed. Item P_5 is the next to be placed and to avoid collisions; it is placed at its CFR boundary. Its translation parameter has a value of 0.5. Both figures have P_4 placed at different positions, and consequently P_5 is also placed in different positions although the translation parameter is the same.

The wasted area that represents the cost function assumes only discrete values, while its variables (the rotations for each item) are continuous. To solve this type of problem, Martins & Tsuzuki [10] proposed a simulated quenching with a new heuristic to determine the next candidate that managed to solve this type of problem.

3.1. Scale factor

The objective function is the wasted space in the container and is discrete, depending on which items have been placed. In order to improve the sensibility of the cost function, intermediate levels can be generated by scaling one of the unplaced items, and attempting to insert the reduced version of the item into the layout. Hence, for each unplaced item, a scale factor between $[0, 1]$ is applied, and the algorithm attempts to place the item, if it fits, the scaled area of the item is subtracted from the objective function. Scale factor was determined by a finite fixed depth binary search, restricted to the interval $[0, 1]$.

4. Adaptive neighborhood

The proposed algorithm is shown in Fig. 3. The main modification is shown in the inner loop, where the choice is to swap two items in the placement sequence (discrete parameters) or to modify the rotation or translation of an item (continuous parameter).

The main concept is that rejected solutions do not contribute to the progress of the optimization process. Therefore, the distribution of the step size for each individual continuous parameter is adapted in order to increase the number of accepted solutions. This

```
x_0 ← <Initial random solution>;
k ← 0;
while <Global stop condition not satisfied> do
    T_{k+1} ← T_k * α;
    k ← k + 1;
    while <Local stop condition not satisfied> do
        u ← random(0, 1);
        if u < 1/3 then
            <Modify placement sequence>;
            flag ← DiscreteParameter;
        else
            if u < 2/3 then
                <Select one vertex from the CFR to place the item>;
                flag ← DiscreteParameter;
            else
                i ← random(0, 1) · n;
                x_{k+1} ← x_k + (1/c_i) Σ_1^{c_i} random(−1/2, 1/2) · e_i;
                flag ← ContinuousParameter;
        ΔE = F(x_{k+1}) − F(x_k);
        if ΔE < 0 then
            x_k ← x_{k+1};
            if flag = ContinuousParameter then
                c_i ← <positive feedback>;
        else
            if random(0, 1) < e^{−ΔE/kT} then
                x_k ← x_{k+1};
                if flag = ContinuousParameter then
                    c_i ← <positive feedback>;
            else
                if flag = ContinuousParameter then
                    c_i ← c_i + 1;
```

Figure 3. The proposed algorithm. Different types of positive feedbacks are studied in this work.

is accomplished by the adoption of a feedback on the proposed algorithm. The next candidate is generated by the modification of *a single parameter*, adding to it a summation of c_i random numbers with a uniform distribution.

$$\mathbf{x}_{k+1} = \mathbf{x}_k + \frac{1}{c_i}\sum_{1}^{c_i} random(-1/2, 1/2) \cdot \mathbf{e}_i \tag{1}$$

where i is the index of the modified parameter and c_i is its crystallization factor. The resulted modification follows a Bates distribution [8, sec. 26.9] centered on 0 with amplitude $1/2$. Its standard deviation is given by $1/2\sqrt{3c_i}$.

For $c_i = 1$, as all operations on parameters are performed in modulus 1; the modification is the equivalent of taking a completely new parameter uniformly distributed in the interval $[0, 1[$. As c_i increases, the expected amplitude of the modification decreases. When at a given iteration, the modification applied to a parameter leads to a rejected solution; the probability distribution (crystallization factor) for that specific parameter is modified in order to have its standard deviation reduced (resulting in lower modification amplitude), this is the negative feedback. When the modification leads to an accepted solution, the distribution (crystallization factor) for that parameter is modified to increase its standard deviation (resulting in larger modification amplitude), this is the positive feedback. Different positive feedbacks are studied in this work (see Table 1). As can be seen, the higher the crystallization factor for a given parameter, the smaller the modification this parameter will receive during the proposed algorithm. The parameter is said to be crystallized.

5. Computational experiments

Crystallization factor c_i controls the standard deviation of the Bates distribution. When a solution is rejected, a negative feedback is applied and the corresponding c_i is increased, causing a decrease in the parameter standard deviation. Accordingly, positive feedback is applied when a solution is accepted, increasing c_i. In the studied problems, placement was restricted to vertexes of the CFR and thus the only continuous parameter is the rotation. Adopted negative feedback consists of incrementing the crystallization factor. For the positive feedback, the four different strategies in Table 1 were tested.

Feedback Method	Positive Feedback	Negative Feedback
A	$CF_i \to CF_i - 1$	$CF_i \to CF_i + 1$
B	$CF_i \to CF_i/2$	$CF_i \to CF_i + 1$
C	$CF_i \to CF_i/4$	$CF_i \to CF_i + 1$
D	$CF_i \to 1$	$CF_i \to CF_i + 1$

Table 1. Feedback strategies. CF_i: Crystallization factor for item i.

The convergence of the algorithm is reached when, at a given temperature, all accepted solutions are equivalent to the best found. This is the global stop condition of the algorithm in Fig. 3. Although a solution as good as the final one is found in less iterations, allowing the algorithm to reach the global convergence is the only generic way to ensure that a solution is the best. The local stop condition shown in Fig. 3 is reached when a predefined number of solutions are accepted.

5.1. Problem instances

All problem instances studied here have a solution in which all items can be fitted in the container. Two puzzles cases were considered: Tangram and Larger First Fails (LF Fails). Tangram is a classic problem and LF Fails consists of a problem which cannot be solved using the larger first heuristic. This heuristic determines that the larger items are placed always ahead of the smaller ones. Fig. 4 shows possible solutions to these problems.

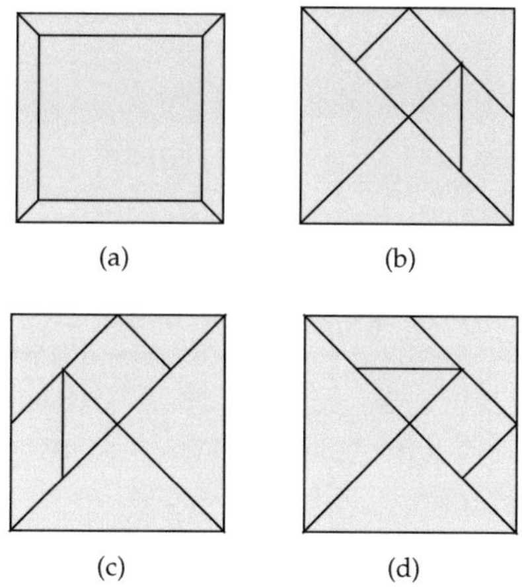

Figure 4. (a) Unique solution for problem LF Fails. (b)-(d) Solutions for the Tangram problem.

6. Results and discussion

The algorithm was implemented in C++ and compiled by GCC 4.4.4. Computational tests were performed using an i7 860 processor with 4GB RAM. Each case was evaluated 100 times. The proposed algorithm is a simulated quenching algorithm which has the following parameters:

- T_0: Initial temperature.
- α: geometric cooling schedule factor.
- N_{acc}: Number of accepted solutions at a given temperature.

Value of T_0 is calculated such that the number of rejected solutions at initial temperature is approximately 10% of the total number of generated solutions. Parameter α is set to 0.99 and N_{acc} is 800.

6.1. Influence of the feedback strategy

Table 2 shows results obtained using each of the proposed feedback strategy, for each problem instance. For the Tangram problem, it can be observed that strategy A has a low convergence percentage, when compared to other feedback strategies, 0.09 less than the rate obtained using the feedback C method. In the case of the LF fails puzzle, results showed similar performance and convergence rate.

Fig. 5 shows the minimum, maximum and average costs explored by the proposed algorithm loop for the LF Fails, for all feedback strategies. The cost function discrete behavior is observable, and it is possible to notice that the global minimum is reached only at low temperatures. In all graphics, the optimum layout was found. One can note that, in Fig. 5.(b) and Fig. 5.(c), the best solution (cost equals zero) was found before reaching convergence. Variation of cost is shown in Fig. 6 and all graphs are very similar independently of the used positive feedback. The rotation crystallization factor for the largest item is displayed in Fig. 7. Possibility of accepting a higher cost solution is lower at low temperatures. As temperature

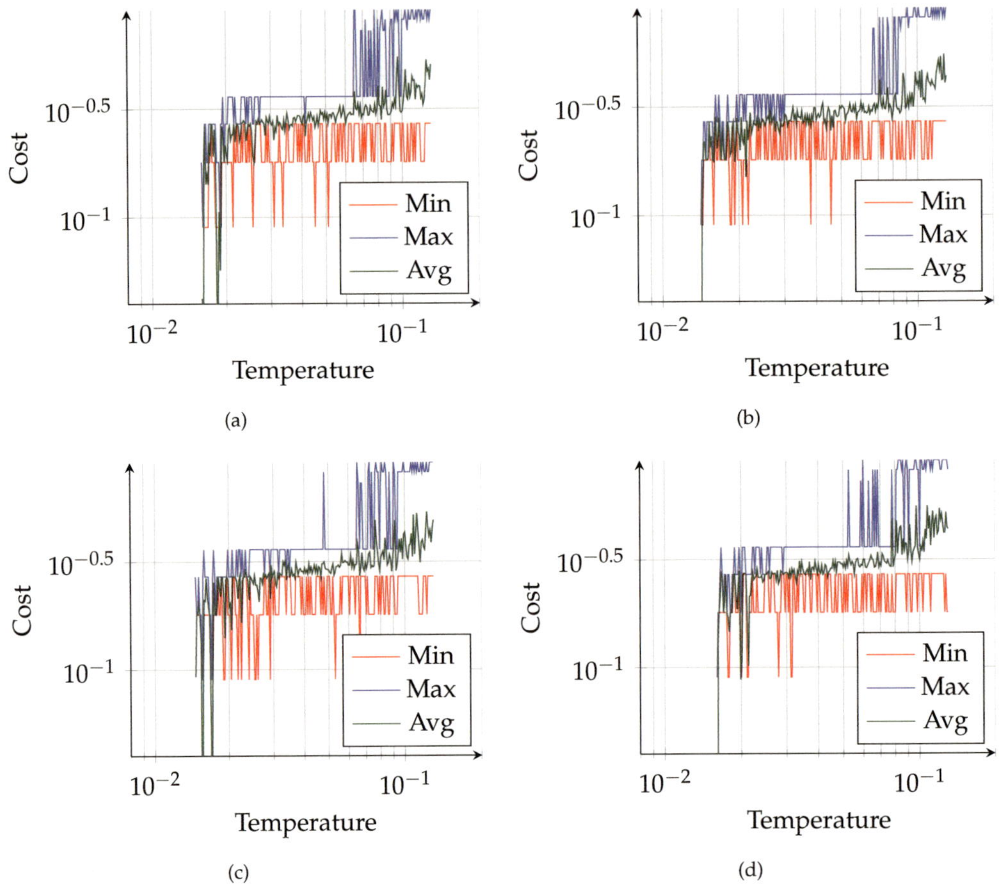

Figure 5. Minimum, maximum and average costs for the LF Fails with different feedbacks. (a) Feedback A. (b) Feedback B. (c) Feedback C. (d) Feedback D.

Problem	Feedback Method	N_{conv}	N_{min}	T_{conv}	P_{conv}
LF fails	A	228935	188814	17.48	1.00
	B	235986	197038	17.78	0.99
	C	235595	195377	17.64	1.00
	D	235481	194394	17.67	1.00
Tangram	A	303517	255611	64.33	0.56
	B	315019	268996	69.07	0.65
	C	319440	270484	69.27	0.62
	D	317403	267057	71.09	0.61

Table 2. Statistics for the LF fails and Tangram puzzles. The columns respectively represent the adopted problem instance, the feedback method, number of iterations to converge, number of iterations to reach the minimum, time in seconds to converge, and the percentage of runs that converged to the global optimum.

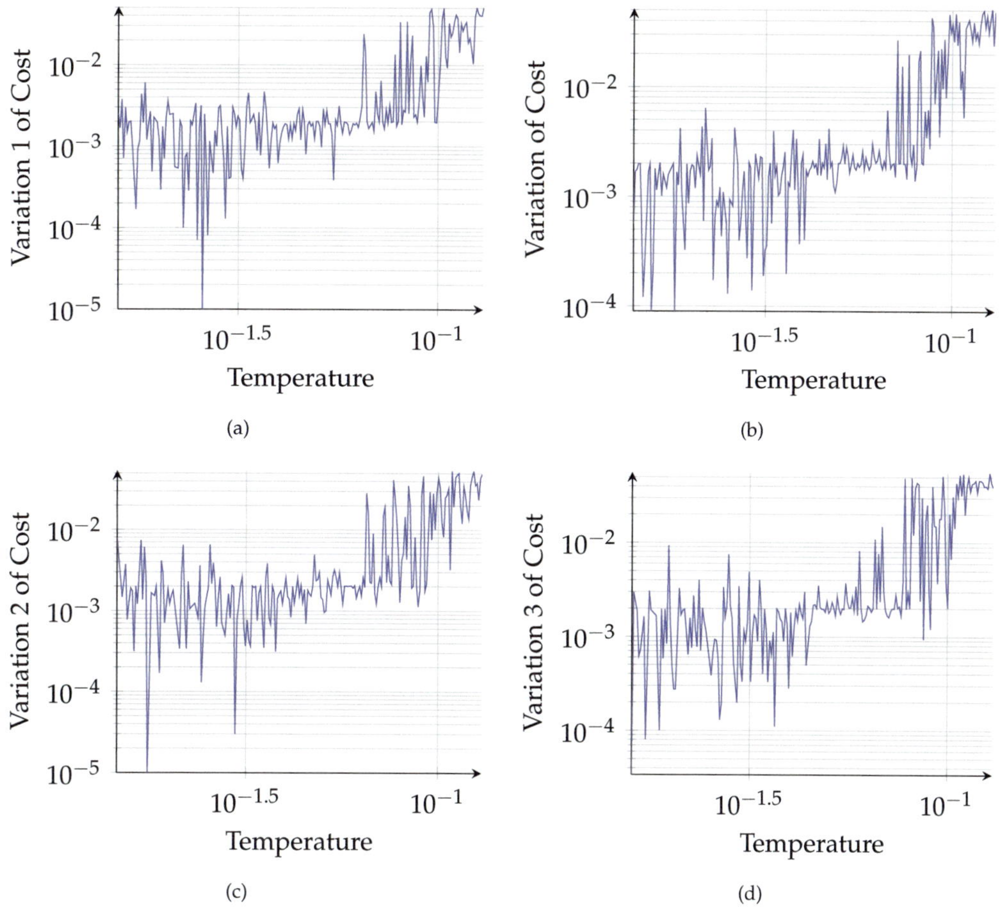

Figure 6. Variation of cost for the LF Fails with different feedbacks. (a) Feedback A. (b) Feedback B. (c) Feedback C. (d) Feedback D.

decreases, the crystallization factor is expected to increase, which is confirmed by the graphics in Fig 7. Positive feedback A is very stable, showing that it is less exploratory. Because of the small number of items, it was not necessary to use the scale factor. Fig. 8 shows the specific heat for each case considered. The specific heat is calculated as [14]

$$C_H(T) = \frac{\sigma^2(T)}{T^2 k_B^2} \tag{2}$$

where T is temperature, $\sigma^2(T)$ is the variation of cost, k_B is a constant. A phase transition occurs at a temperature at which specific heat is maximum, and this triggers the change in state ordering. In several processes, it represents the transition from the exploratory to the refining phase. However, in this specific case, this transition is not observable.

For the Tangram problem, the minimum, maximum and average costs explored by the algorithm in one execution are shown in Fig. 9. The increase in allowable cost function values can be observed. In each of these executions the global minimum was reached only at low

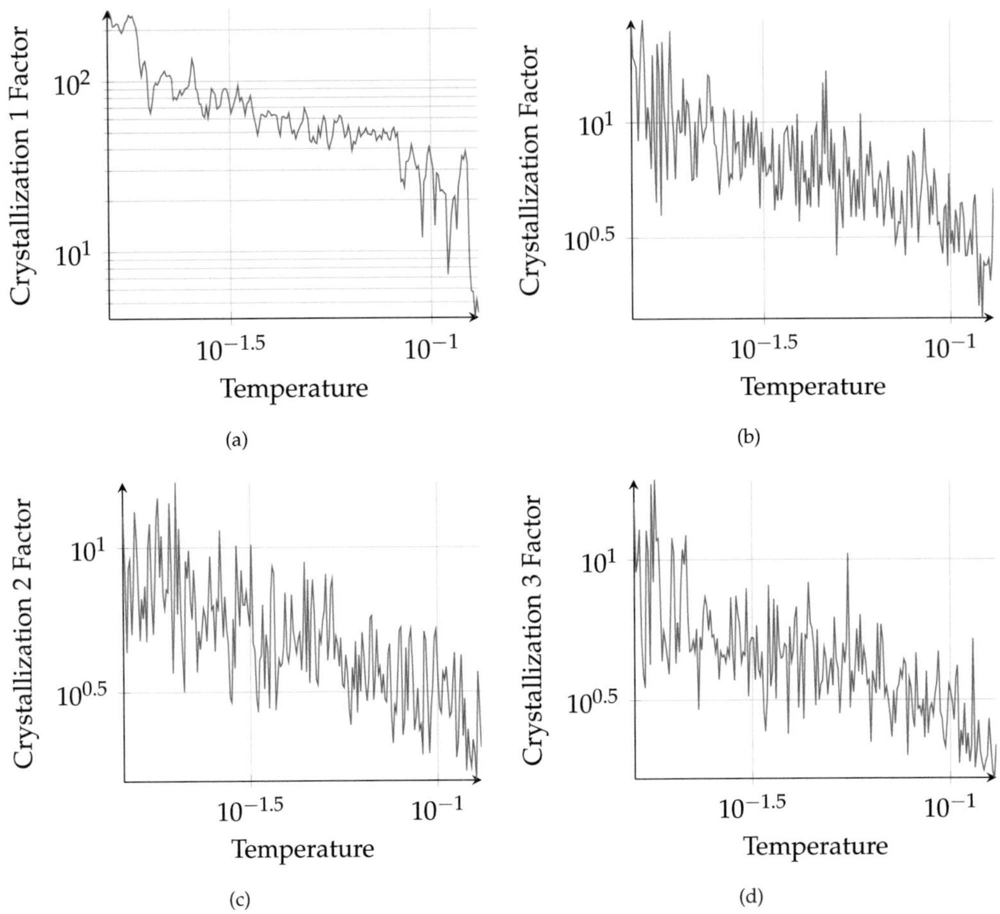

Figure 7. Crystallization factor for the largest item of the LF Fails problem, with different feedbacks. (a) Feedback A. (b) Feedback B. (c) Feedback C. (d) Feedback D.

temperatures. Independently of the used positive feedback, the proposed algorithm reached local minima at lower temperatures, but successfully escaped from them. Cost variance is displayed in Fig. 10. The rotation crystallization factor evolution is shown in Fig. 11, for one of the large triangles. It is possible to observe, when adopting feedback strategies A and C, that there are two distinguishable levels. The final level, at lower temperatures, is very high, indicating that the rotation parameter of the item is crystallized. Again, feedback A is more stable when compared to the others, showing that is less exploratory. As the convergence rate is very poor, the scale factor should be used. Fig. 12 shows the specific heats obtained. In the Tangram casem, it seems that a peak is present. However, further investigations need to be done.

6.2. Influence of the binary search

Binary search is used to improve the sensibility of the discrete objective function, aiming to obtain a higher percentage of convergence for puzzle problems. Its application is not necessary in the case of the LF Fails problem, as almost all executions converged. As a

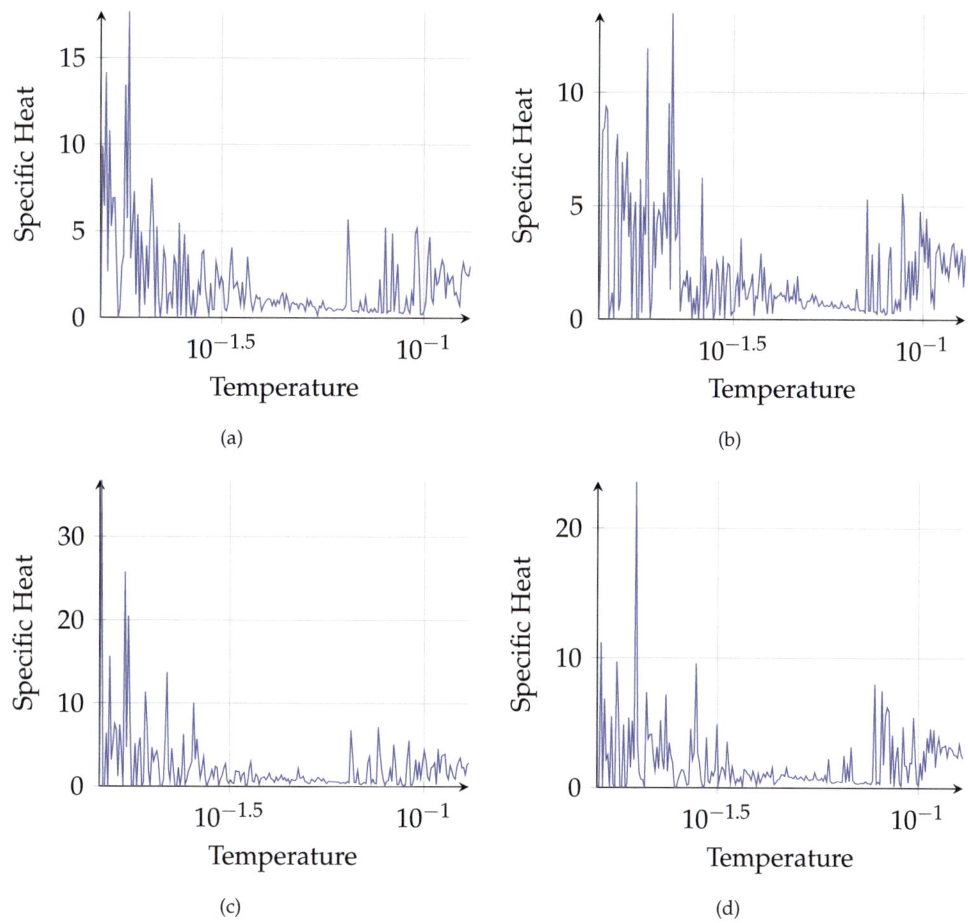

Figure 8. Specific heat for the LF Fails with different feedbacks. (a) Feedback A. (b) Feedback B. (c) Feedback C. (d) Feedback D.

consequence, the binary search was employed only in the Tangram problem. The fixed search depth was set to 1. Table 3 shows the results of the tests. Comparing with the results obtained in Table 2, the convergence rate is observed to be considerably higher when binary search is adopted, reaching 98% in the best case. Drawback is the large number of iterations needed to converge, resulting in longer execution times, approximately 3.5 times higher. As with the previous tests, feedback strategy A obtained less optimum solutions.

The behavior of the optimization process is illustrated through cost function (energy) histograms of the search while the temperature diminishes. For a given temperature, a gray-level histogram of the distribution of the cost function at that temperature is plotted. The resulting graph shows a plot of cost histograms (horizontal bars) and temperature (dots) versus the number of iterations. Darker horizontal bars in the histogram, indicate a higher frequency of occurrence of a particular level of energy at a given temperature. Fig. 13 shows the histogram of the objective function value during the course of the algorithm, without the use of the binary search. Fig. 14 shows the same type of histogram employing the binary search with depth 1. Observing both graphics, one can note the extra levels of energy which

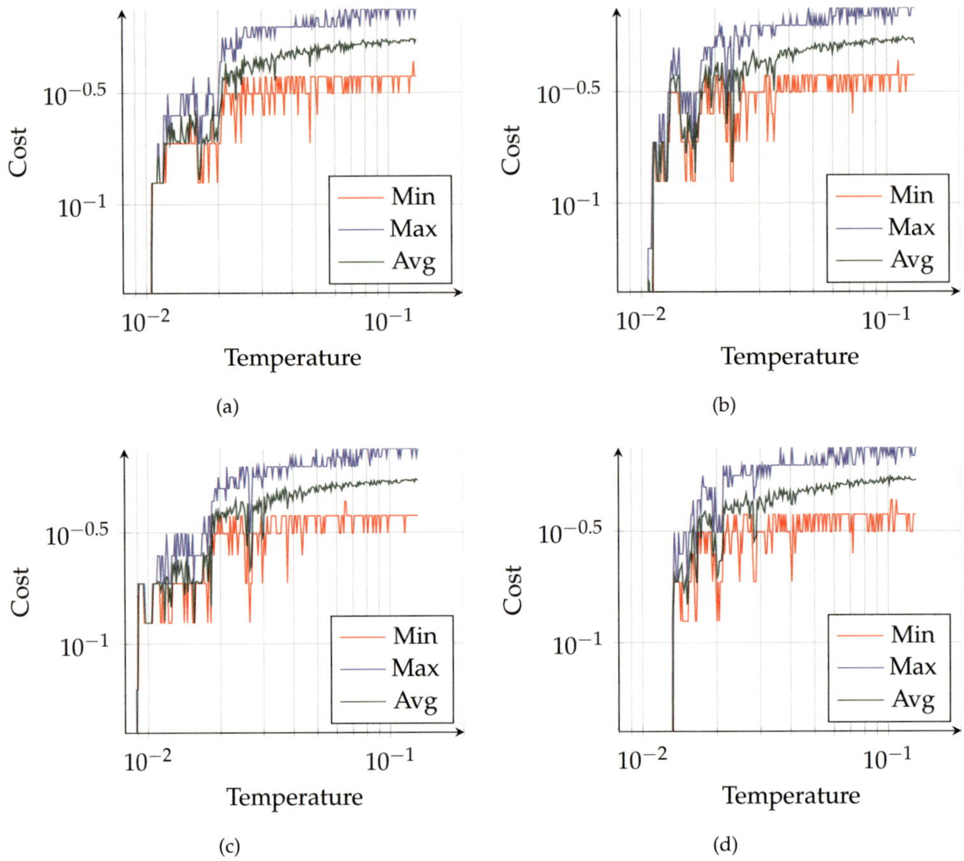

Figure 9. Minimum, maximum and average costs for the Tangram with different feedbacks. (a) Feedback A. (b) Feedback B. (c) Feedback C. (d) Feedback D.

Feedback Method	N_{conv}	N_{min}	T_{conv}	P_{conv}
A	370667	290044	222.54	0.78
B	351141	299052	227.39	0.91
C	343652	327037	228.40	0.98
D	338394	312867	213.91	0.97

Table 3. Statistics for the Tangram puzzles using a binary search with unitary depth. The columns respectively represent the feedback method, number of iterations to converge, number of iterations to reach the minimum, time in seconds to converge, and the percentage of runs that converged to the global optimum.

appears in Fig. 14. Higher values of the search depth were tested, however the convergence rate deteriorates.

From the studied problems, it is possible to observe that positive and negative feedbacks must not be opposites. The negative feedback increases the crystallization factor by a unit and the positive feedback needs to decrease the crystallization factor at a faster speed. If this is not the case, the parameters might get crystallized.

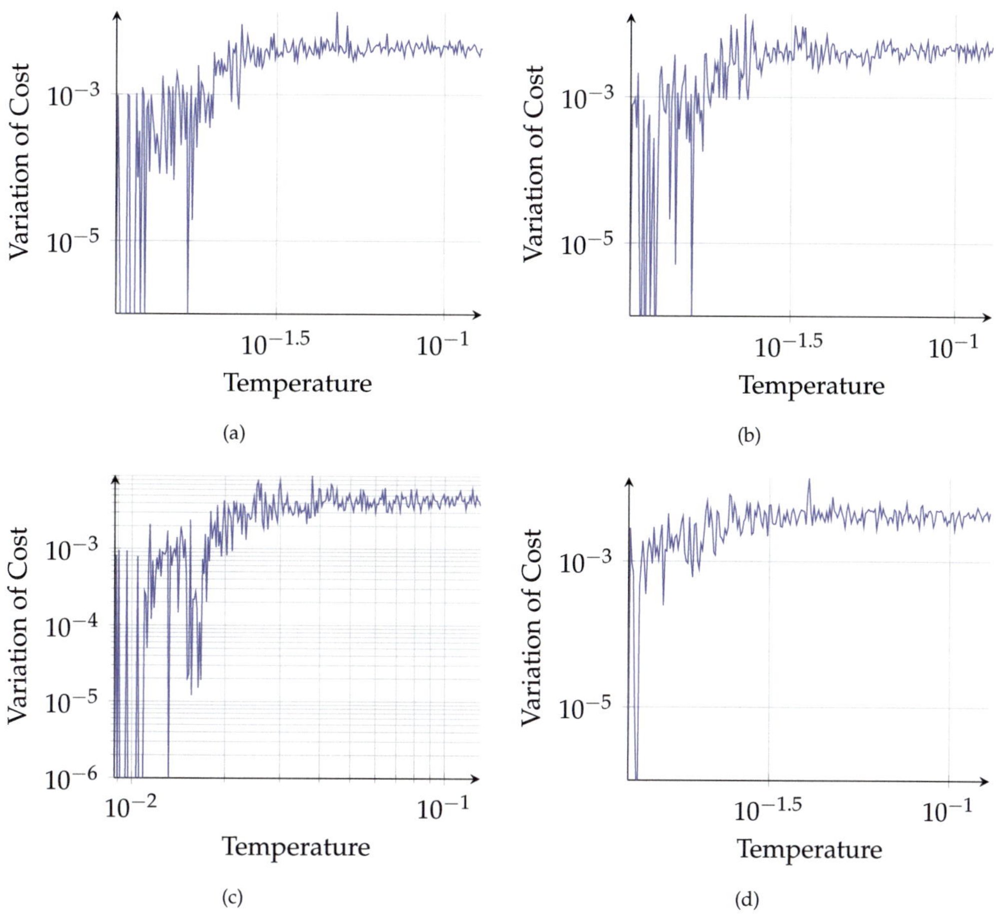

Figure 10. Variation of cost for the Tangram with different feedbacks. (a) Feedback A. (b) Feedback B. (c) Feedback C. (d) Feedback D.

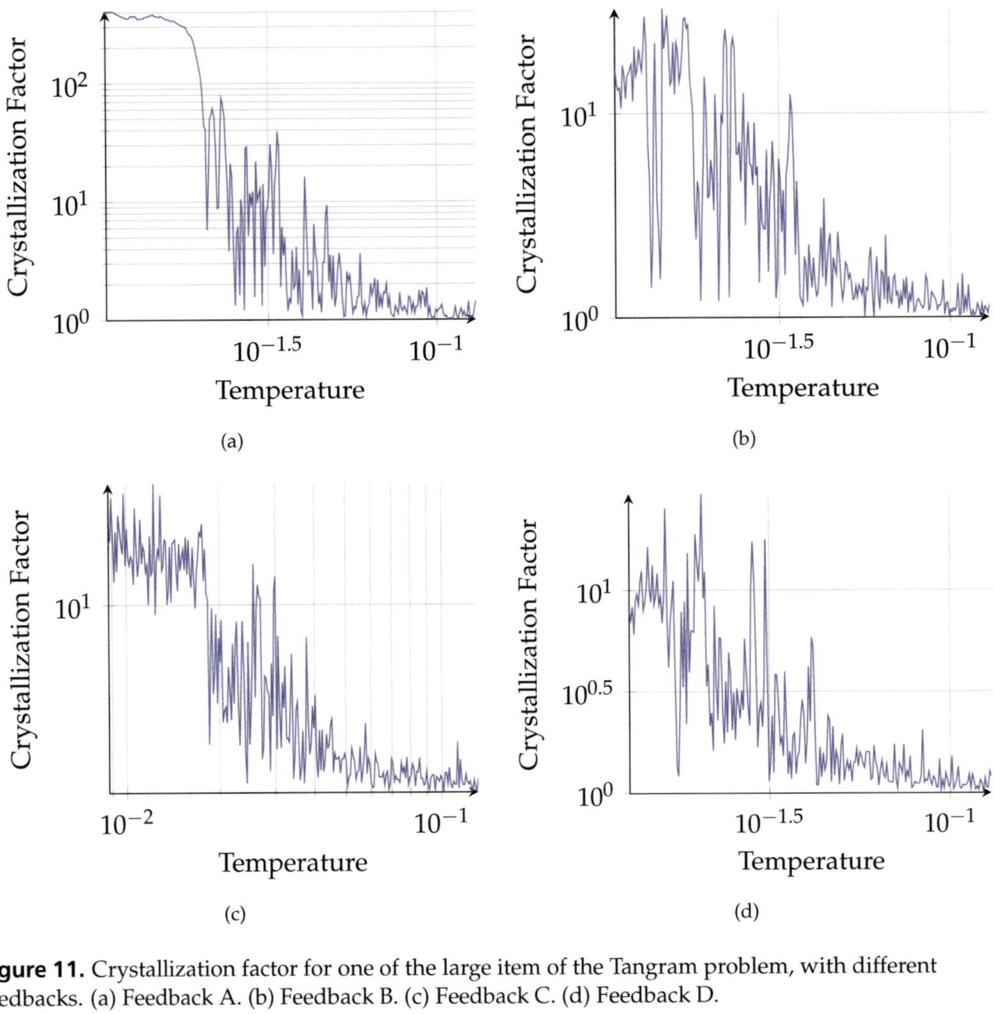

Figure 11. Crystallization factor for one of the large item of the Tangram problem, with different feedbacks. (a) Feedback A. (b) Feedback B. (c) Feedback C. (d) Feedback D.

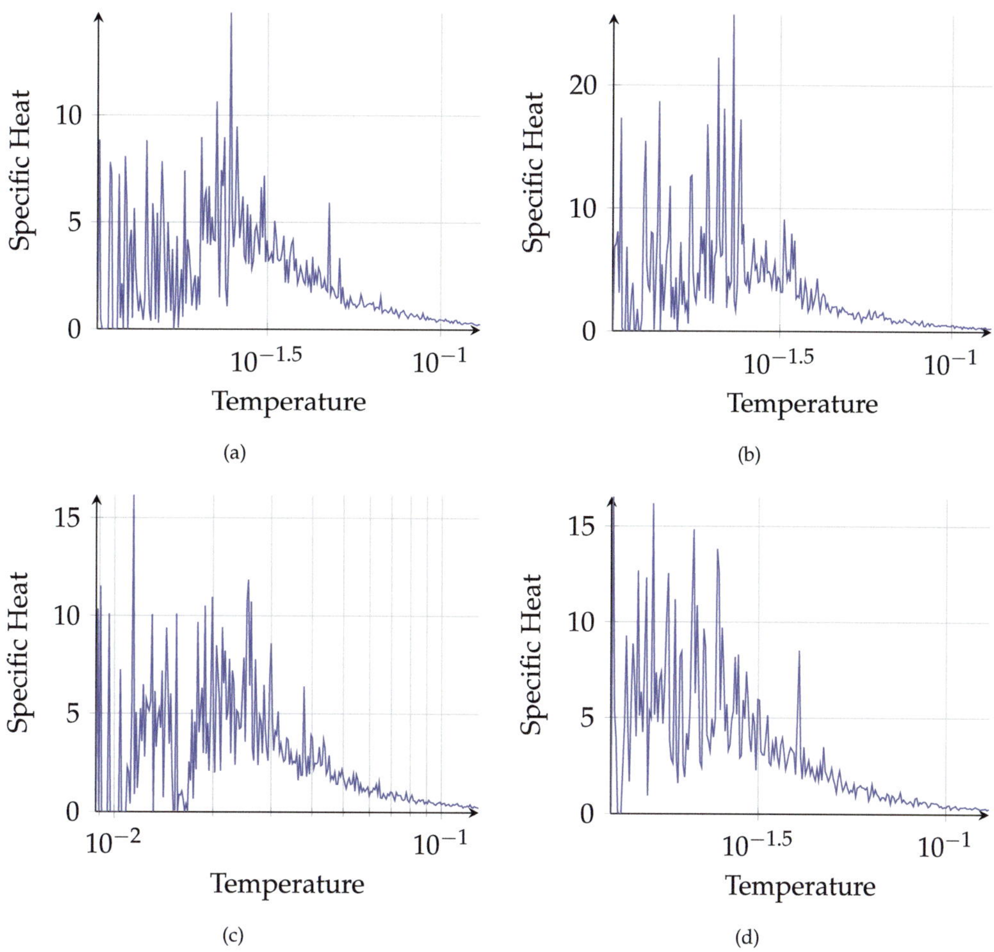

Figure 12. Specific heat for the Tangram with different feedbacks. (a) Feedback A. (b) Feedback B. (c) Feedback C. (d) Feedback D.

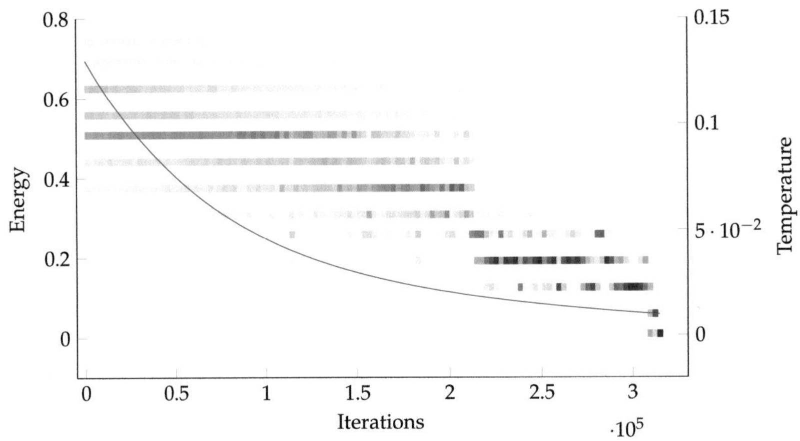

Figure 13. Histogram for the Tangram, not employing binary search.

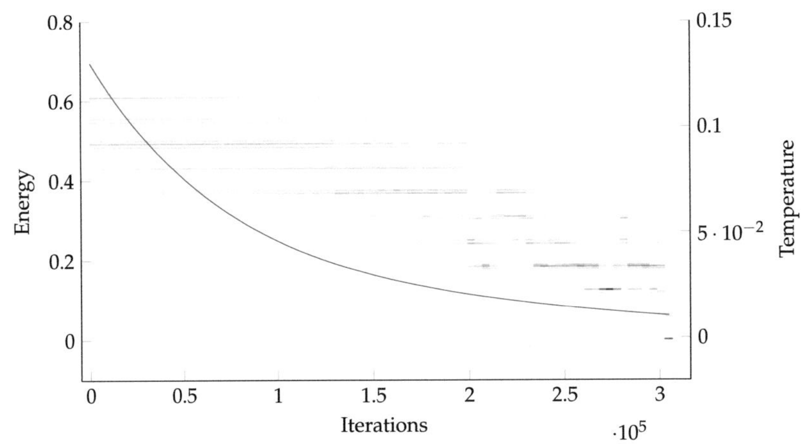

Figure 14. Histogram for the Tangram, employing binary search with fixed depth equal to 1.

7. Conclusion

This work proposed a new simulated quenching algorithm with adaptive neighborhood, in which the sensibility of each continuous parameter is evaluated at each iteration increasing the number of accepted solutions. The proposed simulated quenching was successfully applied to other types of problems: robot path planning [14] and electrical impedance tomography [12]. The placement of an item is controlled by the following simulated quenching parameters: rotation, translation and sequence of placement.

8. Acknowledgements

AK Sato was supported by FAPESP (Grant 2010/19646-0). MSG Tsuzuki was partially supported by the CNPq (Grant 309.570/2010–7). This research was supported by FAPESP (Grants 2008/13127–2 and 2010/18913–4).

Author details

T.C. Martins, A.K.Sato and M.S.G. Tsuzuki
Computational Geometry Laboratory - Escola Politécnica da USP, Brazil

9. References

[1] Bohachevsky, I. O., Johnson, M. E. & Stein, M. L. [1986]. Generalized simulated annealing for function optimization, *Technometrics* 28(3): pp. 209–217.
[2] Brooks, D. G. & Verdini, W. A. [1988]. Computational experience with generalized simulated annealing over continuous variables, *Am. J. Math. Manage. Sci.* 8(3-4): 425–449.
[3] Chen, S. & Luk, B. [1999]. Adaptive simulated annealing for optimization in signal processing applications, *Signal Processing* 79(1): 117 – 128.
[4] Corana, A., Marchesi, M., Martini, C. & Ridella, S. [1987]. Minimizing multimodal functions of continuous variables with the simulated annealing algorithm, *ACM Trans. Math. Softw.* 13(3): 262–280.
[5] Fowler, R. J., Paterson, M. & Tanimoto, S. L. [1981]. Optimal packing and covering in the plane are np-complete, *Inf. Process. Lett.* pp. 133–137.
[6] Heckmann, R. & Lengauer, T. [1995]. A simulated annealing approach to the nesting problem in the textile manufacturing industry, *Annals of Operations Research* 57: 103–133.
[7] Ingber, L. [1996]. Adaptive simulated annealing (asa): Lessons learned, *Control and Cybernetics* 25: 33–54.
[8] Johnson, N. [1994]. *Continuous univariate distributions*, Wiley, New York.
[9] Kirkpatrick, S., Gelatt, C. D. & Vecchi, M. P. [1983]. Optimization by simulated annealing, *Science* 220: 671–680.
[10] Martins, T. C. & Tsuzuki, M. S. G. [2009]. Placement over containers with fixed dimensions solved with adaptive neighborhood simulated annealing, *Bulletin of the Polish Academy of Sciences Technical Sciences* 57: 273–280.
[11] Martins, T. C. & Tsuzuki, M. S. G. [2010]. Simulated annealing applied to the irregular rotational placement of shapes over containers with fixed dimensions, *Expert Systems with Applications* 37: 1955–1972.
[12] Martins, T. C., Camargo, E. D. L. B., Lima, R. G., Amato, M. B. P. & Tsuzuki, M. S. G. [2012]. Image reconstruction using interval simulated annealing in

electrical impedance tomography, *IEEE Transactions on Biomedical Engineering*, URL: *http://dx.doi.org/10.1109/TBME.2012.2188398*.

[13] Miki, M., Hiroyasu, T. & Ono, K. [2002]. Simulated annealing with advanced adaptive neighborhood, *In Second international workshop on Intelligent systems design and application*, Dynamic Publishers, Inc. ISBN, pp. 113–118.

[14] Tavares, R. S., Martins, T. C. & Tsuzuki, M. S. G. [2011]. Simulated annealing with adaptive neighborhood: A case study in off-line robot path planning, *Expert Systems with Applications* 38(4): 2951–2965.

Chapter 2

A Simulated Annealing Algorithm for the Satisfiability Problem Using Dynamic Markov Chains with Linear Regression Equilibrium

Felix Martinez-Rios and Juan Frausto-Solis

Additional information is available at the end of the chapter

http://dx.doi.org/10.5772/46175

1. Introduction

Since the appearance of Simulated Annealing algorithm it has shown to be an efficient method to solve combinatorial optimization problems such as Boolean Satisfiability problem. New algorithms based on two cycles: one external for temperatures and other internal, named Metropolis, have emerged. These algorithms usually use the same Markov chain length in the Metropolis cycle for each temperature. In this paper we propose a method based on linear regression to find the Metropolis equilibrium. Experimentation shows that the proposed method is more efficient than the classical one, since it obtains the same quality of the final solution with less processing time.

Today we have a considerable interest for developing new and efficient algorithms to solve hard problems, mainly those considered in the complexity theory (NP-complete or NP-hard) [8]. The Simulated Annealing algorithm proposed by Kirkpatrick et al. [18] and Cerny [5, 6] is an extension of the Metropolis algorithm [23] used for the simulation of the physical annealing process and is specially applied to solve NP-hard problems where it is very difficult to find the optimal solution or even near-to-optimum solutions.

Efficiency and efficacy are given to Simulated Annealing algorithm by the cooling scheme which consists of initial (c_i) and final (c_f) temperatures, the cooling function $(f(c_k))$ and the length of the Markov chain (L_k) established by the Metropolis algorithm. For each value of the control parameter (c_k) (temperature), Simulated Annealing algorithm accomplishes a certain number of Metropolis decisions. In this regard, in order to get a better performance of the Simulated Annealing algorithm a relation between the temperature and Metropolis cycles may be enacted [13].

The Simulated Annealing algorithm can get optimal solutions in an efficient way only if its cooling scheme parameters are correctly tuned. Due this, experimental and analytical parameters tuning strategies are currently being studied; one of them known as ANDYMARK [13] is an analytical method that has been shown to be more efficient. The objective of these methods is to find better ways to reduce the required computational resources and to increment the quality of the final solution. This is executed applying different accelerating techniques such as: variations of the cooling scheme [3, 27], variations of the neighborhood scheme [26] and with parallelization techniques [12, 26].

In this chapter an analytic adaptive method to establish the length of each Markov chain in a dynamic way for Simulated Annealing algorithm is presented; the method determines the equilibrium in the Metropolis cycle using Linear Regression Method (LRM). LRM is applied to solve the satisfiability problems instances and is compared versus a classical ANDYMARK tune method.

2. Background

In complexity theory, the satisfiability problem is a decision problem. The question is: given the expression, is there some assignment of TRUE and FALSE values to the variables that will make the entire expression true? A formula of propositional logic is said to be satisfiable if logical values can be assigned to its variables in a way that makes the formula true.

The propositional satisfiability problem, which decides whether a given propositional formula is satisfiable, is of critical importance in various areas of computer science, including theoretical computer science, algorithmics, artificial intelligence, hardware design, electronic design automation, and verification. The satisfiability problem was the first problem refered to be as NP complete [7] and is fundamental to the analysis of the computational complexity of many problems [28].

2.1. Boolean satisfiability problem (SAT)

An instance of SAT is a boolean formula which consists on the next components:

- A set S of n variables $x_1, x_2, x_3, ..., x_n$.
- A set L of literals; a literal l_i, is a variable x_i or its negation $\widetilde{x_i}$.
- A set of m clauses: $C_1, C_2, C_3, ..., C_m$ where each clause consists of literals l_i linked by the logical connective OR ($\vee$).

This is:

$$\Phi = C_1 \wedge C_2 \wedge C_3 \wedge ... \wedge C_m \tag{1}$$

where Φ, in Equation 1, is the SAT instance. Then we can enunciate the SAT problem as follows:

Definition 1. *Given a finite set* $\{C_1, C_2, C_3, \ldots, C_m\}$ *of clauses, determine whether there is an assignment of truth-values to the literals appearing in the clauses which makes all the clauses true.*

NP-completeness in SAT problem, only refers to the run-time of the worst case instances. Many of the instances that occur in practical applications can be solved much faster, for example, SAT is easier if the formulas are restricted to those in disjunctive normal form, that is, they are disjunction (OR) of terms, where each term is a conjunction (AND) of literals. Such a formula is indeed satisfiable if and only if at least one of its terms is satisfiable, and a term is satisfiable if and only if it does not contain both x and $\widetilde{x}$ for some variable x, this can be checked in polynomial time.

SAT is also easier if the number of literals in a clause is limited to 2, in which case the problem is called $2 - SAT$, this problem can also be solved in polynomial time [2, 10]. One of the most important restrictions of SAT is HORN-SAT where the formula is a conjunction of Horn clauses (a Horn clause is a clause with at most one positive literal). This problem is solved by the polynomial-time Horn-satisfiability algorithm [9].

The 3-satisfiability (3-SAT) is a special case of k-satisfiability (k-SAT), when each clause contains exactly $k = 3$ literals. 3-SAT is NP-complete and it is used as a starting point for proving that other problems are also NP-hard [31]. This is done by polynomial-time reduction from 3-SAT to the other problem [28].

3. Simulated Annealing algorithm

Simulated Annealing improves this strategy through the introduction of two elements. The first is the Metropolis algorithm [23], in which some states that do not improve energy are accepted when they serve to allow the solver to explore more of the possible space of solutions. Such "bad" states are allowed using the Boltzman criterion: $e^{-\Delta J/T} > rnd(0,1)$, where ΔJ is the change of energy, T is a temperature, and $rnd(0,1)$ is a random number in the interval $[0,1)$. J is called a cost function and corresponds to the free energy in the case of annealing a metal. If T is large, many "bad" states are accepted, and a large part of solution space is accessed.

The second is, again by analogy with annealing of a metal, to lower the temperature. After visiting many states and observing that the cost function declines only slowly, one lowers the temperature, and thus limits the size of allowed "bad" states. After lowering the temperature several times to a low value, one may then "quench" the process by accepting only "good" states in order to find the local minimum of the cost function.

The elements of Simulated Annealing are:

- A finite set S.
- A cost function J defined on S. Let $S^* \subset S$ be the set of global minima of J.
- For each $i \in S$, a set $S(i) \subset S - i$ is called the set of neighbors of i.
- For every i, a collection of positive coefficients $q_{ij}, j \in S(i)$, such that $\sum_{j \in S(i)} q_{ij} = 1$. It is assumed that $j \in S(i)$ if and only if $i \in S(j)$.
- A nonincreasing function $T : N \to (0, \infty)$, called the cooling schedule. Here N is the set of positive integers, and $T(t)$ is called the temperature al time t.
- An initial state $x(0) \in S$.

The Simulated Annealing algorithms consists of a discrete time inhomogeneus Markov chain $x(t)$ [4]. If the current state $x(t)$ is equal to i, chose a neighbor j of i at random; the probability that any particular $j \in S(i)$ is selectec is equal to q_{ij}. Once j is chosen, the next state $x(t+1)$ is determined as follows:

$$\begin{array}{ll} \text{if } J(j) \leq J(i) \text{ then } x(t+1) = j & \\ \text{if } J(j) \leq J(i) \text{ then} & \\ \qquad x(t+1) = j \text{ with probability } e^{-(J(j)-J(i))/T(t)} & \quad (2) \\ \text{else} & \\ \qquad x(t+1) = i & \end{array}$$

In a formal way:

$$P\left[x(t+1) = j | x(t) = i\right] = \begin{cases} q_{ij} e^{\left[-\frac{1}{T(t)} max\{0, J(j)-J(i)\}\right]} & j \neq i, j \in S(i) \\ 0 & j \neq i, j \notin S(i) \end{cases} \quad (3)$$

In Simulated Annealing algorithm we are considering a homogeneus Markov chain $x_T(t)$ wich temperature $T(t)$ is held at a constant value T. Let us assume that the Markov chain $x_{T(t)}$ is irreducible and aperiodic and that $q_{ij} = x_{ji} \forall i, j$, then $x_{T(t)}$ is a reversible Markov chain, and its invariant probability distribution is given by:

$$\pi_T(i) = \frac{1}{Z_T} e^{\left[-\frac{J(i)}{T}\right]} \quad (4)$$

In Equation 4 Z_T is a normalized constant and is evident that as $T \to 0$ the probability π_T is concentrate on the set S^* of global minima of J, this property remains valid if the condition $q_{ij} = q_{ji}$ is relaxed [11].

In the optimization context we can generate an optimal element with high probability if we produce a random sample according to the distribution π_T, known as the Gibbs distribution. When is generated an element of S accomplished by simulating Markov chain $x_T(t)$ until it reaches equilibrium we have a Metropolis algorithm [23].

The Simulated Annealing algorithm can also be viewed as a local search method occasionally moves to higher values of the cost function J, this moves will help to Simulated Annealing escape from local minima. Proof of convergence of Simulated Annealing algorithm can be revised [4].

3.1. Traditional Simulated Annealing algorithms

Figure 1 shows the classic algorithm simulated annealing. In the algorithm, we can see the cycle of temperatures between steps 2 and 5. Within this temperature cycle, are the steps 3 and 4 which correspond to the Metropolis algorithm.

As described in the simulated annealing algorithm, Metropolis cycle is repeated until thermal equilibrium is reached, now we use the formalism of Markov chains to estimate how many times it is necessary to repeat the cycle metropolis of so that we ensure (with some probability) that all solutions of the search space are explored.

Similarly we can estimate a very good value for the initial and final temperature of the temperature cycle. All these estimates were made prior to running the simulated annealing algorithm, using data information SAT problem is solved.

1	Initializing: Initial solution S_i Initial and final temperature: T_i and T_f $T = T_i$
2	Temperatures cycle:
3	Metropolis cycle: Generating S_j from S_i $dif = J(S_j) - J(S_i)$ If $dif < 0$ then $S_i = S_j$ else if $e^{\frac{-dif}{T}} >$ rnd(0,1) $S_i = S_j$
4	Metropolis condition: If thermal equilibrium is reached goto 5 Else goto 3
5	Stop criterion: If the final temperature T_f is reached End Else Update T goto 2

Figure 1. Simulated Annealing algorithm

It is well known that Simulated Annealing requires a well defined neighborhood structure and other parameters as initial and final temperatures T_i and T_f. In order to determine these paratmeters we follow the next method proposed by [30]. So following the analysis made in [30] we give the basis of this method.

Let $P_A(S_j)$ be the accepting probability of one proposed solution S_j generated from a current solution S_i, and $P_R(S_j)$ the rejecting probability. The probability of rejecting S_j can be established in terms of $P_A(S_j)$ as follows:

$$P_R(S_j) = 1 - P_A(S_j) \tag{5}$$

Accepting or rejecting S_j only depends on the cost deterioration size that this change will produce to the current solution, that means:

$$P_A(S_j) = g[J(S_i) - J(S_j)] = g(\Delta J_{ij}) \tag{6}$$

In Equation 6, $J(S_i)$ and $J(S_j)$ are the cost associated to S_i and S_j respectively, and $g(\Delta J_{ij})$ is the probability to accept the cost difference $\Delta J_{ij} = J(S_i) - J(S_j)$.

The solution selected from S_i may be any solution S_j defined by the next neighborhood scheme:

Definition 2. *Let* $\{\forall S_i \in S, \exists$ *a set* $V_{S_i} \subset S | V_{S_i} = V : S \longrightarrow S\}$ *be the neighborhood of a solution* S_i*, where* V_{S_i} *is the neighborhood set of* S_i*,* $V : S \longrightarrow S$ *is a mapping and* S *is the solution space of the problem being solved.*

It can be seen from the Definition 2 that neighbors of a solution S_i only depends on the neighborhood structure V established for a specific problem. Once V is defined, the maximum and minimum cost deteriorations can be written as:

$$\Delta J_{V_{\max}} = \max[J(S_i) - J(S_j)], \forall S_j \in V_{S_i}, \forall S_i \in S \tag{7}$$

$$\Delta J_{V_{\min}} = \min[J(S_i) - J(S_j)], \forall S_j \in V_{S_i}, \forall S_i \in S \tag{8}$$

where $\Delta J_{V_{\max}}$ and $\Delta J_{V_{\min}}$ are the maximum and minimum cost deteriorations of the objective function through J respectively.

3.2. Markov Chains and Cooling Function

The Simulated Annealing algorithm can be seen like a sequence of homogeneous Markov chains, where each Markov chain is constructed for descending values of the control parameter $T > 0$ [1]. The control parameter is set by a cooling function like:

$$T_{k+1} = f(T_k) \tag{9}$$

and T_k must satisfy the next property:

$$\begin{aligned} &\lim_{k\to\infty} T_k = 0 \\ &T_k \geq T_{k+1} \quad \forall k \geq 1 \end{aligned} \tag{10}$$

At the beginning of the process T_k has a high value and the probability to accept one proposed solution is high. When T_k decreases this probability also decreases and only good solutions are accepted at the end of the process. In this regard every Markov chain makes a stochastic walk in the solution space until the stationary distribution is reached. Then a strong relation between the Markov chain length (L_k) and the cooling speed of Simulated Annealing exists: when $T_k \to \infty, L_k \to 0$ and when $T_k \to 0, L_k \to \infty$.

Because the Markov chains are built through a neighborhood sampling method, the maximum number of different solutions rejected at T_f when the current solution S_i is the optimal one, is the neighborhood size $|V_{S_i}|$. In this regard the maximum Markov chain length is a function of $|V_{S_i}|$. In general L_k can be established as:

$$L_k \leq L_{\max} = g(|V_{S_i}|) \tag{11}$$

In Equation 11, $L_{\max}$ is the Markov chain length when $T_k = T_f$, and $g(|V_{S_i}|)$ is a function that gives the maximum number of samples that must be taken from the neighborhood V_{S_i} in order to evaluate an expected fraction of different solutions at T_f . The value of $L_{\max}$ only depends on the number of elements of V_{S_i} that will be explored at T_f.

Usually a Simulated Annealing algorithm uses a uniform probability distribution function $G(T_k)$given by a random replacement sampling method to explore V_{S_i} at any temperature T_k, where $G(T_k)$ is established as follows:

$$G(T_k) = \begin{cases} \frac{1}{|V_{S_i}|} & \forall S_j \in V_{S_i} \\ 0 & \forall S_j \notin V_{S_i} \end{cases} \tag{12}$$

In this regard, the probability to get the solution S_j in N samples is:

$$P_A(S_j) = 1 - e^{-\frac{N}{|V_{S_i}|}} \tag{13}$$

Notice in Equation 13 that $P_A(S_j)$ may be understood as the expected fraction of different solutions obtained when N samples are taken. From Equation 13, N can be obtained as:

$$N = -\ln(1 - P_A(S_j))\,|V_{S_i}| \tag{14}$$

In Equation 14, we define:

$$= -\ln(1 - P_A(S_j)) = -\ln(P_R(S_j)) \tag{15}$$

You can see that $P_R(S_j) = 1 - P_A(S_j)$, $P_R(S_j)$ is the rejection probability. Constant C establishes the level of exploration to be done In this way different levels of exploration can be applied. For example: if a 99% of the solution space is going to be explored, the rejection probability will be $P_R(S_j) = 0.01$, so, from Equation 15 we obtain $C = 4.60$.

Definition 3. *The exploration set of the search space, Φ_C, is defined as follows:*

- *Given the set of probability of acceptance* $\Phi_{P_A} = \{70, 75, 80, 85, 90, 95, 99, 99.9, 99.99, 99.999, ...\}$
- *Using Equation 15:* $\Phi_C = \{1.20, 1.39, 1.61, 1.90, 2.30, 3.00, 4.61, 6.91, 9.21, 11.51, ...\}$

Then in any Simulated Annealing algorithm the maximum Markov chain length (when $T_k = T_f$) may be set as:

$$L_{max} = N = C|V_{S_i}| \tag{16}$$

Because a high percentage of the solution space should be explored, C varies from $1 \leq C \leq 4.6$ which guarantees a good level of exploration of the neighborhood at T_f.

When the process is at the beginning the temperature T_i is very high. This is because in the Boltzman distribution the acceptance probability is directly related with the cost increment $P_A = e^{-(\Delta J/T_k)}$; where T_k is the temperature parameter, therefore:

$$T_k = -\frac{\Delta J}{\ln(P_A)} \tag{17}$$

At the beginning of the process, P_A is close to one (normally 0.99, [21]) and the temperature is extremely high. Almost any solution is accepted at this temperature; as a consequence the stochastic equilibrium of a Markov cycle is reached with the first guess solution. Similarly, when the process is ending the acceptance probability (tipically 0.01) and the temperature closer to zero but the Metropolis cicle is very long.

For instance SAT values $\Delta J_{V_{max}}$ and $\Delta J_{V_{min}}$ in the energy of different states can be estimated at the beginning of the execution on the simulated annealing algorithm. To estimate these values, we can count the maximum number of Clauses containing any of the variables of the problem, the largest number of clauses that can change when we change the value of a variable, is an upper bound to change maximum of Energy and:

$$T_i = -\frac{\Delta J_{V_{\max}}}{\ln(P_A)} = -\frac{\text{max number of clauses}}{\ln(0.99)} \tag{18}$$

Similarly, the minimum of change Energy can be estimated by counting the clauses that are changed when creating a new neighbor and obtain the lowest of these values:

$$T_f = -\frac{\Delta J_{V_{\min}}}{\ln(P_A)} = -\frac{\text{min number of clauses}}{\ln(0.01)} \tag{19}$$

Some criticisms about Simulated Annealing are about the long time of execution of standard Boltzmann-type Simulated Annealing, has many times driven these projects to utilize a temperature schedule too fast to satisfy the sufficiency conditions required to establish a true ergodic search. In this chapter we use a logarithmic an exponential temperature schedule that is consistent with the Boltzmann algorithm follow:

$$T_k = T_0 e^{[(\alpha-1)k]}, 0 < \alpha < 1 \tag{20}$$

From Equation 20 we can obtain:

$$\frac{\Delta T}{\Delta k} = T_k(\alpha - 1), k \gg 1 \tag{21}$$

and

$$\Delta T = T_k(\alpha - 1)\Delta k, k \gg 1 \tag{22}$$

if in the previous expression Δk is equal to 1 then obtain the equation for two successive values of the temperature

$$T_{k+1} = \alpha T_k, 0 < \alpha < 1, k \gg 1 \tag{23}$$

where T_k is the "temperature," k is the "time" index of annealing [16, 17].

3.3. Simulated Annealing algorithm with the Markov chain Lenght dynamically

In [13, 20, 21] authors shown a strong relation between the cooling function and the length of the Markov chain exists. For the Simulated Annealing algorithm, the stationary distribution for each Markov chain is given by the Boltzmann probability distribution, which is a family of curves that vary from a uniform distribution to a pulse function.

At the very beginning of the process (with $T_k = T_i$), Simulated Annealing has a uniform distribution, henceforth any guess would be accepted as a solution. Besides any neighbor of the current solution is also accepted as a new solution. In this way when Simulated Annealing is just at the beginning the Markov chain length is really small, $L_k = Li \approx 1$. When running the temperature cycle of simulated annealing, for values of k greater than 1, the value of T_k is decremented by the cooling function [16], until the final temperature is reached ($T_k = T_f$):

$$T_{k+1} = \alpha T_k \tag{24}$$

In Equation 24 α is normally in the range of $[0.7, 0.99][1]$.

In this regard the length of each Markov chain must be incremented at any temperature cycle in a similar but in inverse way that T_k is decremented. This means that L_k must be incremented until L_{max} is reached at T_f by applying an increment Markov chain factor (β). The cooling function given by Equation 24 is applied many times until the final temperature T_f is reached. Because Metropolis cycle is finished when the stochastic equilibrium is reached, it can be also modeled as a Markov chain as follows:

$$L_{k+1} = \beta L_k \tag{25}$$

In previous Equation 25, L_k represents the length of the current Markov chain at a given temperature, that means the number of iterations of the Metropolis cycle for a T_k temperature. So L_{k+1} represents the length of the next Markov chain. In this Markov Model, β represents an increment of the number of iterations in the next Metropolis cycle.

If the cooling function given by Equation 24 is applied over and over, n times, until $T_k = T_f$, the next geometrical function is easily gotten:

$$T_f = \alpha^n T_i \tag{26}$$

Knowing the initial (T_i) and the final (T_f) temperature and the cooling coefficient (α), the number of times that the Metropolis cycle is executed can be calculated as:

$$n = \frac{\ln T_f - \ln T_i}{\ln \alpha} \tag{27}$$

If we make a similar process for increasing the equation of the Markov chain length, another geometrical function is obtained:

$$L_{\max} = \beta^n L_1 \tag{28}$$

Once n is known by Equation 27, the value of the increment coefficient (β) is calculated as:

$$\beta = e^{\left(\frac{\ln L_{\max} - \ln L_1}{n}\right)} \tag{29}$$

Once L_{max} (calculated form Equation 16), L_1 and β are known, the length of each Markov chain for each temperature cycle can be calculated using Equation 27. In this way L_k is computed dynamically from $L_1 = 1$ for T_i until L_{max} at T_f. First we can obtain T_i from Equation 18 and T_f from Equation 27, with both values and Equation 29 algorithm can calculate β [30].

In Figure 2 we can see the simulated annealing algorithm modifications using Markov chains described above. Below we will explain how we will use the linear regression for the simulated annealing algorithm run more efficiently without losing quality in the solution.

4. Linear Regresion Method (LRM)

We explain, in Section 3.2, how to estimate the initial and final temperature for SAT instances that will be provided to the simulated annealing algorithm to determine if it is satisfiable or not.

As shown in the Figure 3, the algorithm found Metropolis various configurations with different energy at a given temperature.

The typical behavior of the energy for a given temperature can be observed in Figure 3. We set out to determine when the cycle of Metropolis reaches the equilibrium although not all

1	Initializing: Initial solution S_i Initial and final temperature: T_i and T_f Calculate $n, \beta, L_{\max}$ $T = T_i$
2	Temperatures cycle: $L = L_1$
3	Metropolis cycle: Generating S_j from S_i $dif = J(S_j) - J(S_i)$ If $dif < 0$ then $S_i = S_j$ else if $e^{\frac{-dif}{T}} >$ rnd(0,1) $S_i = S_j$
4	Metropolis condition: $L = \beta L$ if $L = L_{max}$ goto 5 Else goto 3
5	Stop criterion: If $T_k = T_f$ End Else $T_k = \alpha T_k$ goto 2

Figure 2. Simulated Annealing algorithm with dinamically Markov chain

of the iterations required by Markov have been executed. In order to determine this zone in adaptive way, we will fit by least squares a straight line and will stop the Metropolis cycle if the slope of this line is equal or smaller than zero. This Linear Regression Method LRM is a well known method but never was applied to detect Metropolis equilibrium in Simulated Annealing.

Suppose that the data set consists of the points:

$$(x_i, y_i), i = 1, 2, 3, ..., n \tag{30}$$

We want to find a function f such that $f(x_i) \approx y_i$. To attain this goal, we suppose that the function f is of a particular form containing some parameters $(a_1, a_2, a_3, ..., a_m)$ which need to be determined.

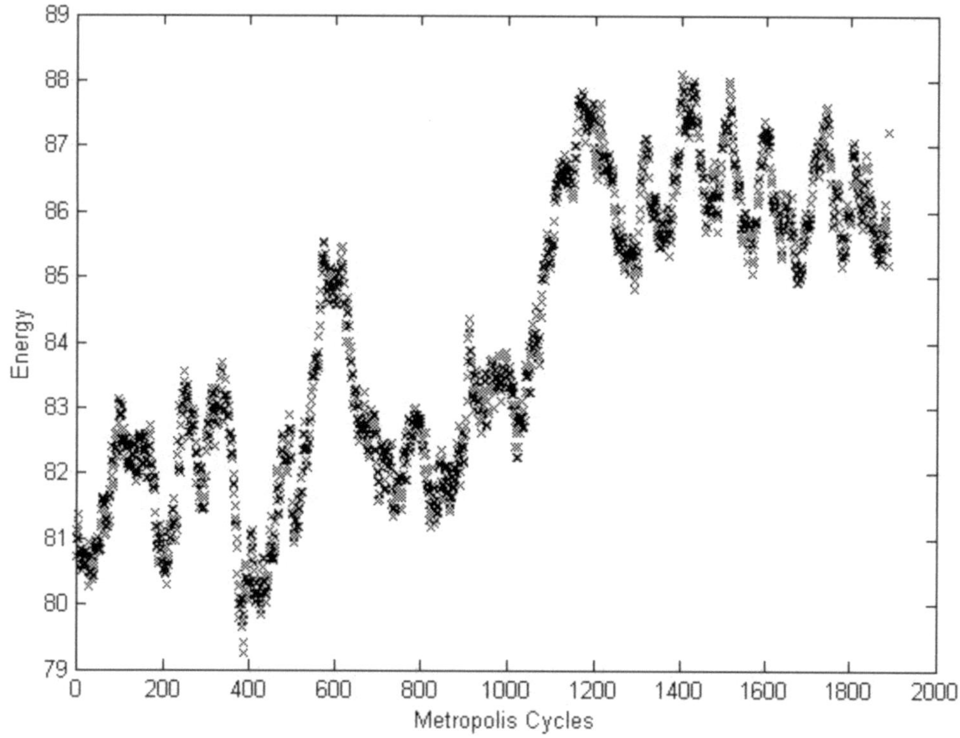

Figure 3. Energy of different states, explored in the Metropolis cycle, for a fixed temperature

In our problem:

$$y_i \approx f(x_i, a, b) = ax_i + b \tag{31}$$

In Equation 31 a and b are not yet known. In our problem $f(x_i, a, b) = f(i, a, b) = J_i$.

As usual, we now seek the values of a and b, that minimize the sum of the squares of the residuals as follows:

$$S = \sum_{i=1}^{n} [y_i - (ax_i + b)]^2 \tag{32}$$

As it is well known regression equations are obtained by differentiating S in Equation 32 with respect to each parameter a and b, and we obtain this system of linear equations:

$$a\sum_{i=1}^{n} x_i^2 + b\sum_{i=1}^{n} x_i = \sum_{i=1}^{n} x_i y_i \tag{33}$$

$$a\sum_{i=1}^{n} x_i + b\sum_{i=1}^{n} 1 = \sum_{i=1}^{n} y_i \tag{34}$$

In Equation 33 and Equation 34 we can define the following constants:

$$A = \sum_{i=1}^{n} x_i^2 \, , \; B = \sum_{i=1}^{n} x_i \, , \; C = \sum_{i=1}^{n} x_i y_i \, , \; D = \sum_{i=1}^{n} y_i \tag{35}$$

Then the system of equations (Equation 33 and 34) can be rewritten as:

$$aA + bB = C \tag{36}$$

$$aB + bn = D \tag{37}$$

We recall that parameter a, the slope of Equation 31 is:

$$a = \frac{Cn - BD}{An - B^2} \tag{38}$$

In our data $x_i = 1, 2, 3, ..., n$, then we can write:

$$A = \sum_{i=1}^{n} x_i^2 = \sum_{i=1}^{n} i^2 = \frac{n(n+1)(2n+1)}{6} \tag{39}$$

and

$$B = \sum_{i=1}^{n} x_i = \sum_{i=1}^{n} i = \frac{n(n+1)}{2} \tag{40}$$

in the same way:

$$C = \sum_{i=1}^{n} iJ_i \tag{41}$$

and

$$D = \sum_{i=1}^{n} J_i \tag{42}$$

By substitution of equations: 39, 40, 41 and 42; in Equation 38 finally we get the equation

$$a = \frac{Cn - BD}{n^3 - n} \tag{43}$$

In order to apply LRM to traditional Simulated Annealing, we apply the following strategy:

1. Metropolis cycle is running as usual, just as explained in the Section 3.3, using the maximum value of Markov chain length calculated by the Equation 28 L_{max}^{C}, C is calculated $P_A^i \in \Phi_{P_A}$
2. When the repeats of metropolis, L, are equal to L_{max}^{C-1} (L_{max}^{C-1} is calculated by Equation 28 with $P_A^{i-1} \in \Phi_{P_A}$).
3. If the value of the slope a, in the equation of the line (Equation 31), found by Equation 43 is close to zero, then stop the cycle of Metropolis although this has not reached the value L_{max}.

Notice from Equation 43 and for Figure 4, that the computation of LRM is $O(n)$ where n is the number of points taken to compute the slope. So the complexity of Simulated Annealing with LRM is not affected [19, 22].

5. Experimental results

In order to prove LRM algorithm we used the SAT instances in Table 1 and Table 2. Some of these instances were generated using the programs proposed by Horie et al. in 1997 [15] and

1	Initializing: Initial solution S_i Initial and final temperature: T_i and T_f Calculate $n, \beta, L_{\max}$ $T = T_i$
2	Temperatures cycle: $L = L_1$

3	Metropolis cycle:
	Generating S_j from S_i
	$dif = J(S_j) - J(S_i)$
	If $dif < 0$ then
	$\quad S_i = S_j$
	else if $e^{\frac{-dif}{T}} >$ rnd(0,1)
	$\quad S_i = S_j$
	If $L \geq L_{max}^{C-1}$ then
	$\quad$ calculate a
	..........If $a \approx 0$ then
	$\quad\quad L = L_{max}$
	..................goto 4

4	Metropolis condition: $L = \beta L$ if $L = L_{max}$ $\quad$ goto 5 Else $\quad$ goto 3
5	Stop criterion: If $T_k = T_f$ $\quad$ End Else $\quad T_k = \alpha T_k$ $\quad$ goto 2

Figure 4. Simulated Annealing algorithm with dinamically Markov chain and LRM

other are in SATLIB [14]. We generated several instances that had the same relation of clauses and variables σ [24, 25].

The measurement of efficiency of this algorithm was based on the execution time and we also obtained a solution quality measure (SQM), SQM is taken as the number of "true" clauses in an instance at the end of the program execution.

SAT problem	Id	Variables	Clauses	σ	SAT?
aim-100-1_6-yes1-1	a1	100	160	1.60	Yes
aim-50-1_6-yes1-3	a2	50	80	1.60	Yes
aim-200-1_6-no-1	a7	200	320	1.60	No
aim-50-1_6-no-2	a8	50	80	1.60	No
g2_V100_C200_P2_I1	g1	100	200	2.00	Yes
aim-50-2_0-no-4	a10	50	100	2.00	No
aim-50-2_0-yes1-1	a3	50	100	2.00	Yes
aim-50-2_0-no-3	a9	50	100	2.00	No
dubois21	d2	63	168	2.67	No
dubois26	d1	78	208	2.67	No
dubois27	d3	81	216	2.67	No
BMS_k3_n100_m429_161	b1	100	283	2.83	Yes
g2_V300_C900_P3_I1	g15	300	900	3.00	Yes
g2_V50_C150_P3_I1	g17	50	150	3.00	Yes
BMS_k3_n100_m429_368	b2	100	308	3.08	Yes
hole6	h2	42	133	3.17	No
par8-1	p1	350	1149	3.28	Yes
aim-50-3_4-yes1-2	a4	50	170	3.40	Yes
hole7	h3	56	204	3.64	No
par8-3-c	p2	75	298	3.97	Yes
par8-5-c	p3	75	298	3.97	Yes

Table 1. SAT instances for testing algorithms

Both algorithms: Simulated Annealing Algorithm with the Markov Chain Lenght dynamically (*SA_C*) and Simulated Annealing with Linear Regresion Method (*SA_LRM*), were implemented in Dell Lattitude with 1 Gb of Ram memory and Pentium 4 processor running at 2.13 GHz.

5.1. Experiment design

The experiments with these algorithms require a considerable run-time, because each instance SAT, is solved several times to take average values of performance.

Another important element to consider is to guarantee that the conditions of execution on the various algorithms are similar (because we measure time of execution on). In this regard, the first thing we did was on the Evaluation of a set of computers with similar hardware and software conditions.

To run this task, there were used programs available from the Internet, that perform different computers test and give us the result of this evaluation [29, 32, 33].

In Table 3, MOS means: million operations per second, MSS million strings per second and MBTS represent millions of bytes transferred per second. As we can see Table 3 the differences

SAT problem	Id	Variables	Clauses	σ	SAT?
g2_V100_C400_P4_I1	g3	100	400	4.00	Yes
hole8	h4	72	297	4.13	No
uuf225-045	u4	225	960	4.27	No
RTI_k3_n100_m429_150	r1	100	429	4.29	Yes
uf175-023	u1	175	753	4.30	Yes
uuf100-0789	u8	100	430	4.30	No
uf50-01	u7	50	218	4.36	Yes
uuf50-01	u9	50	218	4.36	No
ii8a2	i3	180	800	4.44	Yes
g2_V50_C250_P5_I1	g19	50	250	5.00	Yes
hole10	h1	110	561	5.10	No
ii32e1	i2	222	1186	5.34	Yes
anomaly	a6	48	261	5.44	Yes
aim-50-6_0-yes1-1	a5	50	300	6.00	Yes
g2_V50_C300_P6_I1s	g20	50	800	6.00	Yes
jnh201	j1	100	800	8.00	Yes
jnh215	j3	100	800	8.00	No
medium	m1	116	953	8.22	Yes
jnh301	j2	100	900	9.00	Yes

Table 2. SAT instances for testing algorithms, continuation

Computer	1	2	3	4	5	6	7	8	9
Floating Point Math (MOS)	168.1	168.3	168.0	168.0	168.2	168.1	167.9	168.2	168.2
Integer Maths (MOS)	33.61	33.56	33.60	33.59	33.65	33.62	33.56	33.65	33.62
Search for prime numbers (MOS)	126.6	126.6	126.4	126.5	126.7	126.7	126.3	126.6	126.6
String Sorting (MSS)	415.6	416.0	415.3	408.3	415.9	415.8	415.7	415.6	415.7
Memory blocks transfer (MbTS)	470.9	476.6	475.9	465.7	466.4	467.4	468.7	469.0	459.7
Cache memory read (MbTS)	1141	1142	1141	1141	1142	1141	1141	1142.	1088
Non cache memory read (MbTS)	831.8	831.7	832.0	831.2	830.7	831.1	830.6	831.6	831.2
Memory write (MbTS)	377.3	379.7	378.2	378.3	378.0	379.1	378.1	378.1	377.0
Overall calculation speed	232.5	232.6	232.1	232.0	232.4	232.5	232.1	232.5	232.6
Overall memory speed	209.9	210.8	210.4	209.6	209.5	210.0	209.8	209.8	209.9

Table 3. Performance of computers used for experiments

between computers are at most equal to 1.6 percent, so we can infer that we obtain similar results in one or another computer.

Each SAT instance was executed 100 times with a slow cooling function (0.99 in Equation 20 and we obtained the average time of the executions and the average quality of the solution.

The SQM is established by the next expression:

$$SQM = \frac{\text{clauses true}}{\text{total clauses}} \times 100 \quad (44)$$

Both results, SA_C and SA_LRM, were compared using two quotients which we denominated time improvement Q_{time} and quality improvement $Q_{quality}$ defined by:

$$Q_{time} = \frac{AverageTime_{SA\_LRM}}{AverageTime_{SA\_C}} \times 100 \quad (45)$$

$$Q_{quality} = \frac{SQM_{SA\_LRM}}{SQM_{SA\_C}} \times 100 \quad (46)$$

If $Q_{quality}$ is close to 100% this means that both algorithm found good solutions, however $Q_{quality}$ factor must decrease, which implies that the new algorithm SA_LRM is faster than SA_C.

	$L^C_{max} = 4.61$ $L^{C-1}_{max} = 3.00$		$L^C_{max} = 3.00$ $L^{C-1}_{max} = 2.30$		$L^C_{max} = 2.30$ $L^{C-1}_{max} = 1.96$	
Instance	$Q_{quality}$	Q_{time}	$Q_{quality}$	Q_{time}	$Q_{quality}$	Q_{time}
a1	99.4	13.0	99.3	12.9	99.2	11.7
a10	99.0	11.0	99.2	11.6	99.6	9.7
a2	99.5	11.4	98.7	11.2	99.0	9.8
a3	99.9	11.3	99.4	11.6	99.2	9.6
a4	99.2	9.8	99.6	10.4	98.9	8.8
a5	99.5	10.1	99.1	10.6	99.5	9.0
a6	99.0	34.5	99.3	32.1	99.2	27.8
a7	98.7	13.3	99.1	13.9	98.8	11.8
a8	99.6	11.3	99.2	12.2	99.4	9.8
a9	99.2	10.9	98.7	11.6	99.4	9.5
b1	99.6	54.0	99.7	51.0	99.9	46.0
b2	100.2	57.7	99.8	53.5	99.9	49.2
d1	99.2	11.8	99.4	11.6	99.2	10.1
d2	99.5	11.6	99.5	11.5	99.4	10.1
d3	100.0	11.5	99.1	11.3	99.1	9.5
g1	99.8	71.7	99.9	69.7	99.8	64.7
g15	99.7	28.7	99.7	28.3	99.9	29.0
g17	99.3	11.7	99.2	12.7	99.5	10.5
g19	99.6	11.5	98.6	11.5	99.3	9.8.0
g20	99.5	10.5	99.5	11.1	99.4	9.5.0

Table 4. Experimentals results

	$L^{C}_{max} = 4.61$ $L^{C-1}_{max} = 3.00$		$L^{C}_{max} = 3.00$ $L^{C-1}_{max} = 2.30$		$L^{C}_{max} = 2.30$ $L^{C-1}_{max} = 1.96$	
Instance	$Q_{quality}$	Q_{time}	$Q_{quality}$	Q_{time}	$Q_{quality}$	Q_{time}
g3	99.8	20.8	99.6	21.2	99.6	18.4
h1	100.1	54.5	99.6	53.5	99.9	47.1
h2	99.7	22.0	99.3	23.2	99.5	19.1
h3	100.1	30.5	99.7	29.7	99.5	24.9
h4	98.9	41.0	99.3	38.5	99.0	32.7
i2	99.1	55.0	98.9	54.4	99.7	45.7
i3	100.0	65.0	100.2	65.4	100.3	58.7
j1	99.7	11.6	99.7	12.8	99.6	11.1
j2	99.7	11.6	99.8	12.3	99.8	10.4
j3	99.5	11.3	99.7	12.3	99.8	10.2
m1	100.5	74.5	99.9	70.9	99.7	66.8
p1	99.9	66.0	99.9	65.4	99.9	59.7
p2	99.4	13.8	99.0	13.3	99.3	11.8
p3	99.3	14.0	99.5	12.4	99.5	11.0
r1	99.4	24.0	99.7	22.6	99.5	21.1
u1	99.6	49.6	100.0	49.8	100.0	44.5
u4	99.7	53.3	100.0	50.9	99.3	48.1
u7	99.9	14.7	100.0	15.2	99.5	13.1
u8	99.7	26.0	100.0	23.9	99.6	20.0
u9	99.1	13.8	99.5	14.0	99.2	11.9

Table 5. Experimentals results, continuation

From Table 4 and Table 5 experimental results we can obtain the average values for the magnitudes Q_{time} and $Q_{quality}$, as shown in the following Table 6.

As you can see, in Table 6, the quality factor of the solutions, $Q_{quality}$ is very close to 100%, which implies that the SA_LRM algorithm finds solutions as good as the SA_C algorithm, it is important to note that 37% of SAT instances used for the experiments, are not-SAT, which implies that their respective SQM can not be equal to 100% and therefore the quality factor must be less than 100%.

Also in Table 6, we see that the factor Q_{time} diminishes values less than 30%, showing that our algorithm, SA_LRM, is 70% faster than the SA_C algorithm but maintains the same quality of the solution.

As shown in Figure 5 are some instances in which the reduction of run time is only 25% while other reducing runtime up to 90%.

L_{max}^{C}	L_{max}^{C-1}	$Q_{time}(\%)$	$Q_{quality}(\%)$
4.61	3.00	99.6	27.3
3.00	2.30	99.5	26.8
2.30	1.96	99.5	23.8

Table 6. Q_{time} and $Q_{quality}$ averages for all instances tested

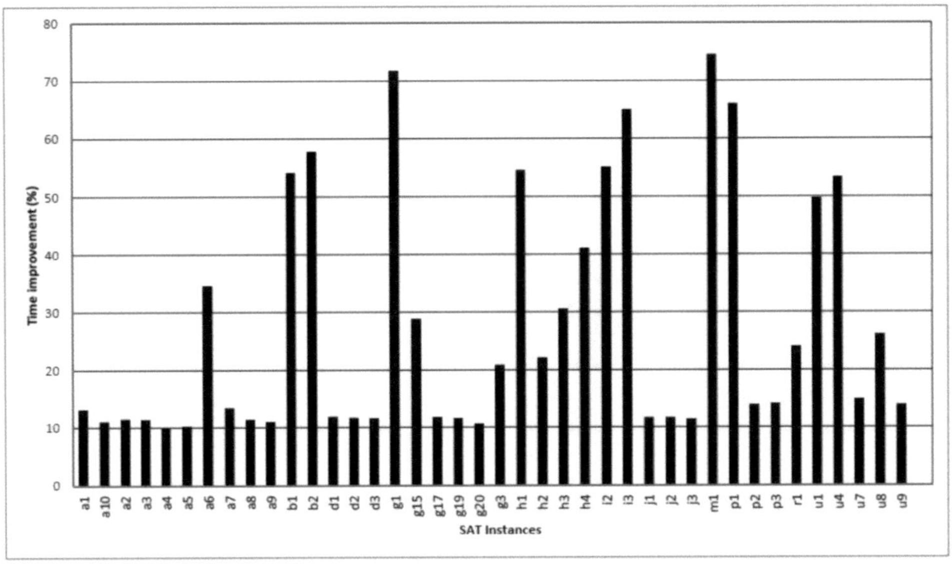

Figure 5. Q_{time} for SAT instances

6. Conclusions

In this paper a new adaptive Simulated Annealing algorithm named SA_LRM that uses least squares method as a way to find the equilibrium zone in Metropolis cycle is presented. When this zone is found our algorithm abort the Metropolis cycle, although the iterations calculated with the dynamic chains of Markov have not been completed. After experimentation we show that SA_LRM is more efficient than those tuned using only an analytical method.

Author details

Felix Martinez-Rios
Universidad Panamericana, México

Juan Frausto-Solis
UPMOR, México

7. References

[1] Aarts, E. & Korst, J. [1989]. *Simulated annealing and Boltzman machines: An stochastic approach to combinatorial optimization and neural computing*, John Wiley and Sons.

[2] Aspvall, B. & Tarjan, M. F. P. R. E. [1979]. Alinear-time algorithm for testing the truth of certain quantified boolean formulas, *Information Processing Letters* 8(3).

[3] Atiqullah, M. [2004]. An efficient simple cooling schedule for simulated annealing, 3045: 396–404.

[4] Bertsimas, D. & Tsitsiklis, J. [1993]. Simulated annealing, *Statistical Science* 8: 10–15.

[5] Cerny, V. [1982]. A thermodynamical approach to the travelling salesman problem: An efficient simulation algorithm, *Comenius University* .

[6] Cerny, V. [1985]. Thermodynamical approach to the traveling salesman problem: an efficient simulation algorithm, *Journal of Optimazation Theory and Applications* 45(1).

[7] Cook, S. A. [1971]. *The complexity of theorem proving procedures*, Proceedings on the third annual ACM symposium on theory of computing.

[8] Crescenzi, P. & Kann, V. [1998]. How to find the best approximation results - a follow-up to garey and johnson, *ACM SIGACT News* 29(4): 90–97.

[9] Dowling, W. F. & Gallier, J. H. [1984]. Linear-time algorithms for testing the satisfiability of propositional horn formulae, *Journal of Logic Programming* 1(3): 267–284.

[10] Even, S., Itai, A. & Shamir, A. [1976]. On the complexity of timetable and multicommodity flow problems, *SIAM Journal on Computing* 5(4): 691–703.
URL: *http://link.aip.org/link/?SMJ/5/691/1*

[11] Faigle, U. & Kern, W. [1991]. Note on the convergence of simulated annealing algorithms, *SIAM journal on control and optimization* 29(1): 153–159.
URL: *http://e-archive.informatik.uni-koeln.de/67/*

[12] Fleischer, M. A. [1996]. Cybernetic optimization by simulated annealing: Accelerating convergence by parallel processing and probabilistic feedback control, *Journal of Heuristics* 1: 225–246.

[13] Frausto-Solis, J., Sanvicente, H. & Imperial, F. [2006]. Andymark: An analytical method to establish dynamically the length of the markov chain in simulated annealing for the satisfiablity problem, *Springer Verlag* .

[14] Hoos, H. H. & Stutzle, T. [2000]. Satlib: An online resource for research on sat, *SAT 2000* pp. 283–292. Disponible en http://www.satlib.org/.

[15] Horie, S. & Watanabe, O. [1997]. Hard instance generation for sat, *ISAAC '97: Proceedings of the 8th International Symposium on Algorithms and Computation*, Springer-Verlag, pp. 22–31.

[16] Ingber, L. [1993]. Simulated annealing: Practice versus theory, *Mathematical and Computer Modelling* 18(11): 29 – 57.

[17] Ingber, L. [1996]. Adaptive simulated annealing (asa): Lessons learned, *Control and Cybernetics* 25: 33–54.

[18] Kirkpatrick, S., Gelatt, C. D. & Vecchi, M. P. [1983]. Optimization by simulated annealing, *Science* (4598)(220): 671–680.

[19] Martinez-Rios, F. & Frausto-Solis, J. [2007]. A hybrid simulated annealing threshold accepting algorithm for satisfiability problems using dynamically cooling schemes, *Electrical and Computer Engineering Series WSEAS* pp. 282–286.

[20] Martinez-Rios, F. & Frausto-Solis, J. [2008a]. Golden annealing method for job shop scheduling problem, *Mathematics and Computers in Science and Engineering, ISSN 1790-2769* .

[21] Martinez-Rios, F. & Frausto-Solis, J. [2008b]. Golden ratio annealing for satisfiability problems using dynamically cooling schemes, *Lecture Notes in Computer Science* 4994: 215–224.

[22] Martinez-Rios, F. & Frausto-Solis, J. [2008c]. Simulated annealing for sat problems using dynamic markov chains with linear regression equilibrium, *MICAI 2008, IEEE* pp. 182–187.

[23] Metropolis, N., Rosenbluth, A. W. R. M. N. & Teller, A. H. [1953]. Equation of state calculations by fast computing machines, *The journal of Chemicla Physics* 21: 1087–1092.

[24] Mezard, M., Parisi, G. & Zecchina, R. [2002]. Analytic and algorithmic solution of random satisfiability problems, *Science* 297(5582): 812–815.

[25] Mezard, M. & Zecchina, R. [2002]. The random k-satisfiability problem: from an analytic solution to an efficient algorithm, *Phys. Rev.* E66(056126).

[26] Miki, M., Hiroyasu, T. & Ono, K. [2002]. Simulated annealing with advanced adaptive neighborhood, *Second international workshop on Intelligent systems design and application*, Dynamic Publishers, Inc., pp. 113–118.

[27] Munakata, T. & Nakamura, Y. [2001]. Temperature control for simulated annealing, *PHYSICAL REVIEW E* 64.

[28] Papadimitriou, C. H. [1994]. *Computational Complexity*, Addison-Wesley.

[29] PassMark [2007]. Cpu burnintest. http://www.passmark.com/download/index.htm. URL: *http://www.passmark.com/download/index.htm*

[30] Sanvicente-Sanchez, H. & Frausto-Solis, J. [2004]. A method to establish the cooling scheme in simulated annealing like algorithms, *Lecture Notes in Computer Science* .

[31] Schaefer, T. J. [1978]. The complexity of satisfiability problems, *Proceedings of the tenth annual ACM symposium on Theory of computing*, STOC '78, ACM, pp. 216–226. URL: *http://doi.acm.org/10.1145/800133.804350*

[32] van Wandelen, C. J. [2007]. Cpubench. http://cpubench.softonic.com/eie/21660. URL: *http://cpubench.softonic.com/eie/21660*

[33] Vorte, B. [2007]. Cpumathmark. http://files.aoaforums.com/code.php?file=931. URL: *http://files.aoaforums.com/code.php?file=931*

Chapter 3

Optimization by Use of Nature in Physics Beyond Classical Simulated Annealing

Masayuki Ohzeki

Additional information is available at the end of the chapter

http://dx.doi.org/10.5772/50636

1. Introduction

We prefer to find the most appropriate choice in daily life for convenience and efficiency. When we go to a destination, we often use a searching program to find the fastest way, the minimum-length path, or most-reasonable one in cost. In such a searching problem, we mathematically design our benefit as a multivariable function (cost function) depending on many candidates and intend to maximize it. Such a mathematical issue is called the optimization problem. Simulated annealing (SA) is one of the generic solvers for the optimization problem [14]. We design the lowest-energy state in a physical system, which corresponds to the minimizer/maximizer of the cost function. The cost function to describe the instantaneous energy of the system is called as the Hamiltonian $H_0(\sigma_1, \sigma_2, \cdots, \sigma_N)$, where σ_i is the degrees of freedom in the system and N is the number of components related with the problem size. The typical instance of the Hamiltonian is a form of the spin glass, which is the disordered magnetic material, since most of the optimization problems with discrete variables can be rewritten in terms of such a physical system,

$$H_0(\sigma_1, \sigma_2, \cdots, \sigma_N) = -\sum_{\langle ij \rangle} J_{ij}\sigma_i\sigma_j, \tag{1}$$

where σ_i indicates the direction of the spin located at the site i in the magnetic material as $\sigma_i = \pm 1$. The summation is taken over all the connected bonds (ij) through the interaction J_{ij}. The configuration of J_{ij} depends on the details of the optimization problem.

Then we introduce an artificial design of stochastic dynamics governed by the master equation.

$$\frac{d}{dt}P(\sigma;t) = \sum_{\sigma'} M(\sigma|\sigma';t)P(\sigma';t), \tag{2}$$

where $P(\sigma;t)$ is the probability with a specific configuration of σ_i simply denoted as σ at time t. The transition matrix is written as $M(\sigma'|\sigma;t)$, which obeys the conservation of probability $\sum_\sigma M(\sigma|\sigma';t) = 1$ and the detailed balance condition

$$M(\sigma|\sigma';t)P_{\rm eq}(\sigma';t) = M(\sigma'|\sigma;t)P_{\rm eq}(\sigma;t). \tag{3}$$

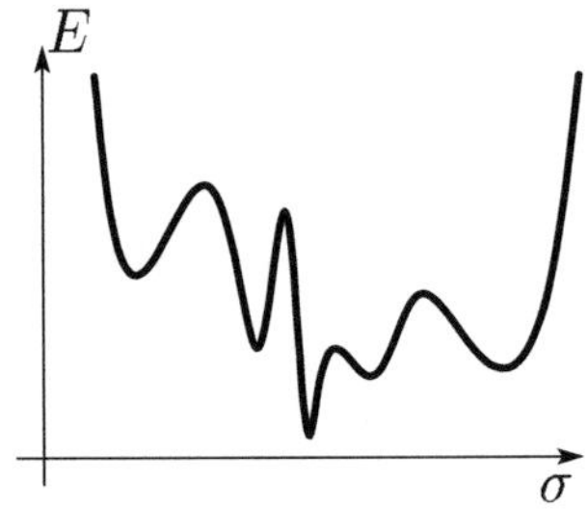

Figure 1. Energy structure of a spin-glass system. We map the possible 2^N configurations to the one-dimensional horizontal axis for simplicity. The vertical axis represents the value of the energy for each configuration.

Here we denote the instantaneous equilibrium distribution (Gibbs-Boltzmann distribution) as

$$P_{\rm eq}(\sigma;t) = \exp(-\beta(t)E(\sigma;t))/Z_t, \tag{4}$$

where the instantaneous energy $E(\sigma;t)$ is the value of the Hamiltonian H_0 and Z_t denotes a normalization factor termed as the partition function. In order to satisfy these conditions, we often use the transition matrix with Metropolis rule as

$$M(\sigma|\sigma';t) = \min(1, \exp(-\beta\Delta E(\sigma|\sigma';t))), \tag{5}$$

where

$$\Delta E(\sigma|\sigma';t) = E(\sigma;t) - E(\sigma';t), \tag{6}$$

or heat-bath rule as

$$M(\sigma|\sigma';t) = \delta_1(\sigma,\sigma')\frac{\exp\left(-\frac{\beta}{2}\Delta E(\sigma|\sigma';t)\right)}{2\cosh\left(\frac{\beta}{2}\Delta E(\sigma|\sigma';t)\right)}, \tag{7}$$

where

$$\delta_1(\sigma,\sigma') = \delta(2, \sum_{i=1}^{N}(1-\sigma_i\sigma_i')). \tag{8}$$

The master equation simulates behavior of relaxation toward a specific distribution associated with the energy of the system. If we evolve the system for a long time with a virtual parameter $\beta(t)$ being a constant β, which is the inverse temperature in context of physics, the probability distribution converges to the equilibrium distribution. To generate lower energy state, let us set $\beta \gg 1$. Unfortunately, the spin-glass system in the low-temperature often exhibits the extremely long time for equilibration of the system. The most relevant reason is on the complicated structure of the energy of the spin-glass system as schematically depicted in Fig. 1. There are barriers between the valleys to avoid hopping from state to state in the low-temperature, where the energy effect is dominant. Therefore it is difficult to reach the equilibrium distribution by a direct simulation with a constant temperature. Instead, by tuning a virtual parameter $\beta(t)$ from zero to a large number, we perform stochastic searching while keeping to trace the instantaneous equilibrium distribution. The mathematical guarantee to converge to the instantaneous equilibrium state by gradually changing the inverse temperature has been proved by Geman and Geman based on the classical inhomogenious (time-dependent) Markov chain representing nonequilibrium

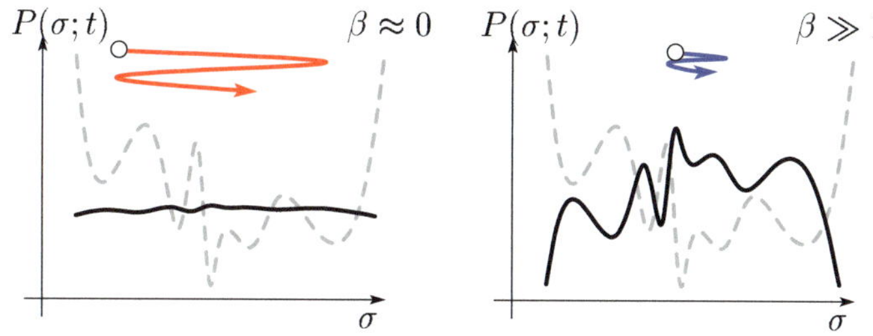

Figure 2. Behavior of the instantaneous state in SA. The left panel shows the stochastic searching in SA in the high-temperature region $\beta \approx 0$, while the right one describes that in the low-temperature one. The dashed curves express the structure of the energy as in Fig. 1, while the thick ones denote the probability for realization of each configuration schematically.

processes [6]. The convergence theorem states that we reach the equilibrium distribution, if we obey the following schedule or slower rate to change the inverse temperature as

$$\beta(t) = \frac{1}{pN} \log(\alpha t + 1), \tag{9}$$

where α is exponentially small in N and p is a constant independence of N. An intuitive way to understand the performance of SA is as follows. The inverse temperature controls the range of searching, roughly speaking. The instantaneous state keeps hopping from state to state in a relatively high-temperature region as depicted in Fig. 2. By gradually decrease of the inverse temperature, we narrow the range of searching. The lower energy state means its realization with a higher possibility following the instantaneous equilibrium distribution as in Fig. 2. Demand of a sufficiently slow control of the inverse temperature implies that we need enough time to find the states with relatively lower energies by the stochastic searching before the barrier avoids globally searching for the lower energy state.

Basically, SA is based on the behavior closely to the instantaneous equilibrium state. Therefore we need to perform the change of the inverse temperature with a sufficiently slow control. In order to improve the performance, in particular to shorten the necessary time, we need to consider the protocol away from the equilibrium state, that is nonequilibrium process.

In this chapter, we show a novel direction to solve efficiently the optimization problem by use of the nature in nonequilibrium physics. In statistical physics, the interest of researchers in nonequilibrium dynamical behavior has increased. Among several remarkable developments, the Jarzynski equality (JE), which is known as a generalization of the second law of thermodynamics, might be possible to change the paradigm in optimization problem by use of the physical nature. The Jarzynski equality relates an average over all realizations during a predetermined nonequilibrium process with an expectation in an equilibrium state. As seen later, the mathematical structure of JE does not depend on the schedule and the rate of changing the external parameter. It means that, if we implement JE to solve the optimization problem, we do not need to demand slow control of the driver of the system. The challenge of the implementation of JE have been performed in several researchers. Although not yet have been studied the performance in the actual application to the optimization problem, we show the possibility of the novel method from several analyses.

2. Population annealing

We introduce a couple of theories in nonequilibrium statistical physics in short, before we show the actual application. They provide the supplement to make the protocol of SA faster. The Jarzynski equality is the most important key.

2.1. Jarzynski equality

Among several recent developments in nonequilibrium statistical mechanics, we take JE as an attempt to improve the performance of SA. The Jarzynski equality relates quantities at two different thermal equilibrium states with those of nonequilibrium processes from $t = 0$ to $t = T$ connecting these two states as [10, 11]

$$\left\langle e^{-\beta W} \right\rangle_{0\to T} = \frac{Z_T}{Z_0}, \tag{10}$$

where the partition functions appearing in the ratio on the right-hand side are for the initial ($t = 0$) and final Hamiltonians ($t = T$). The quantity on the right-hand side can be represented by the exponentiated difference of the free energy $\exp(-\beta\Delta F)$ between the initial and final conditions. The brackets on the left-hand side express the nonequilibrium average over all the instantaneous realizations of the degrees of freedom, for instance spin configurations, during the nonequilibrium process with the following path probability defined as

$$P_{0\to T}(\{\sigma_t\}) = \prod_{k=0}^{n-1} \left\{ e^{\delta t M(\sigma_{k+1}|\sigma_k;t_k)} \right\} P_{\rm eq}(\sigma_0; t_0). \tag{11}$$

It implies that the observations of the nonequilibrium behavior can estimate the equilibrium quantity represented by the partition functions, that is the free energy. This equality is regarded as a generalization of the well-known inequality, the second law of thermodynamics $\langle W \rangle_{0\to T} \geq \Delta F$, which can be reproduced by the Jensen inequality. One of the important features is that JE holds independently of the pre-determined schedule of the nonequilibrium process.

In order to consider the improvement of SA, let us apply the nonequilibrium process with change of the temperature. We then have to employ the pseudo work instead of the ordinary performed work due to the energy difference as

$$Y(\sigma; t_k) = (\beta_{k+1} - \beta_k)\, E(\sigma), \tag{12}$$

where we use discrete time expressions as $t_0 = 0$ and $t_n = T$ for simplicity and we assume that the instantaneous energy does not depend on the time as $E(\sigma)$. The Jarzynski equality holds formally in the case with change of temperature,

$$\left\langle e^{-Y} \right\rangle_{0\to T} = \frac{Z_T}{Z_0}. \tag{13}$$

We show a simple proof of JE for the particular dynamics in SA. Let us consider a nonequilibrium process in a finite-time schedule governed by the master equation. The left-hand side of JE is written as

$$\left\langle e^{-Y} \right\rangle_{0\to T} = \sum_{\{\sigma_k\}} \prod_{k=0}^{n-1} \left\{ e^{-Y(\sigma_{k+1};t_k)} e^{\delta t M(\sigma_{k+1}|\sigma_k;t_k)} \right\} P_{\rm eq}(\sigma_0; t_0). \tag{14}$$

where we use the formal solution of the master equation by the exponentiated transition matrix. We take the first product of the above equation as,

$$\sum_{\sigma_0}\left\{e^{-Y(\sigma_1;t_0)}e^{\delta t M_0(\sigma_1|\sigma_0;t_0)}\right\}P_{\rm eq}(\sigma_0;t_0)$$

$$= P_{\rm eq}(\sigma_1;t_1)\frac{Z_{t_1}}{Z_0}. \tag{15}$$

Repetition of the above manipulation in Eq. (14) yields the quantity in the right-hand side of JE as,

$$\sum_{\sigma_n}P_{\rm eq}(\sigma_n;t_n)\prod_{k=0}^{n-1}\frac{Z_{t_{k+1}}}{Z_{t_k}} = \frac{Z_T}{Z_0}. \tag{16}$$

2.2. Fluctuation theorem

The Jarzynski equality is a consequence from the fluctuation theorem [2–4], which relates the probability $P_{0\to T}(\{\sigma_t\})$ with that of the inverse process $P_{T\to 0}(\{\sigma_t\})$ as

$$\frac{P_{0\to T}(\{\sigma_t\})}{P_{T\to 0}(\{\sigma_t\})}e^{-Y} = \frac{Z_T}{Z_0}. \tag{17}$$

This leads to the more generic result, for an observable $O(\{\sigma_t\})$ depending on the instantaneous spin configurations,

$$\left\langle O(\{\sigma_t\})e^{-Y}\right\rangle_{0\to T} = \langle O_{\rm r}(\{\sigma_t\})\rangle_{T\to 0}\frac{Z_T}{Z_0}, \tag{18}$$

where $O_{\rm r}$ denotes the observable which depends on the backward process $T \to 0$. The brackets with subscript $0 \to T$ express the average with the weight $P_{0\to T}(\{\sigma_t\})$ over possible realizations $\{\sigma_t\}$. For $O = O_{\rm r} = 1$, Eq. (18) reduces to JE.

If we choose an observable depending only on the final state, which is denoted as O_T, instead of $O(\{\sigma_t\})$, $O_{\rm r}$ reads an observable at the initial state in the backward process. Then $\langle O_{\rm r}\rangle_{T\to 0}$ equals to the ordinary thermal average at the initial equilibrium state with β_T represented by $\langle\cdots\rangle_{\beta_T}$, and Eq. (18) leads to

$$\left\langle O_T e^{-Y}\right\rangle_{0\to T} = \langle O\rangle_{\beta_T}\frac{Z_T}{Z_0}. \tag{19}$$

By looking over the above calculations, we can understand the roll of the exponentiated pseudo work. The resultant distribution after SA is given by $P_{0\to T}(\{\sigma_t\})$. The biased distribution with the exponentiated pseudo work, $P_{0\to T}(\{\sigma_t\})\exp(-Y)$ always gives the equilibrium one. In this sense, the exponentiated pseudo work plays a roll to fill the gap between the equilibrium distribution and the resultant one after SA. Therefore, if we skillfully use the exponentiated pseudo work to keep the instantaneous distribution close to the equilibrium one, we can invent the improved version of SA.

2.3. Population annealing

We introduce an improvement of SA by use of the property of JE. Let us consider to implement JE in numerical simulation. We parallelize the instantaneous spin configurations $\{\sigma_t\}$ as $\{\sigma_t\}_{i=1,\cdots,C}$. Each of the configuration independently will be evolved by the master equation. We regard the exponentiated pseudo work on the left-hand side of JE, Eq. (14), as the weight for each realization of the configuration. By computing the pseudo work for each realization and multiplying the weight given by the exponentiated pseudo work, we simultaneously perform the stochastic dynamics governed by the master equation. At the last stage of repetition of the above procedure, we obtain the ratio of the partition function as in the right-hand side of JE. In order to calculate its value, we estimate the empirical average as, after parallel computing of the master equation,

$$\frac{1}{C}\sum_{i=1}^{C}\exp\left(-\sum_{k=1}^{n}Y(t_k;\sigma)\right). \tag{20}$$

While estimating the ratio of the partition functions by JE, implementation of Eq. (19) gives the thermal average of the observable through their ratio. This is the typical implementation of JE in a numerical simulation, which is called as population annealing (PA) [7, 9, 17] as depicted in Fig. 3,

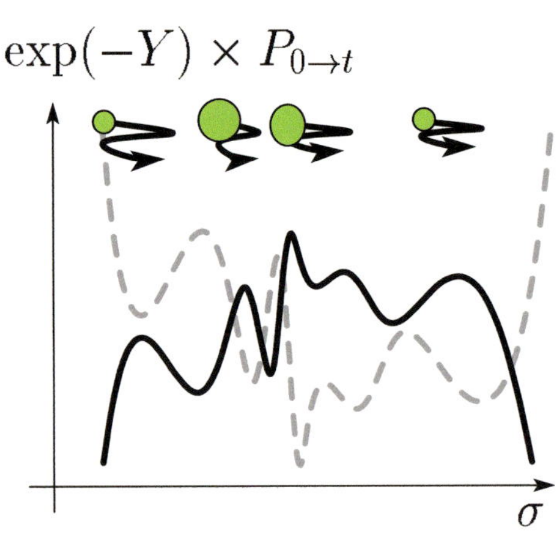

Figure 3. Schematic picture of the process of PA by $C = 4$. The size of the circles denotes the weight given by the multiplication of the exponentiated pseudo work during PA.

We remark that, as proposed in the literatures [7, 9], we have to employ a skillful technique, resampling, to efficiently generate the relevant copies to estimate the nonequilibrium average and maintain the stability of the method. The population annealing with resampling method indeed shows outstanding performance comparable to a standard technique to equilibrate the spin-glass system known as the exchange Monte Carlo method [8]. If we successfully generate the equilibrium distribution in the low-temperature region, we efficiently find the lowest energy state, which corresponds to the optimal solution in context of the optimization problem. Therefore PA is also relevant for the improvement of SA as a solver for the optimization problem. The advantage of PA is cutting the computational time compared to SA, since PA follows the property of JE. It means that we find the optimal solution by use of PA faster than SA.

Below, we propose an ambitious use of PA to evaluate the equilibrium property in the low-temperature in spin glasses by use of the special symmetry [21].

2.4. Spin glass

At first we briefly review a useful analysis in several spin-glass systems, which provides a powerful technique to discuss the possibility of PA. Let us consider a simple model of spin glasses, the $\pm J$ Ising model, on an arbitrary lattice. The Hamiltonian is the same form as Eq. (1) as

$$H = -\sum_{\langle ij\rangle} J\tau_{ij}\sigma_i\sigma_j, \tag{21}$$

where we extract the sign of the interaction τ_{ij} following the distribution function as

$$P(\tau_{ij}) = p\delta(\tau_{ij}-1) + (1-p)\delta(\tau_{ij}+1). \tag{22}$$

The partition function, which is the most important quantity through the free energy, is defined as

$$Z(K;\{\tau_{ij}\}) = \sum_{\{\sigma_i\}}\prod_{\langle ij\rangle}\exp(K\tau_{ij}\sigma_i\sigma_j). \tag{23}$$

The free energy is then given by

$$-\beta F(K;\{\tau_{ij}\}) = \log Z(K;\{\tau_{ij}\}), \tag{24}$$

where the product βJ is rewritten as K. Both of the above quantities depend on the specific configuration $\{\tau_{ij}\}$. In order to evaluate the physical property of spin glasses in equilibrium, we strive the difficult task to deal with the free energy depending on the non-uniform interactions. Instead of the direct manipulation, the averaged quantity over all the possible configurations of $\{\tau_{ij}\}$ may be considered based on the self-averaging property as, in the large-limit N,

$$\frac{1}{N}F(K;\{\tau_{ij}\}) \to \frac{1}{N}\left[F(K;\{\tau_{ij}\})\right], \tag{25}$$

where the square bracket denotes the average over all the combinations of $\{\tau_{ij}\}$ (configurational average). The self-averaging property is valid for other observables, which can be obtained from the free energy per site like the internal energy.

2.5. Gauge transformation

Here let us define a local transformation by the simultaneous change of the interactions and spin variables as, by the binary variables $\epsilon_i = \pm 1$ [18, 19]

$$\tau_{ij} \to \epsilon_i\epsilon_j\tau_{ij} \tag{26}$$

$$\sigma_i \to \epsilon_i\sigma_i. \tag{27}$$

This is called as the gauge transformation. Notice that the gauge transformation does not alter the value of the physical quantity given by the double average over τ_{ij} and σ_i since it changes only the order of the summations. The Hamiltonian can not change its form after the gauge transformation since the right-hand side is evaluated as

$$-\sum_{\langle ij \rangle} J\tau_{ij}\epsilon_i\epsilon_j\epsilon_i\sigma_i\epsilon_j\sigma_j = H, \tag{28}$$

As this case, if the physical quantity is invariant under the gauge transformation (gauge invariant), we can evaluate its exact value even for finite-dimensional spin glasses. The key point of the analyses by the gauge transformation is on the form of the distribution function. Before performing the gauge transformation, the distribution function can take the following form as

$$P(\tau_{ij}) = \frac{\mathrm{e}^{K_p\tau_{ij}}}{2\cosh K_p}, \tag{29}$$

where $\exp(-2K_p) = (1-p)/p$. The gauge transformation changes this into

$$P(\tau_{ij}) = \frac{\mathrm{e}^{K_p\tau_{ij}\epsilon_i\epsilon_j}}{2\cosh K_p}. \tag{30}$$

Let us evaluate the internal energy by aid of the gauge transformation here. The thermal average of the Hamiltonian is given by

$$\langle H \rangle_K = \sum_{\{\sigma_i\}} \frac{1}{Z(K;\{\tau_{ij}\})} H \prod_{\langle ij \rangle} \exp\left(K\tau_{ij}\sigma_i\sigma_j\right) \tag{31}$$

$$= -J\frac{d}{dK}\log Z(K;\{\tau_{ij}\}). \tag{32}$$

We can use the self-averaging property here and thus take the configurational average as

$$[\langle H \rangle_K]_{K_p} = \sum_{\{\tau_{ij}\}} \prod_{\langle ij \rangle} \frac{\exp(K_p\tau_{ij})}{2\cosh K_p} \times \langle H \rangle_K. \tag{33}$$

where $[\cdots]_{K_p}$ denotes the configurational average by the distribution function (29) with K_p. Then we perform the gauge transformation, which does not change the value of the internal energy due to gauge invariance,

$$[\langle H \rangle_K]_{K_p} = \sum_{\{\tau_{ij}\}} \prod_{\langle ij \rangle} \frac{\exp(K_p\tau_{ij}\epsilon_i\epsilon_j)}{2\cosh K_p} \times \langle H \rangle_K. \tag{34}$$

Therefore we here take the summation over all the possible configurations of $\{\sigma_i\}$ and divide it by 2^N (the number of configurations) as

$$[\langle H \rangle_K]_{K_p} = \frac{1}{2^N} \sum_{\{\epsilon_i\}} \sum_{\{\tau_{ij}\}} \prod_{\langle ij \rangle} \frac{\exp(K_p\tau_{ij}\epsilon_i\epsilon_j)}{2\cosh K_p} \times \langle H \rangle_K. \tag{35}$$

We take the summation over $\{\epsilon_i\}$ in advance of that over $\{\tau_{ij}\}$ and then find the partition function with K_p instead of K.

$$[\langle H\rangle_K]_{K_p} = \frac{1}{2^N}\sum_{\{\tau_{ij}\}}\frac{Z(K_p;\{\tau_{ij}\})}{(2\cosh K_p)^{N_B}} \times \langle H\rangle_K. \tag{36}$$

where N_B is the number of bonds. Going back to the definition (31), both of the partition functions on the denominator and numerator can be cancelled when $K_p = K$ as

$$\begin{aligned}[\langle H\rangle_K]_K &= \frac{-J}{2^N(2\cosh K_p)^{N_B}}\sum_{\{\sigma_i\}}\sum_{\{\tau_{ij}\}}\frac{d}{dK}\exp\left(K\tau_{ij}\sigma_i\sigma_j\right)\\ &= -N_B J\tanh K.\end{aligned} \tag{37}$$

Similarly, we can evaluate the rigorous upper bound on the specific heat. The condition $K_p = K$ confines the special subspace in which we can perform the exact analysis for spin glasses. This subspace is called as the Nishimori line (NL) [18, 19].

2.6. Jarzynski equality for spin glasses

By use of the gauge transformation as above introduced briefly, let us consider the application of the relations (18) and (19) to spin glasses, namely PA for such a complicated system in a tricky way. We analyze JE for the spin-glass model for several interesting quantities below.

2.6.1. *Gauge-invariant quantities like internal energy*

We apply Eq. (19) to a gauge-invariant quantity $G(\{\tau_{ij}\})$ for the purpose of evaluation of the equilibrium quantity in spin glasses. The configurational average for Eq. (19) yields

$$\left[\left\langle G_T(\{\tau_{ij}\})\mathrm{e}^{-Y}\right\rangle_{K_0\to K_T}\right]_{K_p} = \left[\langle G(\{\tau_{ij}\})\rangle_{K_T}\frac{Z(K_T;\{\tau_{ij}\})}{Z(K_0;\{\tau_{ij}\})}\right]_{K_p}. \tag{38}$$

The quantity on the left-hand side is the configurational and nonequilibrium averages of the observable $G(\{\tau_{ij}\})$ at final stage of PA, that is after the protocol $K_0 \to K_T$ with the factor $e^{-\beta W}$. On the other hand, $\langle G(\{\tau_{ij}\})\rangle_{K_T}$ on the right-hand side expresses the configurational and thermal averages of the equilibrium state for the final Hamiltonian.

The gauge transformation $\sigma_i \to \sigma_i\epsilon_i$, $\tau_{ij} \to \tau_{ij}\epsilon_i\epsilon_j$ $(\forall i,j)$ leads us to

$$\left[\langle G(\{\tau_{ij}\})\rangle_{K_T}\frac{Z(K_T;\{\tau_{ij}\})}{Z(K_0;\{\tau_{ij}\})}\right]_{K_p} = \sum_{\{\tau_{ij}\}}\frac{\langle G(\{\tau_{ij}\})\rangle_{K_T}\prod_{\langle ij\rangle}\mathrm{e}^{K_p\tau_{ij}\epsilon_i\epsilon_j}}{(2\cosh K_p)^{N_B}}\frac{Z(K_T;\{\tau_{ij}\})}{Z(K_0;\{\tau_{ij}\})}. \tag{39}$$

All the quantities in this equation are invariant under the gauge transformation. The summation over $\{\epsilon_i\}$ and division by 2^N gives

$$\left[\langle G(\{\tau_{ij}\})\rangle_{K_T}\frac{Z(K_T;\{\tau_{ij}\})}{Z(K_0;\{\tau_{ij}\})}\right]_{K_p} = \sum_{\{\tau_{ij}\}}\frac{\langle G(\{\tau_{ij}\})\rangle_{K_T}Z(K_p;\{\tau_{ij}\})}{2^N(2\cosh K_p)^{N_B}}\frac{Z(K_T;\{\tau_{ij}\})}{Z(K_0;\{\tau_{ij}\})}. \tag{40}$$

On the other hand, let us evaluate the quantity $\left[\langle G(\{\tau_{ij}\})\rangle_{K_T}\right]_{K_p}$. Similarly to the above calculation, the following identity can be obtained by the gauge transformation,

$$\left[\langle G(\{\tau_{ij}\})\rangle_{K_T}\right]_{K_p} = \sum_{\{\tau_{ij}\}} \frac{\langle G(\{\tau_{ij}\})\rangle_{K_T} Z(K_p;\{\tau_{ij}\})}{2^N(2\cosh K_p)^{N_B}}. \tag{41}$$

By Setting $K_p = K_0$ in Eq. (40) and $K_p = K_T$ in the above equation, we reach the following nonequilibrium relation,

$$\left[\langle G_T(\{\tau_{ij}\})\mathrm{e}^{-Y}\rangle_{K_0\to K_T}\right]_{K_0} = \left[\langle G(\{\tau_{ij}\})\rangle_{K_T}\right]_{K_T}\left(\frac{\cosh K_T}{\cosh K_0}\right)^{N_B}. \tag{42}$$

If we set $G_T(\{\tau_{ij}\}) = 1$ in the resultant equation, the Jarzynski equality for spin glass is obtained,

$$\left[\left\langle \mathrm{e}^{-Y}\right\rangle_{K_0\to K_T}\right]_{K_0} = \left(\frac{\cosh K_T}{\cosh K_0}\right)^{N_B}. \tag{43}$$

Equation (43) leads to the lower bound on the pseudo work, using Jensen's inequality for the average of e^{-Y},

$$[\langle Y\rangle_{K_0\to K_T}]_{K_0} \geq -N_B\log\left(\frac{\cosh K_T}{\cosh K_0}\right). \tag{44}$$

By substituting $G_T(\{\tau_{ij}\}) = H$ into Eq. (42), we obtain

$$\left[\left\langle H\mathrm{e}^{-\beta W}\right\rangle_{K_0\to K_T}\right]_{K_0} = [\langle H\rangle_{K_T}]_{K_T}\left(\frac{\cosh K_T}{\cosh K_0}\right)^{N_B}. \tag{45}$$

This equation shows that the internal energy after the cooling as in SA or heating process starting from a temperature on NL, which is in the present case $K_0 = K_p$, is proportional to the internal energy in equilibrium on NL corresponding to the final temperature $K_T = K_p$ as in Fig. 4. We here assume to perform PA and take an average over all results after many repetitions. The nonequilibrium process starts from NL $(1/K_0, 1/K_0)$ and ends at the point away from NL $(1/K_0, 1/K_T)$. Notice that the ordinary procedure in PA gives the estimation of the equilibrium quantity at the last condition as $(1/K_0, 1/K_T)$. However the corresponding internal energy is at the different point but on NL as $(1/K_T, 1/K_T)$. It means that we can obtain the equilibrium quantities in the different amount of the randomness from the initial condition through the configurational average of the results from PA.

2.6.2. *Gauge-non-invariant quantities*

In statistical physics, it is important to detect the order of the instantaneous spin configuration in the system. For instance, as in Fig. 4, there are several phases, ferromagnetic, paramagnetic and spin-glass ones, involved in the spin-glass model. They have the characteristic quantities to distinguish themselves, termed as the order parameter. The order parameter to identify the phase boundary between the ferromagnetic and paramagnetic phases is the magnetization defined as

$$m = \frac{1}{N}\sum_{i=1}^{N}\sigma_i \tag{46}$$

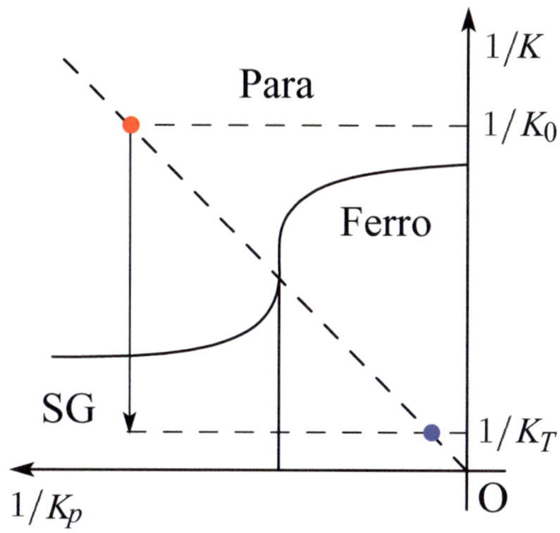

Figure 4. The nonequilibrium process as in SA/PA on the left-hand side as in Eq. (45) drawn as an arrow on the phase diagram. The corresponding equilibrium state as on the right-hand side of Eq. (45) is at the point on NL. The solid curves are the phase boundaries and the dashed line of 45° represents NL.

Therefore it is important to observe the behavior of the first momentum of spin variable in equilibrium. For this purpose, we choose $\sigma_i(T)$ for O in Eq. (19) and consider the possibility of the application of PA. After the configurational average, we obtain

$$\left[\left\langle \sigma_i(T)\mathrm{e}^{-Y}\right\rangle_{K_0\to K_T}\right]_{K_p} = \left[\langle\sigma_i\rangle_{K_T}\frac{Z(K_T;\{\tau_{ij}\})}{Z(K_0;\{\tau_{ij}\})}\right]_{K_p}. \tag{47}$$

Gauge transformation for the right-hand side in this equation yields

$$\left[\langle\sigma_i\rangle_{K_T}\frac{Z(K_T;\{\tau_{ij}\})}{Z(K_0;\{\tau_{ij}\})}\right]_{K_p} = \sum_{\{\tau_{ij}\}}\frac{Z(K_T;\{\tau_{ij}\})}{Z(K_0;\{\tau_{ij}\})}\langle\sigma_i\rangle_{K_T}\epsilon_i\prod_{\langle ij\rangle}\frac{\mathrm{e}^{K_p\tau_{ij}\epsilon_i\epsilon_j}}{2\cosh K_p}. \tag{48}$$

We again sum both sides of this equation over all the possible configurations of $\{\epsilon_i\}$ and divide the obtained quantity by 2^N to find

$$\left[\langle\sigma_i\rangle_{K_T}\frac{Z(K_T;\{\tau_{ij}\})}{Z(K_0;\{\tau_{ij}\})}\right]_{K_p} = \sum_{\{\tau_{ij}\}}\frac{Z(K_p;\{\tau_{ij}\})}{2^N(2\cosh K_p)^{N_B}}\langle\sigma_i\rangle_{K_T}\langle\epsilon_i\rangle_{K_p}\frac{Z(K_T;\{\tau_{ij}\})}{Z(K_0;\{\tau_{ij}\})}. \tag{49}$$

The following relation can also be obtained in a similar manipulation,

$$[\langle\sigma_i\rangle_{K_0}]_{K_p} = \sum_{\{\tau_{ij}\}}\frac{Z(K_p;\{\tau_{ij}\})}{2^N(2\cosh K_p)^{N_B}}\langle\sigma_i\rangle_{K_0}\langle\epsilon_i\rangle_{K_p}. \tag{50}$$

By setting $K_p = K_0$ in Eq. (49) and $K_p = K_T$ in Eq. (50), we reach a relation

$$\left[\langle\sigma_i\rangle_{K_T}\frac{Z(K_T;\{\tau_{ij}\})}{Z(K_0;\{\tau_{ij}\})}\right]_{K_0} = [\langle\sigma_i\rangle_{K_0}]_{K_T}\left(\frac{\cosh K_T}{\cosh K_0}\right)^{N_B}. \tag{51}$$

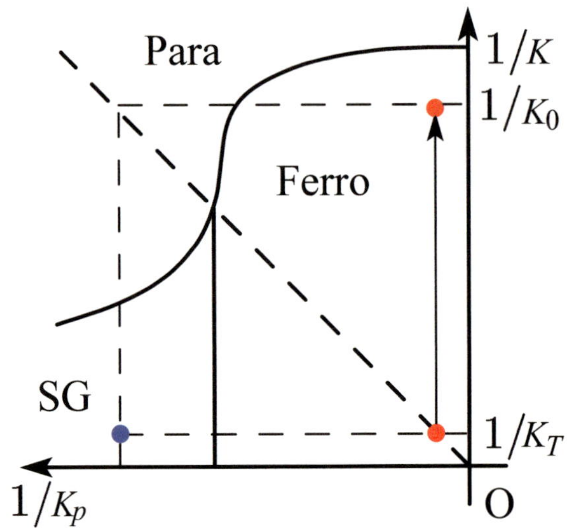

Figure 5. Equations (52) and (53) relate equilibrium quantities at the point $(1/K_0, 1/K_T)$ (the lower-left dot) with other physical quantities estimated during the nonequilibrium process shown in the arrow. The lower-left dot is in the spin glass phase whereas the corresponding arrow is in the ferromagnetic phase.

As a result, we obtain a nonequilibrium relation ,

$$\left[\langle\sigma_i(T)\mathrm{e}^{-Y}\rangle_{K_0\to K_T}\right]_{K_0} = [\langle\sigma_i\rangle_{K_0}]_{K_T}\left(\frac{\cosh K_T}{\cosh K_0}\right)^{N_B}. \tag{52}$$

The same method yields another relation for the correlation functions to similarly measure the magnitude of order in the system

$$\left[\langle\sigma_0(T)\sigma_r(T)\mathrm{e}^{-Y}\rangle_{K_0\to K_T}\right]_{K_0} = [\langle\sigma_0\sigma_r\rangle_{K_0}]_{K_T}\left(\frac{\cosh K_T}{\cosh K_0}\right)^{N_B}. \tag{53}$$

The obtained relations (52) and (53) relate the equilibrium physical quantities away from NL (the right-hand sides) with other quantities measured during the nonequilibrium process from a point on NL to another point away from NL (the left-hand sides) as depicted in Fig. 5. The spin-glass system in the low-temperature region exhibits the extremely slow relaxation toward equilibrium. This feature hampers to observe the equilibrium behavior of spin glasses. However our results imply that the configurational average of PA would overcome the difficulty. One may attempts the heating process from NL in order to evaluate the low-temperature property through Eqs. (52) and (53) as depicted in Fig. 5. The Jarzynski equality holds irrespectively of the schedule to control the external field. It means that we can investigate the low-temperature behavior for spin glasses without suffering from critical slowing down. Unfortunately, however, the exponentiated pseudo work does not hold the self-averaging property. It means that the sample-to-sample fluctuation between different configurations of $\{\tau_{ij}\}$ remains to be relevant even in a large-N system. Therefore, if we estimate the empirical average of realizations of $\{\tau_{ij}\}$ following the obtained equalities, we

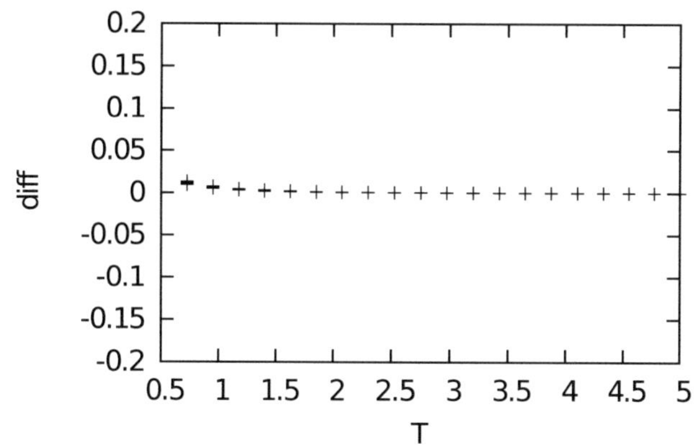

Figure 6. Superimposed plots of Differences between the exact value and the estimation given by PA. The horizontal axis denotes the instantaneous temperature during PA. The vertical axis represents the difference from the exact values given by the transfer matrix method in advance.

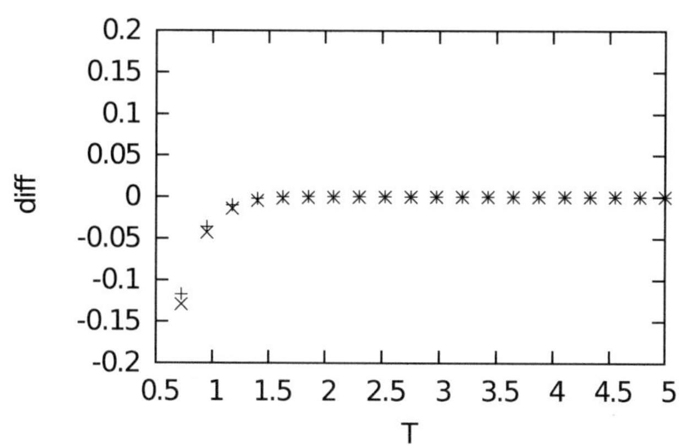

Figure 7. Difference between the exact values and the empirical averages over 100 realizations. The cross points give the deviation of the empirical averages of the exact values of ratios of the partition functions from the exact value on the right-hand side of Eq. (43), while the tilted-cross ones represent that of the empirical average of the estimation given by PA.

do not correctly reproduce the quantity of the right-hand side. We describe the test results for estimation of the ratio of the partition functions as in Eq. (13) by PA for 100 realizations of $\{\tau_{ij}\}$ in Fig. 6. We perform PA for the $\pm J$ Ising model on the square lattice with a linear size $L = 6$, and Monte-Carlo step $R = 1000$. The number of copies is $C = 100$. The population annealing can correctly estimate the ratio of the partition functions for each realization, but their simple average does not coincide with the quantity on the right-hand side of Eq. (43) as in Fig. 7. Both of the results are away from the exact solutions due to the sample-to-sample fluctuation and show nontrivial behavior depending on the linear size. These facts imply lack of self-averaging property. Therefore, if we exploit all the above results given by the configurational average of the exponentiated pseudo work, we have to overcome this violation due to lack of the self-averaging property. This is one of the remaining problem associated with this procedure with PA for spin glasses.

3. Quantum annealing

Observant readers may begin to recognize the possibility to use physical nature to drive the system in searching the lowest energy (ground state) instead of thermal fluctuation controlled by the inverse temperature. We show another strategy to find the ground state recently studied in a field of physics, quantum annealing (QA) [13].

3.1. Quantum adiabatic computation

In quantum-mechanical system, we can use a parallel way to drive all the candidates of the desired solution in optimization problem by use of superposition. Quantum annealing uses quantum fluctuation between superposed states to search for the ground state. One of the successful strategies is to use the adiabatic evolution known as quantum adiabatic computation (QAC) [5]. In QAC, as the procedure of SA, we control to gradually decrease the strength of quantum fluctuations to drive the system. Similarly to SA, the convergence into the optimal solution of QAC (the ground state) is also guaranteed by a mathematical proof [15].

In QAC, we introduce a non-commutative operator to drive the system by quantum nature in addition to the original Hamiltonian H_0, which is designed to represent the optimization problem to be solved, as

$$H(t) = f(t)H_0 + (1 - f(t))\, H_1, \tag{54}$$

where $f(t)$ is assumed to be a monotonically increasing function satisfying $f(0) = 0$ and $f(T) = 1$. For instance, $f(t) = t/T$, where T denotes the computation time for QAC. In order to exemplify the explicit instance, we again assume that H_0 is the spin glass Hamiltonian as

$$H_0 = -\sum_{\langle ij \rangle} J_{ij} \sigma_i^z \sigma_j^z, \tag{55}$$

where σ_i^z is the z component of the Pauli operators defined as

$$\sigma^x = \begin{pmatrix} 0 & 1 \\ 1 & 0 \end{pmatrix}, \quad \sigma^y = \begin{pmatrix} 0 & \mathrm{i} \\ -\mathrm{i} & 0 \end{pmatrix}, \quad \sigma^z = \begin{pmatrix} 1 & 0 \\ 0 & -1 \end{pmatrix}. \tag{56}$$

We take the computational basis of the eigenstates of the z-component of the Pauli matrix (Ising variables) to represent the instantaneous state as $|\Psi(t)\rangle = |\sigma_1^z, \sigma_2^z, \cdots, \sigma_N^z\rangle$. The transverse-field operator is often used as quantum fluctuations for implementing QAC for the spin-glass model

$$H_1 = -\Gamma_0 \sum_{i=1}^{N} \sigma_i^x. \tag{57}$$

where Γ_0 is the strength of the transverse field. The whole Hamiltonian of QAC (although widely used also for QA) thus becomes

$$H(t) = f(t) \sum_{\langle ij \rangle} J_{ij} \sigma_i^z \sigma_j^z + (1 - f(t))\, \Gamma_0 \sum_{i=1}^{N} \sigma_i^x. \tag{58}$$

The quantum adiabatic computation starts from a trivial ground state of H_1. In the present case, the ground state of the transverse-field operator H_1 is simply written by a uniform linear combination as $\sum_{\{\sigma\}} |\sigma_1^z, \sigma_2^z, \cdots, \sigma_N^z\rangle / \sqrt{2}^N$.

The adiabatic theorem guarantees that the instantaneous state at time t, $|\Psi(t)\rangle$, is very close to the instantaneous ground state for a sufficiently large T (implying slow control) as $|0(t)\rangle$, if the instantaneous ground state $|0(t)\rangle$ is non-degenerate. The condition for $|0(t)\rangle$, $\langle 0(t)|\Psi(t)\rangle \approx 1-\delta^2(\delta \ll 1)$ to hold can be written as [15]

$$\frac{\max\left|\langle 1(t)|\frac{dH(t)}{dt}|0(t)\rangle\right|}{\min \Delta^2(t)} = \delta, \tag{59}$$

where $|1(t)\rangle$ is the instantaneous first excited state, and $\Delta(t)$ is the energy gap between the ground state and first excited one. The maximum and minimum should be evaluated between 0 and T. In the present case, since $dH(t)/dt \propto 1/T$, the above adiabatic condition is reduced into

$$T \propto \frac{1}{\delta \min \Delta^2(t)}. \tag{60}$$

It means that if we desire to solve the optimization problems by use of QAC, which one of the specialized version of QA, we take the computational time proportional to the inverse square of the energy gap. If the problem involved with the exponential closure of the energy gap for increasing of N, QAC must take extremely long time to obtain the ground state with a high probability [12, 25]. Interestingly, we can reproduce the convergence theorem of SA (9) from the above adiabatic condition with recourse to a mathematical mapping of the procedure of SA into the quantum dynamics [15, 23]. It implies that the nature of QAC can be understood through that of SA by the mathematical mapping technique.

Below, we would provide a new paradigm to solve faster than the ordinary scheme of SA. A fast sweep of the system yields nonequilibrium behavior. Although we have not yet understood deeply the nonequilibrium phenomena, there are a few well-established theories which rises to applications to the optimization problem. One possibility is PA for the quantum system by use of JE and its alternatives [20]. Here we again employ JE to give another scheme of QA while considering the nonequilibrium behavior.

3.2. Jarzynski equality for isolated quantum system

In order to consider the nonequilibrium behavior away from the adiabatic dynamics of QAC, we shortly review JE for an isolated quantum system [1, 24].

To directly use JE in the protocol to find the ground state of the spin-glass Hamiltonian H_0 as in Eq. (55), we prepare the dynamical quantum system following the time-dependent Hamiltonian (58). In addition, we pick up a state from the canonical ensemble for $H(0) = H_1 = -\Gamma_0 \sum_i \sigma_i^x$ at the initial stage of the procedure and then let it evolve following the time-dependent Schrödinger equation. We measure the performed work in the isolated quantum system as $W = E_m(T) - E_n(0)$, which is simply given by the difference between the outputs of projective measurements of the initial and final energies. Here m and n denote the indices of the instantaneous eigenstates as $H(T)|m(T)\rangle = E_m(T)|m(T)\rangle$ and $H(0)|n(0)\rangle = E_n(0)|n(0)\rangle$, respectively. The time-evolution operator is given by the following unitary operator as

$$U_T = \mathcal{T} \exp\left(\mathrm{i}\int_0^T dt H(t)\right), \tag{61}$$

where $\mathcal{T}$ denotes the time ordered product. Therefore the transition probability between the initial and final stages is given as

$$P_{m,n}(0 \to T) = |\langle \Psi_m(T)|U_T|\Psi_n(0)\rangle|^2. \tag{62}$$

The path probability for the nonequilibrium process starting from the equilibrium ensemble is then evaluated as

$$P_{m,n}(0 \to T)\frac{\exp(-\beta E_n(0))}{Z_0(\beta;\{J_{ij}\})}, \tag{63}$$

where we express the instantaneous partition function for the instantaneous Hamiltonian at each time t as $Z_t(\beta;\{J_{ij}\})$.

By directly evaluating the nonequilibrium average of the exponentiated work but for the isolated system, we reach JE applicable to non-adiabatic version of QA. We define the nonequilibrium average of the exponentiated work as

$$\left\langle e^{-\beta W}\right\rangle_{\mathrm{QA}} = \sum_{m,n} e^{-\beta W} P_{m,n}(0 \to T)\frac{\exp(-\beta E_n(0))}{Z_0(\beta;\{J_{ij}\})},$$

which becomes the left-hand side of JE. The quantity defined here is evaluated as

$$\begin{aligned}\left\langle e^{-\beta W}\right\rangle_{\mathrm{QA}} &= \sum_{m,n}\frac{e^{-\beta E_m(T)}}{Z_0(\beta;\{J_{ij}\})}P_{m,n}(0 \to T)\\ &= \sum_{m}\frac{e^{-\beta E_m(T)}}{Z_0(\beta;\{J_{ij}\})}\\ &= \frac{Z_T(\beta;\{J_{ij}\})}{Z_0(\beta;\{J_{ij}\})},\end{aligned} \tag{64}$$

where we used the fact that the performed work W was a classical number and

$$\begin{aligned}\sum_n P_{m,n}(0 \to T) &= \sum_n \langle \Psi_m(T)|U_T|\Psi_n(0)\rangle\langle \Psi_n(0)|U_T^\dagger|\Psi_i(T)\rangle\\ &= \sum_m \langle \Psi_m(T)|U_T U_T^\dagger|\Psi_m(T)\rangle = 1.\end{aligned} \tag{65}$$

If we measure the physical observable $\hat{O}_T$ at the last of the nonequilibrium process, we obtain another equation as, similarly to the classical version,

$$\langle \hat{O}_T e^{-\beta W}\rangle_{\mathrm{QA}} = \langle \hat{O}\rangle_\beta \frac{Z_T(\beta;\{J_{ij}\})}{Z_0(\beta;\{J_{ij}\})}, \tag{66}$$

where the subscript on the square brackets in the right-hand side denotes the thermal average in the last equilibrium state with the inverse temperature β. The ratio of Eqs. (64) and (66) gives

$$\frac{\langle \hat{O}_T e^{-\beta W}\rangle_{\mathrm{QA}}}{\langle e^{-\beta W}\rangle_{\mathrm{QA}}} = \langle \hat{O}\rangle_\beta. \tag{67}$$

The resultant equation suggests that we can estimate the thermal average under the Hamiltonian H_0 on the right-hand side through the left-hand side of JE. This fact may be useful in the evaluation of equilibrium average, since the left-hand side is evaluated without slow adiabatic processes. Differently from QAC, we must sweep the quantum system repeatedly to correctly estimate the nonequilibrium average as in JE, but in short time. We propose such a procedure as the alternative of QAC, the non-adiabatic quantum annealing (NQA) based on the property of JE as above established. First we discuss the possibility as a solver of the optimization problem below.

4. Non-adiabatic quantum annealing

In order to investigate the property of the ground state, we tune the inverse temperature into a very large value $\beta \gg 1$. The nonequilibrium average on the left-hand side of JE involves a non-extensive quantity, the exponentiated work, whose value fluctuates significantly from process to process. Therefore the average on the left-hand side must be calculated by many trials of annealing processes. Thus, rare events with large values of the exponentiated work (i.e. $\beta|W| \gg \Gamma_0$) would contribute to the average significantly, and we have to repeat the annealing process very many, typically exponentially, times in order to reach the correct value of the average. This property would be a bottleneck of the simple implementation of NQA, instead of long time involved by the closure of the energy gap in QAC. What about PA in the classical counter part? In order to generate the relevant contributions in PA, we use the biased distribution with the exponentiated pseudo work through resampling technique [7, 9]. Without resampling technique, we cannot efficiently reproduce the prediction given by JE. In this fact, if we implement the biased distribution in the quantum system, we would use NQA without suffering from rare events. It means that we can perform NQA in order to find the ground state in a short time with several repetitions. So far, it is the very difficult task to realize the biased distribution in the quantum system. However it is worthwhile to consider its possibility in the future.

4.1. Several analyses of non-adiabatic quantum annealing

Unfortunately, we have not reached any positive answers on the performance of NQA. Instead let us here evaluate several properties in nonequilibrium process as in NQA for the particular spin glasses. We can exactly analyze nonequilibrium behavior by combination of JE with the gauge transformation, although there are few exact results in nonequilibrium quantum dynamical system with many components [22].

Following the prescription of JE, let us consider a repetition of NQA starting from the equilibrium ensemble. The initial Hamiltonian of NQA is given only by the transverse field $H(0) = H_1$. It turns out that the starting point of our analyses is the specialized JE to the case for NQA as

$$\langle e^{-\beta W} \rangle_{\rm QA} = \frac{Z_T(\beta, \{J_{ij}\})}{(2\cosh \beta\Gamma_0)^N}. \tag{68}$$

We assume that the interactions $\{J_{ij}\}$ follow the distribution function for the $\pm J$ Ising model (22), which is better to be rewritten as

$$P(J_{ij}) = \frac{\exp(\beta_p J_{ij})}{2\cosh \beta_p J}, \tag{69}$$

where we do not use $K = \beta J$ for transparency, and $\exp(-2\beta_p J) = (1-p)/p$.

4.2. Gauge transformation for quantum spin systems

For several special spin glasses as the $\pm J$ Ising model, the gauge transformation is available for analyses on the dynamical property even under quantum fluctuations. The time-dependent Hamiltonian as in Eq. (58) is invariant under the following local transformation,

$$\sigma_i^x \to \sigma_i^x,\ \sigma_i^y \to \xi_i \sigma_i^y,\ \sigma_i^z \to \xi_i \sigma_i^z,\ J_{ij} \to J_{ij}\xi_i\xi_j \quad (\forall i,j), \tag{70}$$

where $\xi_i (= \pm 1)$ is called as a gauge variable. This transformation is designed to preserve the commutation relations between different components of Pauli matrices [16].

4.3. Relationship between two different paths of NQA

Below, we reveal several properties inherent in NQA by the gauge transformation. Let us take the configurational average of Eq. (68) over all the realizations of $\{J_{ij}\}$ for the special case with $\beta = \beta_1$ and $\beta_p = \beta_2$ as

$$\left[\langle e^{-\beta_1 W}\rangle_{\mathrm{QA}}\right]_{\beta_2} = \left[\frac{Z_T(\beta_1;\{J_{ij}\})}{(2\cosh\beta_1\Gamma_0)^N}\right]_{\beta_2}. \tag{71}$$

The right-hand side is written explicitly as

$$\left[\langle e^{-\beta_1 W}\rangle_{\mathrm{QA}}\right]_{\beta_2} = \sum_{\{J_{ij}\}} \frac{\exp\left(\beta_2 \sum_{\langle ij\rangle} J_{ij}\right)}{(2\cosh\beta_2 J)^{N_B}} \frac{Z_T(\beta_1;\{J_{ij}\})}{(2\cosh\beta_1\Gamma_0)^N}. \tag{72}$$

Let us here apply the gauge transformation as introduced above. Since the time-dependent Hamiltonian is invariant, we may sum over all possible configurations of the gauge variables $\{\xi_i\}$ and divide the result by 2^N in order to obtain the quantity on the left-hand side,

$$\left[\langle e^{-\beta_1 W}\rangle_{\mathrm{QA}}\right]_{\beta_2} = \sum_{\{J_{ij}\}} \frac{Z_T(\beta_2;\{J_{ij}\}) Z_T(\beta_1;\{J_{ij}\})}{2^N (2\cosh\beta_2 J)^{N_B} (2\cosh\beta_1\Gamma_0)^N}. \tag{73}$$

A similar quantity of the average of the exponentiated work for spin glass with the inverse temperature β_2 and the parameter for the quenched randomness β_1 gives

$$\left[\langle e^{-\beta_2 W}\rangle_{\mathrm{QA}}\right]_{\beta_1} = \sum_{\{J_{ij}\}} \frac{Z_T(\beta_2;\{J_{ij}\}) Z_T(\beta_1;\{J_{ij}\})}{2^N (2\cosh\beta_1 J)^{N_B} (2\cosh\beta_2\Gamma_0)^N}. \tag{74}$$

Comparing Eqs. (73) and (74), we find the following relation between two different non-adiabatic processes,

$$\left[\langle e^{-\beta_1 W}\rangle_{\mathrm{QA}}\right]_{\beta_2} = \left[\langle e^{-\beta_2 W}\rangle_{\mathrm{QA}}\right]_{\beta_1} \left(\frac{\cosh\beta_1 J}{\cosh\beta_2 J}\right)^{N_B} \left(\frac{\cosh\beta_2\Gamma_0}{\cosh\beta_1\Gamma_0}\right)^N. \tag{75}$$

We describe the two different paths of NQA related by this equality in Fig. 8. Setting $\beta_2 = 0$

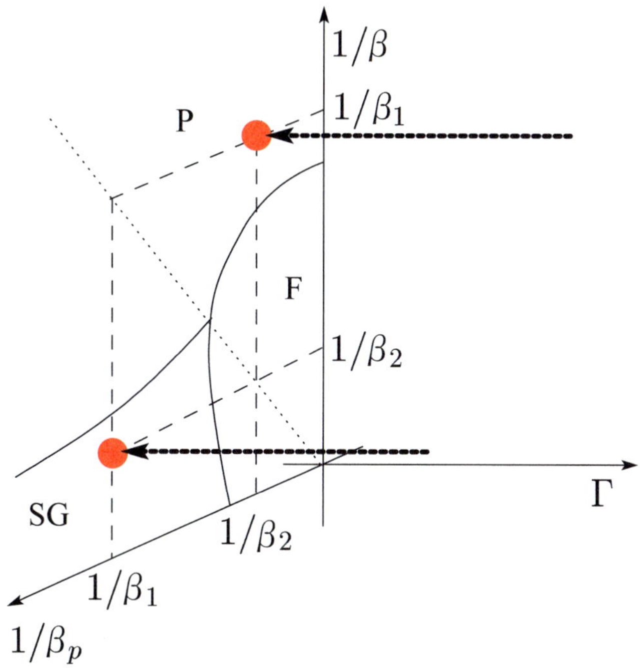

Figure 8. Two different processes of NQA in Eq. (75). The left-hand side of Eq. (75) represents the process toward the upper-right red dot and the right-hand side ends at the lower-left dot. Three phases expressed by the same symbols as in Fig. 4 are separated by solid curves and the dotted line describes NL ($\beta_p = \beta$).

in Eq. (75), (implying $p = 1/2$, the symmetric distribution), we find a nontrivial relation on the performed work during NQA

$$\left[\langle \mathrm{e}^{-\beta_1 W} \rangle_{\mathrm{QA}}\right]_0 = \frac{(\cosh \beta_1 J)^{N_B}}{(\cosh \beta_1 \Gamma_0)^N}. \tag{76}$$

The symmetric distribution ($\beta_2 = 0$ on the left-hand side) makes it possible to reduce the right-hand side to the above trivial expression. It is remarkable that NQA, which involves very complex dynamics, satisfies such a simple identity irrespective of the speed of annealing T. The Jensen inequality for the above equality leads us to the lower bound for the performed work as

$$[\langle W \rangle_{\mathrm{QA}}]_0 \geq -\frac{N}{\beta} \log \left(\frac{(\cosh \beta J)^z}{\cosh \beta \Gamma_0} \right). \tag{77}$$

where z is the coordination number as $z = N_B/N$. Here we generalize the inverse temperature to β from the specific choice β_1. This lower bound is loose, since the direct application of the Jensen inequality to JE for NQA yields, after the configurational average with the symmetric distribution,

$$[\langle W \rangle_{\mathrm{QA}}]_0 \geq \frac{1}{\beta} D(0|\beta) - \frac{N}{\beta} \log \left(\frac{(\cosh \beta J)^z}{\cosh \beta \Gamma_0} \right). \tag{78}$$

where $D(\beta|\beta')$ is the Kullback-Leibler divergence defined as

$$D(\beta|\beta') = \sum_{\{J_{ij}\}} \tilde{P}_\beta(\{J_{ij}\}) \log \frac{\tilde{P}_{\beta'}(\{J_{ij}\})}{\tilde{P}_\beta(\{J_{ij}\})} \tag{79}$$

Here we defined the marginal distribution for the specific configuration $\{J_{ij}\}$ summed over all the possible gauge transformations,

$$\tilde{P}_\beta(\{J_{ij}\}) = \frac{1}{2^N} \sum_{\{\xi_i\}} \prod_{\langle ij \rangle} P(J_{ij}) = \frac{Z_T(\beta; \{J_{ij}\})}{2^N (2\cosh \beta J)^{N_B}}. \tag{80}$$

Since the Kullback-Leibler divergence does not become non-negative, the work performed by the transverse field during a nonequilibrium process in the symmetric distribution (i.e. the left-hand side of Eq. (78)) does not lower below the second quantity on the right-hand side of Eq. (78), namely Eq. (77). This fact means that Eq. (77) is a loose lower bound.

4.4. Exact relations involving inverse statistics

Beyond the above results, we can perform further non-trivial analyses for the nonequilibrium process in the special conditions. Let us next take the configurational average of the inverse of JE, Eq. (68), as

$$\left[\frac{1}{\langle e^{-\beta W} \rangle_{QA}} \right]_{\beta_p} = \left[\frac{(2\cosh \beta \Gamma_0)^N}{Z_T(\beta; \{J_{ij}\})} \right]_{\beta_p}. \tag{81}$$

The application of the gauge transformation to the right-hand side yields

$$\left[\frac{1}{\langle e^{-\beta W} \rangle_{QA}} \right]_{\beta_p} = \sum_{\{J_{ij}\}} \frac{\exp\left(\beta_p \sum_{\langle ij \rangle} J_{ij}\xi_i\xi_j\right)}{(2\cosh \beta_p J)^{N_B}} \frac{(2\cosh \beta \Gamma_0)^N}{Z_T(\beta; \{J_{ij}\})}. \tag{82}$$

Summation of the right-hand side over all the possible configurations of $\{\xi_i\}$ and division of the result by 2^N give

$$\left[\frac{1}{\langle e^{-\beta W} \rangle_{QA}} \right]_{\beta_p} = \sum_{\{J_{ij}\}} \frac{Z_T(\beta_p; \{J_{ij}\})}{2^N (2\cosh \beta_p J)^{N_B}} \frac{(2\cosh \beta \Gamma_0)^N}{Z_T(\beta; \{J_{ij}\})}. \tag{83}$$

This equation reduces to, by setting $\beta_p = \beta$, namely on the Nishimori line,

$$\left[\frac{1}{\langle e^{-\beta W} \rangle_{QA}} \right]_{\beta} = \frac{(\cosh \beta \Gamma_0)^N}{(\cosh \beta J)^{N_B}}. \tag{84}$$

Comparison of Eqs. (76) and (84) gives

$$\left[\langle e^{-\beta W} \rangle_{QA} \right]_0 = \left(\left[\frac{1}{\langle e^{-\beta W} \rangle_{QA}} \right]_{\beta} \right)^{-1}. \tag{85}$$

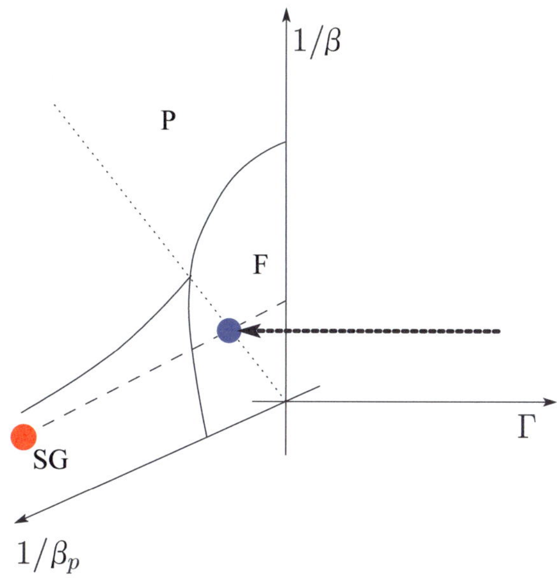

Figure 9. Two different nonequilibrium processes in NQA through Eq. (85). The same symbols are depicted as in Fig. 4. The blue circle represents the determination of the process on the right-hand side of Eq. (85), whereas the red one is for the left-hand side.

As shown in Fig. 9, the resultant equation leads us to a fascinating relationship of the two completely different processes through the inverse statistics. One denotes NQA toward the Nishimori line, while the other expresses for the symmetric distribution.

We can find the exact results through the inverse statics of the inverse statics of Eq. (66). Let us further consider the case for the two-point correlation $O_T = \sigma_i^z \sigma_j^z$. Taking the configurational average of both sides under the condition $\beta_p = \beta$, we find

$$\left[\frac{1}{\langle \sigma_i^z \sigma_j^z \mathrm{e}^{-\beta W} \rangle_{\mathrm{QA}}}\right]_\beta = \frac{(\cosh \beta \Gamma_0)^N}{(\cosh \beta J)^{N_B}} \left[\frac{1}{\langle \sigma_i^z \sigma_j^z \rangle_\beta}\right]_\beta . \tag{86}$$

The quantity on the right-hand side becomes unity by the analysis with the gauge transformation as has been shown in the literatures [18, 19]. We thus reach a simple exact relation

$$\left[\frac{1}{\langle \sigma_i^z \sigma_j^z \mathrm{e}^{-\beta W} \rangle_{\mathrm{QA}}}\right]_\beta = \frac{(\cosh \beta \Gamma_0)^N}{(\cosh \beta J)^{N_B}}, \tag{87}$$

which is another exact identity for processes of NQA.

We obtain several exact nontrivial relations between completely different paths in NQA as shown above by use of the gauge transformation, which is a specialized tool to analyze spin glasses. We should notice that such results are very rare for the nonequilibrium behavior in disordered quantum system. The importance of the above equalities is still not clear. We emphasize that, when we realize the quantum spin systems in experiments, the above results would be a valuable platform to confirm their precisions and conditions. The quantum annealing is originally intended to be a tool implemented in so-called quantum computer. Therefore we need theoretical studies not only on the performance in the application but also how to make good condition to implement the protocol. In this regard, our studies shed light on the future indicator to open the way to realize the generic solver in the quantum computer. We must continue the active studies in this direction.

5. Conclusion

We reviewed several recent active studies on the alternatives of SA with use of the novel substantial progress in statistical physics. The key point was to exploit the nonequilibrium behavior during performing the active control on the system. The Jarzynski equality states the possibility to estimate the equilibrium quantities by the average quantity through the nonequilibrium behavior. It means that we can invent several new strategies by use of JE away from the paradigm of the simulated annealing, which sticks to the quasi-static dynamics. The population annealing is a starting point of the studies in this direction. It is certain that population annealing find out the desired solution of the optimization problem from the property of JE faster than SA. Roughly speaking, the population annealing cuts the computational time (CPU) by use of the parallel dynamics (memory). The remaining problem is to evaluate its qualitative performance of PA for the optimization problem.

Not only the direct use of PA, we propose another type of its application in this chapter. Regarding on this type, we show skillful analyses by use of the special symmetry hidden in spin glasses to give several nontrivial exact relations. The resultant relations are useful to investigate the low-temperature region for spin glasses if we implement them by aid of PA, since we do not suffer from the critical slowing down peculiar in spin glass.

Meanwhile, if we employ a different rule to drive the system, we would be able to find the way to solve the optimization problem as SA. We reviewed QA, which was by use of quantum fluctuations as a driver. The specialized version of QA, QAC, is found to has a crucial bottleneck to solve a part of the optimization problem. Therefore we need to remove this problem while keeping its generality as a solver of the optimization problem. We again considered the application of JE to propose an alternative method, NQA. Although we do not assess its quantitative performance in the application to the optimization problem, our proposal gives a new paradigm to solve the optimization problem through the physical process like SA. We have to emphasize that QA was invented to solve the optimization problem in quantum computer. Therefore we must prepare the quantitative results to verify the precision and conditions in the actual experience on quantum computers. Along this line, we gave several results for the nonequilibrium behavior in the quantum system with gauge symmetry. These studies would be significant in the future development to realize the quantum computation.

Beyond the original version of SA, in order to find the desired solution as fast as possible, we must be away from the quasi-static procedure. The key point is to deal with nonequilibrium behavior. The further understanding of its peculiar behavior in statistical physics would be helpful to invent a genius and generic solver as PA and NQA.

Acknowledgement

The author thanks the fruitful discussions with Prof. Koji Hukushima of University of Tokyo and Prof. Hidetoshi Nishimori of Tokyo Institute of Technology. This work was partially supported by MEXT in Japan, Grant-in-Aid for Young Scientists (B) No.24740263.

Author details

Masayuki Ohzeki
Department of Systems Science, Graduate School of Informatics, Kyoto University, Yoshida-Honmachi, Sakyo-ku, Kyoto, Japan

6. References

[1] Campisi, M., Talkner, P. & Hänggi, P. [2009]. Fluctuation theorem for arbitrary open quantum systems, *Phys. Rev. Lett.* 102: 210401.
URL: *http://link.aps.org/doi/10.1103/PhysRevLett.102.210401*

[2] Crooks, G. E. [1998]. Nonequilibrium measurements of free energy differences for microscopically reversible markovian systems, *J. Stat. Phys.* 90: 1481–1487.

[3] Crooks, G. E. [1999]. Entropy production fluctuation theorem and the nonequilibrium work relation for free energy differences, *Phys. Rev. E* 60: 2721–2726.
URL: *http://link.aps.org/doi/10.1103/PhysRevE.60.2721*

[4] Crooks, G. E. [2000]. Path-ensemble averages in systems driven far from equilibrium, *Phys. Rev. E* 61: 2361–2366.
URL: *http://link.aps.org/doi/10.1103/PhysRevE.61.2361*

[5] Farhi, E., Goldstone, J., Gutomann, S. & Sipser, M. [2000]. Quantum Computation by Adiabatic Evolution, *arXiv:quant-ph/0001106v1* .

[6] Geman, S. & Geman, D. [1984]. Stochastic relaxation, gibbs distributions, and the bayesian restoration of images, *Pattern Analysis and Machine Intelligence, IEEE Transactions on* PAMI-6(6): 721 –741.

[7] Hukushima, K. & Iba, Y. [2003]. Population annealing and its application to a spin glass, *AIP Conference Proceedings* 690(1): 200–206.

[8] Hukushima, K. & Nemoto, K. [1996]. Exchange monte carlo method and application to spin glass simulations, *Journal of the Physical Society of Japan* 65(6): 1604–1608.
URL: *http://jpsj.ipap.jp/link?JPSJ/65/1604/*

[9] Iba, Y. [2001]. Population monte carlo algorithms, *Trans. Jpn. Soc. Artif. Intel.* 16(2): 279–286.

[10] Jarzynski, C. [1997a]. Equilibrium free-energy differences from nonequilibrium measurements: A master-equation approach, *Phys. Rev. E* 56: 5018–5035.
URL: *http://link.aps.org/doi/10.1103/PhysRevE.56.5018*

[11] Jarzynski, C. [1997b]. Nonequilibrium equality for free energy differences, *Phys. Rev. Lett.* 78: 2690–2693.
URL: *http://link.aps.org/doi/10.1103/PhysRevLett.78.2690*

[12] Jörg, T., Krzakala, F., Kurchan, J. & Maggs, A. C. [2008]. Simple glass models and their quantum annealing, *Phys. Rev. Lett.* 101: 147204.
URL: *http://link.aps.org/doi/10.1103/PhysRevLett.101.147204*

[13] Kadowaki, T. & Nishimori, H. [1998]. Quantum annealing in the transverse ising model, *Phys. Rev. E* 58: 5355–5363.
URL: *http://link.aps.org/doi/10.1103/PhysRevE.58.5355*

[14] Kirkpatrick, S., Gelatt, C. D. & Vecchi, M. P. [1983]. Optimization by simulated annealing, *Science* 220(4598): 671–680.
URL: *http://www.sciencemag.org/content/220/4598/671.abstract*

[15] Morita, S. & Nishimori, H. [2008]. Mathematical foundation of quantum annealing, *Journal of Mathematical Physics* 49(12): 125210.
URL: *http://link.aip.org/link/?JMP/49/125210/1*

[16] Morita, S., Nishimori, H. & Yukiyasu, O. [2006]. Gauge theory for quantum spin glasses, *Phys. Rev. E* 61: 2361–2366.
URL: *http://link.aps.org/doi/10.1103/PhysRevE.61.2361*

[17] Neal, R. M. [2001]. Annealed importance sampling, *Statistics and Computing* 11: 125–139. 10.1023/ A:1008923215028.
URL: *http://dx.doi.org/10.1023/A:1008923215028*

[18] Nishimori, H. [1981]. Internal energy, specific heat and correlation function of the bond-random ising model, *Progress of Theoretical Physics* 66(4): 1169–1181.
URL: *http://ptp.ipap.jp/link?PTP/66/1169/*

[19] Nishimori, H. [2001]. *Statistical Physics of Spin Glasses and Information Processing: An Introduction*, Oxford University Press.

[20] Ohzeki, M. [2010]. Quantum annealing with the jarzynski equality, *Phys. Rev. Lett.* 105: 050401.
URL: *http://link.aps.org/doi/10.1103/PhysRevLett.105.050401*

[21] Ohzeki, M. & Nishimori, H. [2010]. Nonequilibrium relations for spin glasses with gauge symmetry, *J. Phys. Soc. Jpn.* 79: 084003.
URL: *http://jpsj.ipap.jp/link?JPSJ/79/084003/*

[22] Ohzeki, M., Nishimori, H. & Hitoshi, K. [2011]. Nonequilibrium work on spin glasses in longitudinal and transverse fields, *J. Phys. Soc. Jpn.* 80: 084002.
URL: *http://jpsj.ipap.jp/link?JPSJ/80/084002/*

[23] Somma, R. D., Batista, C. D. & Ortiz, G. [2007]. Quantum approach to classical statistical mechanics, *Phys. Rev. Lett.* 99: 030603.
URL: *http://link.aps.org/doi/10.1103/PhysRevLett.99.030603*

[24] Tasaki, H. [2000]. Jarzynski Relations for Quantum Systems and Some Applications, *arXiv:0009244* .

[25] Young, A. P., Knysh, S. & Smelyanskiy, V. N. [2010]. First-order phase transition in the quantum adiabatic algorithm, *Phys. Rev. Lett.* 104: 020502.
URL: *http://link.aps.org/doi/10.1103/PhysRevLett.104.020502*

Section 2

SA Applications

Chapter 4

Bayesian Recovery of Sinusoids with Simulated Annealing

Dursun Üstündag and Mehmet Cevri

Additional information is available at the end of the chapter

http://dx.doi.org/10.5772/50449

1. Introduction

The ultimate goal of collecting data is to gain meaningful information about a physical system. However, in many situations, the quantities that we would like to determine are different from the ones which we are able to have measured. If the data we measured depends on the quantities we want, it contains at least some information about them. Therefore, our general interest is to subtract this information from data.

Let the vector $\boldsymbol{\theta}$ contain the parameters to be estimated from the (measurements) vector $\mathbf{D}$, which is the output of the physical system that one wants to be modeled. The physical system is described by a vector function f in the form:

$$y(t) = f(t;\boldsymbol{\theta}), \tag{1}$$

where t represents time. In many experiments, the recorded data $\mathbf{D} = \{d_1, d_2, ..., d_N\}$ are sampled from an unknown function $y(t)$ together with errors $e(t)$ at discrete times $(t_1, t_2, ..., t_N)^T$:

$$d_i = y(t_i) + e(t_i), \quad (i = 1, ..., N). \tag{2}$$

The measurement errors $e(t)$ are generally assumed to be drawn independently from a zero mean Gaussian probability distribution with a standard deviation of σ . On the other hand, different signal models correspond to different choices of signal model function $f(t, \boldsymbol{\theta})$. In this chapter, we restrict our attention to the static[1] sinusoidal model given by

[1] Static refers to that the amplitudes of the sinusoids do not change with time.

$$
\begin{aligned}
f(t;\boldsymbol{\theta}) &= \sum_{j=1}^{\rho} B_j \cos(t\omega_j) + B_{j+\rho} \sin(t\omega_j) \\
&= \sum_{j=1}^{2\rho} G_j(t;\boldsymbol{\omega}) B_j
\end{aligned}
\tag{3}
$$

where B_j's represent the amplitudes of the signal model and

$$
G(t;\boldsymbol{\omega}) = \begin{bmatrix}
\cos(\omega_1 t_1) & \dots & \cos(\omega_\rho t_1) & \sin(\omega_1 t_1) & \dots & \sin(\omega_\rho t_1) \\
\cos(\omega_2 t_2) & \dots & \cos(\omega_\rho t_2) & \sin(\omega_2 t_2) & \dots & \sin(\omega_\rho t_2) \\
\vdots & \vdots & \vdots & \vdots & \vdots & \vdots \\
\cos(\omega_1 t_N) & \dots & \cos(\omega_\rho t_N) & \sin(\omega_1 t_N) & \dots & \sin(\omega_\rho t_N)
\end{bmatrix}. \tag{4}
$$

The goal of data analysis is usually to use the observed data $\mathbf{D}$, to infer the values of parameters $\boldsymbol{\theta} = \{\boldsymbol{\omega}, \mathbf{B}\}$. Besides estimating the values of the parameters, there are two additional important problems. The one is to obtain an indication of the uncertainties associated with each parameter, i.e. some measures of how far they are away from the true parameters. The other we will not consider here is to assess whether or not the model is appropriate for explaining the data.

Sinusoidal parameter estimation in additive white noise within a Bayesian framework has been an important problem in signal processing and still now is an active research area because of its wide variety of applications in multiple disciplines such as sonar, radar, digital communications and biomedical engineering. The purpose of this research is therefore to develop accurate and computationally efficient estimators for sinusoidal parameter, namely, amplitudes and frequencies. In above problem, one may or may not know the number of sinusoids. When it is unknown, it is called model selection (Andrieu and Doucet, 1999; Üstündag, 2011) and is not subject to this chapter. Under an assumption of a known number of sinusoids, several algorithms have already been used in the parameter estimation literature, such as least-square fitting (Press et al., 1995), discrete Fourier transform (Cooley & Tukey, 1964), and periodogram (Schuster, 1905). With least square fitting, the model is completely defined and the question remaining is to find the values of the parameters by minimizing the sum of squared residuals. The discrete Fourier transform has been a very powerful tool in Bayesian spectral analysis since Cooley and Tukey introduced the fast Fourier transform (FFT) technique in 1965, followed by the rapid development of computers. In 1987, Jaynes derived periodogram directly from the principles of Bayesian inference. After his work, researchers in different branches of science have given much attention to the relationship between Bayesian inference and parameter estimation and they have done excellent works in this area for last fifteen years (Bretthorst, 1990; Üstündag et al., 1989, 1991; Harney, 2003; Gregory, 2005; Üstündag & Cevri, 2008, 2011; Üstündag, 2011).

In this chapter, we studied Bayesian recovery of sinusoids using estimation approach proposed by Bretthorst for a general signal model equation and combined it with a simulated annealing (SA) algorithm to obtain a global maximum of the posterior probability density function (PDF) $P(\boldsymbol{\omega}|\mathbf{D},I)$ for frequencies . Unfortunately, conventional algorithms (Press et al., 1995) based on the gradient direction fail to converge for this problem. Even when they converge, there is no assurance that they have found the global, rather than a local maximum. This is because the logarithm of the PDF $P(\boldsymbol{\omega}|\mathbf{D},I)$ is so sharply peaked and highly nonlinear function of frequencies. In this respect, a pattern search algorithm described by Hook-Jevees (Hooke & Jevees, 1962) to overcome this problem has already been used by some researchers in literature of estimation. However, we have found out that this approach does not converge unless the starting point is much closer to the optimum so that we have developed an algorithm in which this Bayesian approach is combined with a simulated annealing (SA) algorithm (Kirkpatrick, et al., 1983; Corana et al., 1987; Goffe et al., 1994; Ingber, 1996), which is a function optimization strategy based on an analogy with the creation of crystals from their melts. This explores the entire surface of the posterior PDF for the frequencies and tries to maximize it while moving both uphill and downhill steps, whose sizes are controlled by a parameter Γ that plays the role of the temperature in a physical system. By slowly lowering the temperature Γ towards zero according to a properly chosen schedule, one can show that the globally optimal solutions are approached asymptotically. Thus, it is largely independent of the starting values, often a critical input in conventional algorithms, and also offers different approach to finding parameter values of sinusoids through a directed, but random, search of the parameter space. In this context, an algorithm of this Bayesian approach is developed and coded in Mathematica programming language (Wellin at al., 2005) and also tested for recovering noisy sinusoids with multiple frequencies. Furthermore, simulation studies on synthetic data sets of a single sinusoid under a variety of signal to noise ratio (SNR) are made for a comparison of its performance with Cramér-Rao lower bound (CRLB), known as a lower limit on variance of any unbiased estimator. The simulations results support its effectiveness.

2. Bayesian parameter estimation

Let us now reconsider above problem within a Bayesian context (Bernardo & Smith, 2000; Bretthorst, 1988; Gregory, 2005; Harney, 2003; Jaynes, 2003; Ruanaidh & Fitzgerald, 1996; Üstündag & Cevri, 2008). As with all Bayesian calculations, the first step is to write down Bayes' theorem for the joint PDF for all of the unknown parameters $\{\boldsymbol{\omega},\mathbf{B},\sigma^2\}$:

$$P\left(\boldsymbol{\omega},\mathbf{B},\sigma^2 \mid \mathbf{D},\mathbf{I}\right)=\frac{1}{P\left(\mathbf{D}\mid\mathbf{I}\right)}P\left(\boldsymbol{\omega},\mathbf{B},\sigma^2 \mid \mathbf{I}\right)P\left(\mathbf{D}\mid \boldsymbol{\omega},\mathbf{B},\sigma^2,\mathbf{I}\right). \tag{5}$$

The quantity $P\left(\boldsymbol{\omega},\mathbf{B},\sigma^2 \mid \mathbf{I}\right)$ is called the prior PDF; it represents prior knowledge of the parameters $\{\boldsymbol{\omega},\mathbf{B},\sigma^2\}$ given the information $\mathbf{I}$. The sampling distribution $P\left(\mathbf{D}\mid \boldsymbol{\omega},\mathbf{B},\sigma^2,\mathbf{I}\right)$ is the likelihood (Edwards, 1972) of the data $\mathbf{D}$, given the model parameters. The probability function $P(\mathbf{D} \mid \mathbf{I})$ is the evidence, which is a constant for parameter estimation and is used here for normalizing the posterior PDF $P\left(\boldsymbol{\omega},\mathbf{B},\sigma^2 \mid \mathbf{D},\mathbf{I}\right)$. Therefore, it can be dropped in Equation (5) so that the joint PDF for the unknown parameters turns out to be the following form:

$$P\left(\boldsymbol{\omega},\mathbf{B},\sigma^2 \mid \mathbf{D},\mathbf{I}\right) \propto P\left(\boldsymbol{\omega},\mathbf{B},\sigma^2 \mid \mathbf{I}\right) P\left(\mathbf{D} \mid \boldsymbol{\omega},\mathbf{B},\sigma^2,\mathbf{I}\right). \tag{6}$$

A key component in Bayes theorem is the likelihood function $P\left(\mathbf{D} \mid \boldsymbol{\omega},\mathbf{B},\sigma^2,\mathbf{I}\right)$ which is proportional to the probability density of the noise. If its standard deviation σ is assumed to be known, then the likelihood function takes on the following form:

$$P\left(\mathbf{D} \mid \boldsymbol{\omega},\mathbf{B},\sigma^2,\mathbf{I}\right) = (2\pi\sigma^2)^{\frac{-N}{2}} exp\left(-\frac{NQ}{2\sigma^2}\right), \tag{7}$$

where the exponent Q is defined as follows

$$Q = \sum_{i=1}^{N} \left(d_i - \sum_{j=1}^{2\rho} B_j G_j(t_i;\boldsymbol{\omega}) \right)^2 . \tag{8}$$

This is equivalent to

$$Q = \overline{\mathbf{D}^2} - \frac{2}{N} \sum_{j=1}^{2\rho} \sum_{i=1}^{N} d_i B_j G_j\left(t_i;\boldsymbol{\omega}\right) + \frac{1}{N} \sum_{j=1}^{2\rho} \sum_{k=1}^{2\rho} \Omega_{jk} B_j B_k, \tag{9}$$

where

$$\overline{\mathbf{D}^2} = \frac{1}{N} \sum_{i=1}^{N} d_i^2 \tag{10}$$

and

$$\Omega_{jk} = \sum_{i=1}^{N} G_j\left(t_i;\boldsymbol{\omega}\right) \otimes G_k\left(t_i;\boldsymbol{\omega}\right), \quad (j,k = 1,\ldots,2\rho). \tag{11}$$

In order to obtain the posterior PDF for $\boldsymbol{\omega}$, Equation (6) can be integrated with respect to the nuisance parameters $\mathbf{B}$ under the knowledge of σ^2:

$$P\left(\boldsymbol{\omega},\sigma \mid \mathbf{D},\mathbf{I}\right) = \int P\left(\boldsymbol{\omega},\mathbf{B},\sigma \mid \mathbf{D},\mathbf{I}\right) d\mathbf{B}. \tag{12}$$

With the choices of an uniform prior PDF or independent Gaussians distributions with mean zero and known standard deviation for the amplitudes, the integral equations in (12) turn out to be a Gaussian integral which can be evaluated analytically (Bretthorst, 1988). To do this it is simply to convert the square matrix $\boldsymbol{\Omega}$ into a special type of matrix- a so called diagonal matrix- that shares the same fundamental properties of the underlying matrix. In other words, it is equivalent to transforming the underlying system of equations into a special set of coordinate axes. Therefore, this diagonalization process (Bretthorst, 1988) effectively introduces new orthonormal model functions,

$$H_j(t;\boldsymbol{\omega}) = \frac{1}{\sqrt{\lambda_j}} \sum_{k=1}^{2\rho} \upsilon_{jk} G_k(t;\boldsymbol{\omega}), \ (j = 1,...,2\rho), \tag{13}$$

and also gives a new expression for the signal model function in Equation (3):

$$f(t;\boldsymbol{\theta}) = \sum_{k=1}^{2\rho} A_k H_k(t;\boldsymbol{\omega}), \tag{14}$$

The new amplitudes A_k 's are related to the old amplitudes B_j 's by

$$A_k = \sqrt{\lambda_k} \sum_{j=1}^{2\rho} B_j \upsilon_{kj}, (k = 1,...,2\rho) \tag{15}$$

where υ_{kj} represents the j th component of the k th normalized eigenvector of Ω_{jk}, with λ_j as the corresponding eigenvalue. Substituting these expressions into Equation (9) and defining

$$h_j = \sum_{i=1}^{N} d_i H_j(t_i;\boldsymbol{\omega}), \quad (j = 1,2,...,2\rho) \tag{16}$$

to be the projection of the data onto the orthonormal model functions $H_j(t;\boldsymbol{\omega})$, we can then proceed to perform the 2ρ integration over A_j in Equation (12) to obtain

$$P(\mathbf{D} \mid \boldsymbol{\omega}, \mathbf{B}, \sigma, I) \propto \sigma^{-N+2\rho} exp\left(-\frac{N\overline{\mathbf{D}^2} - 2\rho\overline{\mathbf{h}^2}}{2\sigma^2} \right) \tag{17}$$

with

$$\overline{\mathbf{h}^2} = \frac{1}{2\rho} \sum_{j=1}^{2\rho} \mathrm{h}_j^2. \tag{18}$$

This represents the mean-square of the observed projections. If σ is known, the joint posterior probability density of the parameters $\boldsymbol{\omega}$, conditional on the data and our knowledge of σ is given by

$$P\left(\boldsymbol{\omega} \middle| \mathbf{D}, \sigma, I\right) \propto exp\left(\frac{\rho\overline{\mathbf{h}^2}}{\sigma^2} \right). \tag{19}$$

If there is no prior information about noise, then σ is known as a nuisance parameter and must also be eliminated by integrating it out. Using Jeffreys prior (Jeffreys, 1961) $\frac{1}{\sigma}$ and integrating Equation (12) over σ we obtain

$$P(\boldsymbol{\omega}|\mathbf{D},I) \propto \left(1-\frac{2\rho\overline{\mathbf{h}^2}}{N\overline{\mathbf{D}^2}}\right)^{\frac{2\rho-N}{2}}. \tag{20}$$

This has the form of the "Student *t*- distribution". As well as determining the values of the $\boldsymbol{\omega}$ parameters for which the posterior PDF for $\boldsymbol{\omega}$ is a maximum, it is also desirable to compute uncertainties associated with them. To do this, let us assume the case where σ is known and let $\hat{\boldsymbol{\omega}}$ represent the estimated values of $\boldsymbol{\omega}$. Following to Bretthorst's work (Bretthorst, 1988), we can expand the function $\overline{\mathbf{h}^2}$ in a Taylor series at the point $\hat{\boldsymbol{\omega}}$, such that

$$P(\boldsymbol{\omega} \mid \mathbf{D},I) \propto exp\left(-\sum_{j=1}^{\rho}\sum_{k=1}^{\rho}\frac{b_{jk}}{2\sigma^2}\Delta_j\Delta_k\right) \tag{21}$$

with b_{jk} defined as $b_{jk} = -\rho\left.\frac{\partial^2\overline{\mathbf{h}^2}}{\partial\omega_j\partial\omega_k}\right|_{\substack{\omega_j=\hat{\omega}_j\\ \omega_k=\hat{\omega}_k}}$ and $\Delta_j = (\hat{\omega}_j - \omega_j)$ for a single frequency case. For an arbitrary model the matrix b_{jk} cannot be calculated analytically; however, it can be evaluated numerically. The calculations of the mean and standard deviations for $\boldsymbol{\omega}$ parameters require for evaluating Gaussian integrals by first changing to the orthogonal variables as was done above with the amplitudes. Let τ_j and u_{kj} represent the jth eigenvalue and eigenvector of the matrix b_{jk}, respectively. Then the new orthogonal variables are given by

$$s_j = \sqrt{\tau_j}\sum_{k=1}^{\rho}\Delta_k u_{kj}, \quad \Delta_j = \sum_{k=1}^{\rho}\frac{s_k u_{jk}}{\sqrt{\tau_k}}. \tag{22}$$

By using these orthogonal variables to perform the ρ Gaussian integrals, the estimate variance for ω_k can be calculated:

$$\sigma_{\omega}^2 = \left\langle\sigma^2\right\rangle\sum_{j=1}^{\rho}\frac{u_{jk}^2}{\tau_j}. \tag{23}$$

Therefore, the approximations for $\boldsymbol{\omega}$ can be implemented in the form:

$$\langle\omega_j\rangle = \hat{\omega}_j \pm \sigma_{\omega_j}, \quad (j = 1,2,3,\ldots,\rho). \tag{24}$$

In agree with Bretthorst, the expectation values of the amplitudes are given by $\langle A_j\rangle = h_j$. From Equation (15) the expected values for the old amplitudes $\mathbf{B}$ becomes

$$\langle B_k\rangle = \sum_{j=1}^{2\rho}\frac{1}{\sqrt{\lambda_j}}h_j\upsilon_{jk}. \tag{25}$$

The uncertainty in A_j is $\pm\sigma$, so that the corresponding uncertainty in $\mathbf{B}$ is

$$\sigma_{\mathbf{B}} = \sigma \sum_{j=1}^{2\rho} \frac{1}{\sqrt{\lambda_j}} \upsilon_{jk} \upsilon_{jk} \quad (k = 1,2,3,...,2\rho) \tag{26}$$

3. Implementation of Simulated Annealing (SA) algorithm

Bayesian approach introduced by Bretthorst is briefly summarized in section 2 but, it is referred to Bretthorst's works (Bretthorst, 1988) for more detail information. Consequently, Bayesian parameter estimation turns into a global optimization problem which is a task to find the best possible solution $\boldsymbol{\omega}$ for Equations (19) or (20) satisfying

$$\max_{\omega \in \Xi_\rho} \left\{ P\left(\boldsymbol{\omega} \middle| \mathbf{D}, I\right) \right\}, \tag{27}$$

in the solution space $\Xi_\rho = \{0, \pi\}$ .Because there is no variation at negative frequencies and the highest possible frequencies corresponds to wave that under goes a complete cycle in two unit intervals, so the lower limit on the range is 0 and all the variation is accounted for by frequencies less than π .

Over last few decades, researchers have developed many computational algorithms to address such type of global optimization problems (Metropolis et al., 1953; Jeffreys, 1961; Kirkpatrick, et al., 1983; Laarhoven & Aarts, 1987; Stefankovic et al., 2009). Although there are numerous algorithms which are suggested to achieve this goal, few of them are capable of locating it effectively. Therefore, we follow the SA algorithm, suggested by Corana (Corana et. al., 1987) and modified by Goffe (Goffe et al., 1994), which is a kind of probabilistic algorithm for finding the global optimum of Equation (27) although its various alternative versions have already been used in statistical applications. A brief review of the most work on many algorithms based on that of SA, together with areas of applications is provided by Ingber and Binder (Binder, 1986; Ingber, 1994).

The algorithm begins with an initial guess of the frequencies $\boldsymbol{\omega}^0$, a trial step-length vector $\mathbf{v}^0$ and a global parameter Γ_0 (called the initial temperature). Each step of the SA algorithm replaces the current frequency with randomly generated new frequency. In other words, the next candidate point $\boldsymbol{\omega}^{k+1}$ is generated by varying one component $j \in \{1,...,\rho\}$ of the current point $\boldsymbol{\omega}^k$ at a time:

$$\omega_j^{k+1} = \omega_j^k + \xi v_j^k, \tag{28}$$

where ξ is a uniformly distributed random number from the interval [-1,1] and $\mathbf{v}^k = \left\{ v_j^k \right\}$ is a step vector. The function value of $P\left(\boldsymbol{\omega}^{k+1} \middle| \mathbf{D}, I\right)$ is then computed. If

$$P\left(\boldsymbol{\omega}^{k+1} \middle| \mathbf{D}, I\right) \geq P\left(\boldsymbol{\omega}^k \middle| \mathbf{D}, I\right) \tag{29}$$

then the point $\boldsymbol{\omega}^{k+1}$ is accepted as the $(k+1)$th iteration point, it is replaced with $\boldsymbol{\omega}^{k}$ and algorithm moves uphill. If $P(\boldsymbol{\omega}^{k+1}|\mathbf{D},I)$ is the largest posterior PDF, denoted as $P_{opt}(\boldsymbol{\omega}|\mathbf{D},I)$ its value and $\boldsymbol{\omega}^{k+1}$ are recorded since this is the best current value of the optimum. This forces the system toward a state corresponding to a local maximum or possibly a global maximum. However, most large optimization problems, like the one given in Equation (27), have many local maxima and optimization algorithm is therefore often trapped in a local maximum. To get out of a local maximum, a decrease of the function value $P(\boldsymbol{\omega}^{k+1}|\mathbf{D},I)$ is accepted with a certain probability. This is accomplished by the Metropolis-criterion (Metropolis et al., 1953) which is based on the changes of obtaining new state with the posterior PDF of frequencies value, defined as

$$p = \frac{\exp\left(-\dfrac{P(\boldsymbol{\omega}^{k+1}|\mathbf{D},I)}{\Gamma^k}\right)}{\exp\left(-\dfrac{P(\boldsymbol{\omega}^{k+1}|\mathbf{D},I)}{\Gamma^k}\right)+\exp\left(-\dfrac{P(\boldsymbol{\omega}^{k}|\mathbf{D},I)}{\Gamma^k}\right)}, \approx \exp\left(-\frac{\Delta P}{\Gamma^k}\right) \tag{30}$$

where ΔP represents difference between the present and previous posterior PDF values of frequencies, e.g., $\Delta P = P(\boldsymbol{\omega}^{k+1}|\mathbf{D},I) - P(\boldsymbol{\omega}^{k}|\mathbf{D},I)$. Whenever

$$P(\boldsymbol{\omega}^{k+1}|\mathbf{D},I) < P(\boldsymbol{\omega}^{k}|\mathbf{D},I) \tag{31}$$

p is computed and compared to p', a uniformly distributed random number from the interval $(0,1)$. If $p > p'$, the new point $\boldsymbol{\omega}^{k+1}$ is accepted and replaced with $\boldsymbol{\omega}^{k}$ and the algorithm moves downhill, i.e. lower temperatures and larger differences in posterior PDF's values. This continues until all ρ components have been altered and thus ρ new points have successively accepted or rejected according to the Metropolis criterion. After this process is repeated n_s times the step vector $\mathbf{v}^k$ is adjusted by the following rule:

$$v_j^{k+1} = \begin{cases} v_j^k\left(1+c_j\dfrac{(n_j-0.6n_s)}{0.4n_s}\right) & if\dfrac{n_j}{n_s}>0.6 \\ \dfrac{v_j^k}{\left(1+c_j\dfrac{(0.4n_s-n_j)}{0.4n_s}\right)} & if\dfrac{n_j}{n_s}<0.4 \\ v_j^k & \text{otherwise} \end{cases} \tag{32}$$

where n_j is the number of accepted moves along the direction j and the parameter c_j, which is initially defined by user, controls the step variation along the jth direction. The aim of these variations in a step length is to maintain average percentage of accepted moves at about one-half of the total number of moves.

An alternative step size of above SA algorithm is given by replacing Equation (28) with

$$\omega_j^{k+1} = \omega_j^k + \Gamma_k \sqrt{\gamma_j^k} N(0,1), \; (j = 1,2,\ldots,\rho). \tag{33}$$

where $N(0,1)$ is a standard Normal distribution, Γ_k represents the temperature at the kth iteration and rescales the step size and ordering ω_j. On the other hand, γ_j^k is the CRLB for the jth component of angular frequencies at the $(k+1)$th iteration (Ireland, 2007) and provides a theoretical lower limit on how precisely parameter ω_j can be extracted from noisy measurements. In this respect, it is defined in the form:

$$\gamma_j = \mathbf{J}^{-1}(\omega_j) \tag{34}$$

where the Fisher information matrix $\mathbf{J}(\boldsymbol{\omega})$ (Kay, 1993), defined by

$$\mathbf{J}(\boldsymbol{\omega}) = E\left[\frac{\partial^2 \ln P(\mathbf{D}|\boldsymbol{\omega}, I)}{\partial \boldsymbol{\omega}^2}\right] = \frac{1}{\sigma^2}\sum_{j=1}^{N} \frac{\partial f_j(\boldsymbol{\omega})}{\partial \boldsymbol{\omega}} \left(\frac{\partial f_j(\boldsymbol{\omega})}{\partial \boldsymbol{\omega}}\right)^T, \tag{35}$$

is an expectation of the second derivatives of the signal function with respect to $\boldsymbol{\omega}$. Assuming that the matrix $\mathbf{J}(\boldsymbol{\theta})$ is diagonal for a large N so that its inversion is straightforward. In this case, the diagonal elements yield the lower bound (Stoica et al., 1989; Lagrange, 2005) for the variance of $\hat{\boldsymbol{\omega}}$ asymptotically and we can write,

$$\gamma_j = \mathrm{var}_{CRLB}\left(\hat{\omega}_j\right) \cong \frac{24\hat{\sigma}^2}{N^3(\hat{B}_j^2 + \hat{B}_{j+\rho}^2)}, \tag{36}$$

where $\hat{\sigma}^2$ represents the estimated variance of the noise and is described in Equation (41). This whole cycle is then repeated n_Γ times, after which the temperature is decreased by a factor $\alpha_\Gamma \in (0,1)$. This process is generally called annealing (or cooling) schedule which is the heart of the algorithm and effects the number of times the temperature is decreased. If a fast cooling takes place then the problem will be trapped in a local maximum. Therefore, there are various annealing schedules suggested by different researchers (Ingber, 1994; Stefankovic, et al., 2009) for lowering the temperature but we choose the following:

$$\Gamma_{k+1} = \alpha_\Gamma \frac{\Gamma_k}{\exp\{(\alpha_\Gamma - 1)k^{\frac{1}{\rho}}\}} \tag{37}$$

Because of being exponential rather than logarithmic, it is sometimes known as simulated quenching (SQ) (Vasan et al., 2009; Aguiare et al., 2012). In case of a well conditioned estimation problem like, say, frequency estimation problem in signal processing, it is clear that the convenience of SA algorithm, together with a need for some global search over local optima, makes a strong practical case for the use of SQ. Therefore, different parameters have different finite ranges and different annealing time- dependent sensitivities. Classical annealing algorithms have distributions that sample infinite ranges and there is no decision for considering differences in each parameter dimension; e.g. different sensitivities might be necessary to use different annealing schedules. This requires the development of a new PDF to embed in the desired features (Ingber, 1994) so that it leads to variants of SA that justifies exponential temperature annealing schedules.

Termination of the algorithm occurs when average function value of the sequences of the points after each $n_s \times n_\Gamma$ step cycle reaches a stable state:

$$\begin{aligned} &\left|P_k\left(\boldsymbol{\omega}\middle|\mathbf{D},I\right)-P_{k-l}\left(\boldsymbol{\omega}\middle|\mathbf{D},I\right)\right|\le\varepsilon\ \ (l=1,\dots,4)\\ &\left|P_k\left(\boldsymbol{\omega}\middle|\mathbf{D},I\right)-P_{opt}\left(\boldsymbol{\omega}\middle|\mathbf{D},I\right)\right|\le\varepsilon \end{aligned} \tag{38}$$

where ε is a small positive number defined by user and l indicates the last four successive iteration values of the posterior PDF of the frequencies that are being stored. Further details of the algorithm initialization are problem-dependent and are given in Section 5.

4. Power spectral density

Before we discuss the computer simulated examples, there is something we need to say about how to display the results. The usual way the result from a spectral analysis is displayed is in the form of a power spectral density that shows the strength of the variation (energy) as a function of frequency. In Fourier transform spectroscopy this is typically taken as the squared magnitude of the discrete Fourier transform of the data. In order to display our results in the form of a power spectral density (Bretthorst, 1988; Gregory, 2005), it is necessary to give an attention to its definition that shows how much power is contained in a unit frequency. According to Bretthorst (Bretthorst, 1988) the Bayesian power spectral density is defined as the expected value of the power of the signals over the joint posterior PDF:

$$S(\boldsymbol{\omega})=\frac{N}{2}\int\sum_{j=1}^{\rho}\left(B_j^2+B_{j+\rho}^2\right)P\left(\boldsymbol{\omega},\mathbf{B}\middle|D,\sigma^2,I\right)dB_1\dots dB_{2\rho}. \tag{39}$$

Performing integrals analytically over $B_1,\dots,B_{2\rho}$ by using these orthonormal model functions defined in section 2, the power spectral density can therefore be approximated as

$$S(\boldsymbol{\omega})=2\left(\sigma^2+\sum_{j=1}^{2\rho}h_j^2(\hat{\boldsymbol{\omega}})\right)\sum_{k=1}^{\rho}\sqrt{\frac{b_{kk}}{2\pi\sigma^2}}\exp\left(-\frac{b_{kk}(\hat{\omega}_k-\omega)^2}{2\sigma^2}\right). \tag{40}$$

This function stresses information about the total energy carried by the signal and about the accuracy of each line. In the next section, we will present some numerical examples how well this technique works.

5. Computer simulated examples

To verify the performance of the algorithm, we generated a simulated data vector according to one, two and with five sinusoids. Here t_i runs over the symmetric time interval $-T$ to T in $(2T+1)=512$ integer steps and the components of the noise e_i are generated from the zero mean Gaussian distribution with a known deviation σ, initially and added to the simulated data. However, one of the interests in an experiment is also to estimate noise power σ^2 so that it is assumed to be unknown. In agreement with Bretthorst, this is given in the following form:

$$\langle\sigma^2\rangle = \frac{\left(N\overline{\mathbf{D}^2} - 2\rho\overline{\mathbf{h}^2}\right)}{N-2\rho-2}\left(1 \pm \sqrt{\frac{2}{N-2\rho-4}}\right). \tag{41}$$

Clearly, it is seen that the accuracy of the estimate depends on how long we sample and the signal-to-noise ratio (SNR), defined as the ratio of the root mean square of the signal amplitude to the noise σ^2. In addition, one may also get the following expression of SNR:

$$\text{SNR} = \sqrt{\frac{2\rho}{N}\left(1 + \frac{\overline{\mathbf{h}^2}}{\sigma^2}\right)}. \tag{42}$$

When the standard deviation of the noise is unknown, an empirical SNR is obtained by replacing σ^2 in Equation (42) with the estimated noise variance in (41).

In our first example, we generate the data set from the following equation:

$$d_i = 0.001 + 0.5403\cos(0.3t_i) - 0.8415\sin(0.3t_i) + e_i \quad (i = 1,\ldots,512). \tag{43}$$

We then carried out the Bayesian analysis of the simulated data, assuming that we know the mathematical form of the model but not the value of the parameters. We first gave starting values to the list of frequencies to begin a multidimensional search for finding a global maximum of the posterior PDF of the frequencies $\boldsymbol{\omega}$ given in Equations (19) or (20). As an initial estimate of the frequencies $\boldsymbol{\omega}^0$ for the maximization procedure, it is possible to take random choices from the interval $(0,\pi)$. However, it is better to start with the locations of the peaks chosen automatically from the Fourier power spectral density graph by using a computer code written in Mathematica.

In agreement with Corana (Corana, et al., 1987), reasonable values of the parameters that control the SA algorithm are chosen as $n_s = 20$, $n_\Gamma = \max(100, 5\rho)$ and $c_j = 2$, $(j = 1,\ldots,\rho)$. Then the global optimization algorithm starts at some high temperature $\Gamma_0 = 100$. Thus the sequence of points is generated until a sort of equilibrium is approached; that is a sequence of points $\boldsymbol{\omega}^0, \boldsymbol{\omega}^1, \boldsymbol{\omega}^2, \ldots$ whose average value of $P(\boldsymbol{\omega}|\mathbf{D}, I)$ reaches a stable value as iteration proceeds. During this phase the step vector $\mathbf{v}$ is periodically adjusted by the rule defined in Equation (32). The best point $\boldsymbol{\omega}$ reached so far is recorded. After thermal equilibration, the temperature Γ is reduced by using the annealing scheduled in Equation (37) with a factor $\alpha_\Gamma = 0.85$ and a new sequence is made starting from this best point $\boldsymbol{\omega}$, until thermal equilibrium is reached again and so on. The SA algorithm first builds up a rough view of the surface by moving large step lengths. As the temperature Γ falls and the step decreases, it is slowly focuses on the most promising area. Therefore it proceeds toward better maximum even in the presence of many local maxima by adjusting the step vector that can allow the algorithm to shape the space within it may move, to that within which ought to be as defined by the PDF $P(\boldsymbol{\omega}|\mathbf{D}, I)$. Consequently, the process is stop at a temperature low enough that no more useful improvement can be expected, according to a stopping criterion in Equation (38).

	Bayesian Parameter estimation	
σ : Known N : 512 3ρ : 3 SNR : 0.725464 $\hat{\sigma}$: 1	Parameters	Estimated Values
	ω_1	0.2998 ± 0.0005
	B_1	0.6035 ± 0.0589
	B_2	-0.8174±0.0626

Table 1. Computer simulations for a single harmonic frequency model

Once the frequencies are estimated, we then carried on calculating the amplitudes and parameter errors approximately using Equations (25), (23) and (26), respectively. However, an evaluation of the posterior PDF at a given point $\boldsymbol{\omega}$ cannot be made analytically. It requires a numerical calculation of projections onto orthonormal model functions, related to Eigen-decomposition of the $(2\rho \times 2\rho)$ dimensional matrix $\Omega(\boldsymbol{\omega})$. Therefore, the proposed algorithm was coded in *Mathematica* programming language (Wellin, P., et al., 2005), that provides a much flexible and efficient computer programming environment. Furthermore, it also contains a large collection of built-in functions so that it results much shorter computer codes than those written in C or FORTRAN programming languages.

The computer program was run on the workstation with four processors, which of each has got Intel Core 2 Quad Central Processing Unit (CPU), in two cases where the standard deviation of noise is known or not. The output of the computer simulations when $\sigma = 1$ is illustrated in Table 1. The results when σ is unknown are almost similar with that of Table 1. Parameter values are quoted as (*value*) ± (*standard deviation*). It can be seen that a single frequency and amplitudes are recovered very well. The estimated value of SNR and the standard deviation of the noise are also shown in Table 1.

In our second example, we consider a signal model with two close harmonic frequencies:

$$d_i = 0.5403\cos(0.3t_i) - 0.8415\sin(0.3t_i) - 0.4161\cos(0.31t_i) - 0.9093\sin(0.31t_i), \quad (i = 1,...,512). \tag{44}$$

In a similar way, we produced the same size data corrupted by the zero mean Gaussian noise with $\sigma = 1$. We run our *Mathematica* code again in the case where the deviation of noise is unknown. The results, shown in Table 2, indicate that all values of the parameters within the calculated accuracy are clearly recovered.

	Bayesian Parameter estimation	
	Parameters	Estimated Values
σ : Unknown N : 512 $3p$: 6 SNR : 1.02936 $\hat{\sigma}$: 0.99639	ω_1	0.3000±0.0009
	ω_2	0.3105±0.0010
	B_1	0.5206±0.0632
	B_2	-0.8698±0.0631
	B_3	-0.3819±0.0634
	B_4	-0.8556±0.0636

Table 2. Computer simulations for two closed harmonic frequency model

In general, we consider a multiple harmonic frequency model signal:

$$d_i = \cos(0.1\, t_i + 1) + 2\cos(0.15t_i + 2) + 5\cos(0.3t_i + 3) + 2\cos(0.31t_i + 4) + 3\cos(t_i + 5) + e_i \quad (i = 1,...,512). \tag{45}$$

The best estimates of parameters are tabulated in Table 3. Once again, all the frequencies have been well resolved, even the third and fourth frequencies which are too closed not to be separated by the Fourier power spectral density shown in Figure 3. Actually with the Fourier spectral density when the separation of two frequencies is less than the Nyquist step, defined as $2\pi / N$, the two frequencies are indistinguishable. This is simply because there are no sample points in between the two frequencies in the frequency domain. If $|\omega_3 - \omega_4| > 2\pi / N$ theoretically the two frequencies can then be distinguished. If $|\omega_3 - \omega_4|$ is not large enough, the resolution will be very poor. Therefore, it is hard to tell where the two frequencies are located. This is just the inherent problem of the discrete Fourier power spectral density. In this example two frequencies are separated by 0.01, which is less than the Nyquist step size. There is no way by using Fourier power spectral density that one can resolve the closed frequencies less than the Nyquist step. However, Bayesian power spectral density shown in Figure 3 gives us very good results with high accuracy. Finally, we constructed the signal model in Equation (3), whose parameters, amplitudes and frequencies, are randomly chosen from uniform distribution in the intervals $[-5,5]$ and $[0,\pi]$, respectively and used it to generate data samples of $N=512$ by adding a zero mean Gaussian random noise with $\sigma=1$. The proposal algorithm was rerun for recovering sinusoids from it and the results are tabulated in Table (4). It can be seen that frequencies are specially recovered with high accuracies. Ten frequencies signal model are shown in Fig.5.

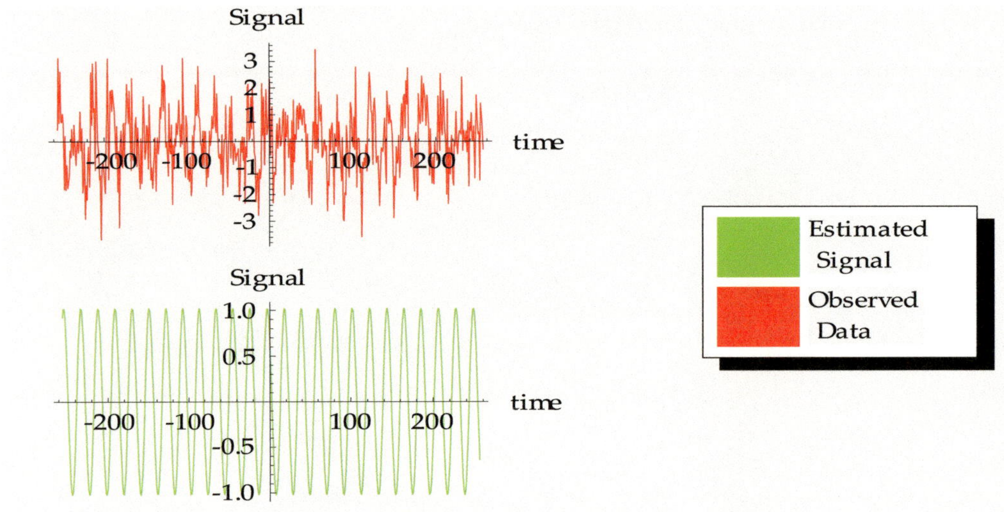

Figure 1. Recovering signal from noisy data produced from a single harmonic frequency signal model.

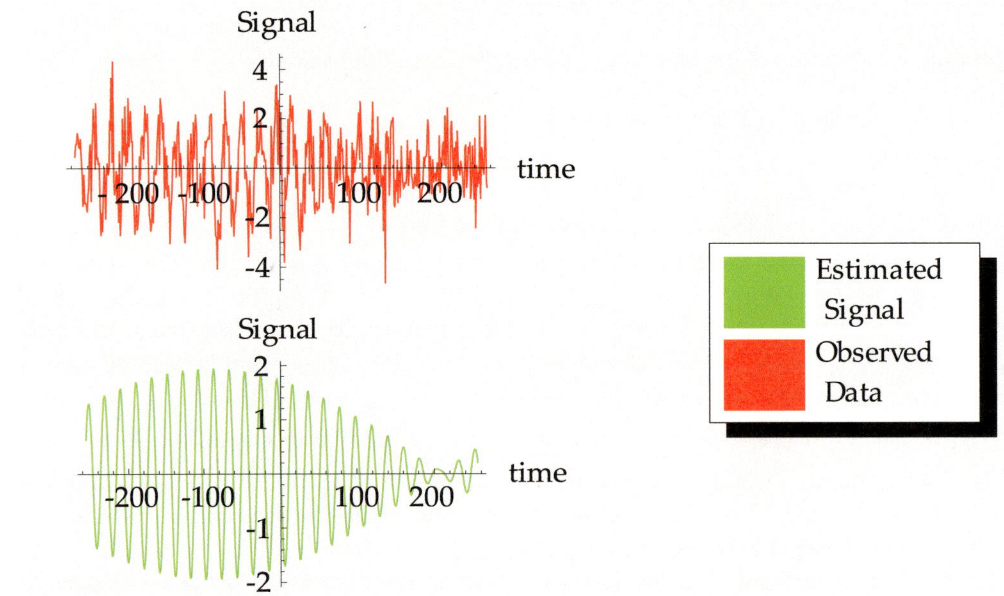

Figure 2. Recovering signals from noisy data produced from two closed harmonic frequency signal model.

Parameters		True values	Estimated values	Bretthorst's results
Frequencies	ω_1	0.10	0.1006 ± 0.0004	0.0998 ± 0.0001
	ω_2	0.15	0.1495 ± 0.0002	0.1498 ± 0.0002
	ω_3	0.30	0.3001 ± 0.0001	0.3001 ± 0.0002
	ω_4	0.31	0.3099 ± 0.0004	0.3120 ± 0.0001
	ω_5	1.00	1.0000 ± 0.0001	0.9999 ± 0.0001
Amplitudes	a_1	1.00	0.9905 ± 0.08	0.99 ∓ 0.08
	a_2	2.00	1.9600 ± 0.08	2.08 ± 0.08
	a_3	5.00	5.1058 ± 0.09	4.97 ± 0.08
	a_4	2.00	1.8199 ± 0.09	1.95 ± 0.08
	a_5	3.00	2.9556 ± 0.08	2.92 ± 0.08

Table 3. Computer simulations for a multiple harmonic frequency model[2]

On the other hand, modifications of this algorithm (Üstündag & Cevri, 2008; 2011) have already been made by generating the next candidate in Equation (33) from normal distribution with a mean of the current estimation whose standard deviation is a square root of the CRLB (Ireland, 2007) given in Equation (36), which is a lower limit to the variance of the measurement of the frequencies, so this generates a natural scale size of the search space around their estimated values. It is expected that better solutions lie close the ones that are already good and so normally distributed step size is used. Consequently, the results we obtained are comparable with or higher than those obtained previously. In addition, all the results discussed so far are also consistent with those of Bretthorst (Bretthorst, 1988) and also demonstrate the advantages of the Bayesian approach together with SA algorithm. Moreover it appears to be very reliable, in the sense that it always converged to neighborhood of the global maximum. The size of this neighborhood can be reduced by altering the control parameters of the SA algorithm, but this can be expensive in terms of CPU consumption. Moreover, we initially assumed that the values of the random noise in data were drawn from the Gaussian density with the mean $\mu = 0$ and the standard deviation σ. Figure 4 shows the exact and estimate probability densities of the random noise in data. It is seen that the estimated (dotted) probability density is closer to the true (solid) probability density and the histogram of the data is also much closer to the true probability density. The histogram is known as a nonparametric estimator of the probability density because it does not depend on specified parameters.

Computer simulations had been carried out to compare the performance of the method with the CRLB. To do this, we generated 64 data samples from a single real tone frequency signal

[2]In order to compare the results with Bretthorst's in this example we converted $a_i = \sqrt{B_i^2 + B^2_{i+\rho}}, (i = 1,\ldots,\rho)$.

model $(\omega = 0.3, B_1 = B_2 = 1)$ and added it to the variety of noise levels. After 50 independent trials under different SNR ratios, the mean square errors (MSE) for the estimated frequencies were obtained and their logarithmic values were plotted with respect to SNR ratios that vary between zero and 20 dB (deci Bell). It can be seen from Figure 6 that the proposed estimator has threshold about 3 dB of the SNR and follows nicely the CRLB after this value. As expected, larger SNR gives smaller MSE. However, many of existing methods in signal processing literature have a MSE that is close to the CRLB when the SNR is more than 20 dB and they usually perform poorly when the SNR is decreased.

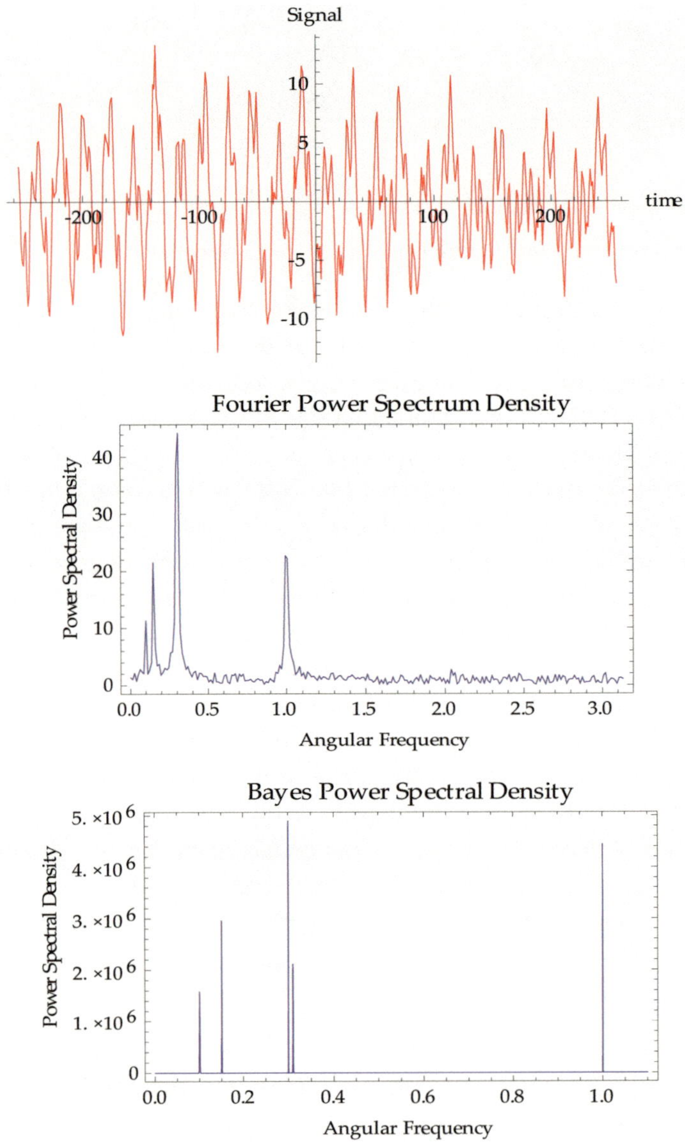

Figure 3. Spectral analysis of multiple frequency signal model

True Frequencies									
ω_1	ω_2	ω_3	ω_4	ω_5	ω_6	ω_7	ω_8	ω_9	ω_{10}
0.36824	0.4109	0.4841	0.4961	0.8146	0.9715	1.0167	1.5978	3.0807	3.1150
True Amplitudes									
B_1	B_2	B_3	B_4	B_5	B_6	B_7	B_8	B_9	B_{10}
-1.838	-2.156	-3.096	2.919	2.765	2.519	-0.185	4.994	-3.862	1.371
B_{11}	B_{12}	B_{13}	B_{14}	B_{15}	B_{16}	B_{17}	B_{18}	B_{19}	B_{20}
4.634	-0.691	-4.947	-4.441	-2.984	-1.409	4.559	1.0755	-3.004	0.206
Estimated Frequencies									
ω_1	ω_2	ω_3	ω_4	ω_5	ω_6	ω_7	ω_8	ω_9	ω_{10}
0.3683 ±0.00002	0.4107 ±0.00003	0.4842 ±0.00002	0.4961 ±0.00004	0.8145 ±0.00003	0.9715 ±0.00002	1.0166 ±0.00002	1.5979 ±0.00002	3.0807 ±0.00002	3.1153 ±0.00007
Estimated Amplitudes									
B_1	B_2	B_3	B_4	B_5	B_6	B_7	B_8	B_9	B_{10}
-1.8644 ±0.06	-2.1858 ±0.06	-3.1385 ±0.06	2.7857 ±0.06	2.7536 ±0.06	2.5492 ±0.06	-0.2409 ±0.06	4.9877 ±0.06	-3.8841 ±0.06	1.2779 ±0.06
B_{11}	B_{12}	B_{13}	B_{14}	B_{15}	B_{16}	B_{17}	B_{18}	B_{19}	B_{20}
4.6103 ±0.06	-0.6783 ±0.06	-5.0206 ±0.06	-4.3572 ±0.06	-3.0533 ±0.06	-1.3511 ±0.06	4.6550 ±0.06	1.1273 ±0.06	-2.9807 ±0.06	0.2479 ±0.06

Table 4. Computer simulations for ten harmonic frequency signal model

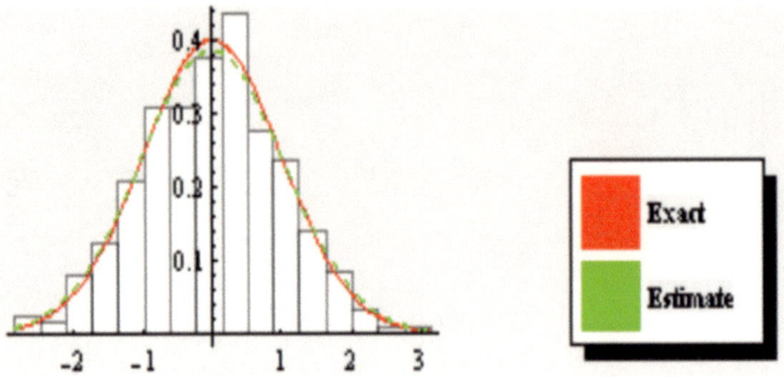

Figure 4. Comparison of exact and estimate probability densities of noise in data $(\sigma = 1)$.

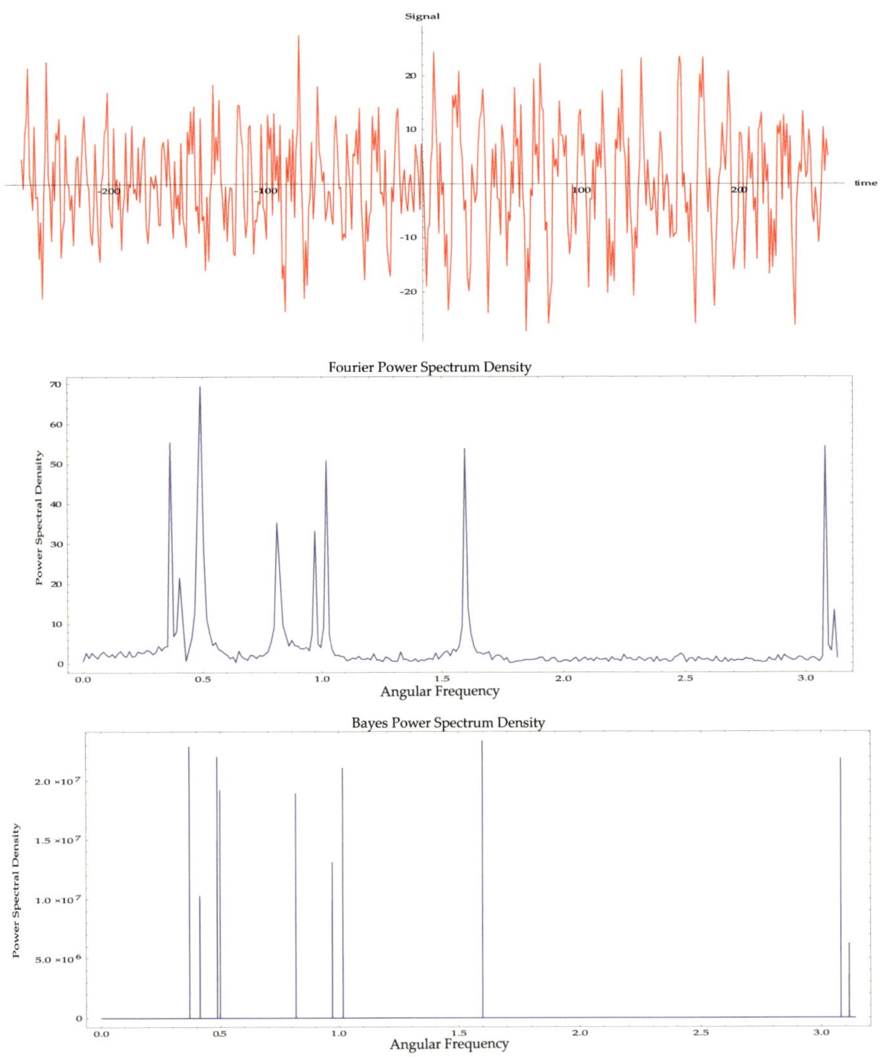

Figure 5. Spectral analysis of ten frequency signal model.

The computational complexity of the algorithm is dependent upon a few parameters such as annealing schedule, data samples and parameters. Under using same annealing schedule, Fig. 7 shows only CPU time of different simulations in a variety of number of data samples. It can be clearly seen that an increase in these numbers causes larger consumption of CPU time. With fixed size of components set and specifically annealing schedule of SA algorithm, the overall execution time of the cooling and decision is almost constant, but the runtime of the first two stages (move and evaluate) mostly depends on complicated design constraints and objective functions. Because the move and the evaluation process in the SA algorithm play an important role in CPU resource usage, improving the calculation ability for these stages will be the most feasible approach for an optimizing SA so that parallel computing is one of best approaches for this goal.

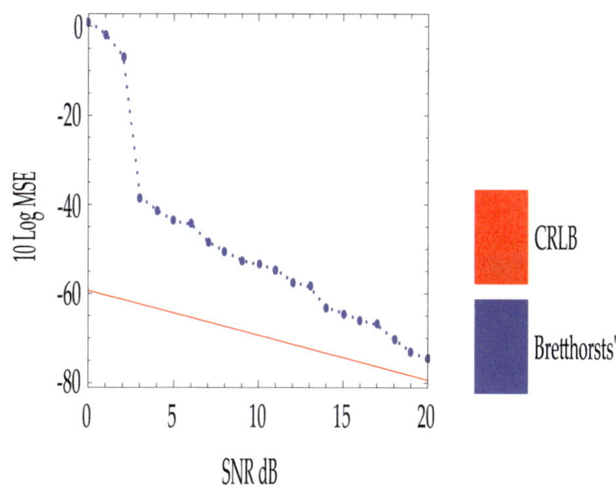

Figure 6. The calculated MSE of the proposed method compared with CRLB versus different SNR with a white noise.

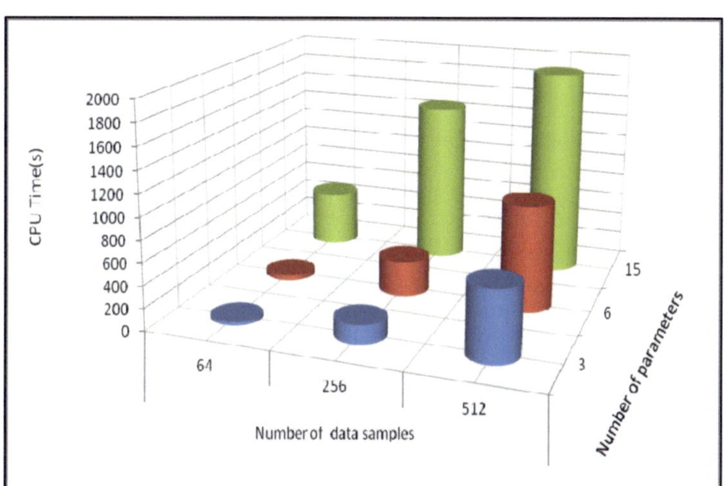

Figure 7. CPU times versus with number of parameters and data samples.

6. Conclusion

In this work we have partially developed a Bayesian approach combined with SA algorithm and applied it to spectral analysis and parameter algorithm for estimating parameters of sinusoids from noisy data. Overall, results presented here show that it provides rational approach for estimating the parameters, namely the frequencies and the amplitudes, can be recovered from the experimental data and the prior information with high accuracy, especially the frequency, which is the most important parameter in spectral analysis. A significant advantage of this approach comes from the very large posterior probabilities, which are sharply peaked in the neighborhood of the best fit. This helps us to simplify the problem of choosing starting values for the iteration and it provides a rational approach for estimating, in an optimal way, values of parameters by performing a random search. On the other hand, for sufficiently high SNR, MSEs of frequencies will attain CRLB so that it justifies the accuracy of the frequency estimation. Although the SA algorithm spends large consumption of CPU time, it is competitive when compared to the multiple runs often used with conventional algorithms to test different starting values. As expected, parallel implementation of SA algorithm reduces CPU resource usage.

Data analysis given in this chapter has also been applied to more complicated models and conditions, such as signals decay, periodic but non-harmonic signals, signals with non-stationary, etc.,. In general, we have also not addressed the problem of model selection which is the part of spectral analysis. In this case, one has enough prior information in a given experiment to select the best model among a finite set of model functions so that Bayesian inference helps us to accomplish it. Therefore, it will deserve further investigations.

Nomenclature

$\mathbf{B}$	: Nuisance parameters
$\{B_j, B_{j+\rho}\}$	: Amplitudes of jth sinusoidal
$P(.)$	: Marginal probability density function
$P(.\|.)$	: Conditional probability density function
b_{jk}	: A matrix defined as $-\rho \frac{\partial^2 \overline{\mathbf{h}^2}}{\partial\omega_j \partial\omega_k}\Big\|_{\substack{\omega_j=\hat{\omega}_j \\ \omega_k=\hat{\omega}_k}}$
c_j	: A parameter which controls the step variation along the jth direction
$\mathbf{D}$	: A set of observed data
$e(t)$	: Measurement errors
ε	: A small positive number defined by user
$E[.]$	: Expectation of a function or random variable
$G(.;.)$	: Model functions that contains sinus and cosines terms
$\mathbf{h}$	: Projections of data onto orthonormal model functions

$H(.;.)$	: Orthonormal model functions
$\overline{\mathbf{h}^2}$	: Mean of square of projections
$\mathbf{I}$	: Prior information
$\mathbf{J}(\cdot)$	: Fisher information matrix
$\mathbf{J}^{-1}(\cdot)$	: Inverse of Fisher information matrix
ρ	: Number of frequencies.
N	: Number of data samples
$N(0,1)$	A standard Normal distribution
n_j	: The number of accepted moves along the direction j.
k	: Number of iterations
Q	: Least square function
$S(.)$	: Bayesian power spectral density
s^2	: Sampling variance
$(t_1, t_2, ..., t_N)^T$	: Discrete time set
σ_ω	: Standard deviation of angular frequency
σ_B	: Standard deviation of amplitude
Ξ_ρ	: Solution space of angular frequencies
τ_j	: jth Eigenvalue of the matrix b_{jk}
u_{kj}	: jth Eigenvector of the matrix b_{jk}
α_Γ	: A factor between 0 and 1.
$\boldsymbol{\theta}$	: Parameters vector
λ_j	: j th component of normalized Eigenvalues of Ω_{jk}
υ_{kj}	: j th component of k th normalized Eigenvector of Ω_{jk}
$\mathbf{v}^k$	: Step-length vector
$\mathbf{v}^0$	: Trial step-length vector
ξ	: Uniformly distributed random number from the interval $[-1,1]$
σ^2	: Variance of noise
$\langle\sigma^2\rangle$	: Expected value of noise variance
$\gamma_j^{(k)}$	: CRLB for j th component of $\boldsymbol{\omega}^k$
$\boldsymbol{\omega}$	: Vector of angular frequency
$\hat{\boldsymbol{\omega}}$	: Estimated angular frequency vector
$\boldsymbol{\omega}^0$	: An initial guess vector for frequencies
Γ_0	: The initial temperature
$\boldsymbol{\Omega}$	: Diagonal matrix
$\lvert\cdot\rvert$	: Absolute value

CRLB	: Cramér-Rao lower bound
PDF	: Probability Density Function
SA	: Simulated Annealing
SNR	: Signal to Noise Ratio
SQ	: Simulated Quenching
MSE	: Mean Squared Errors
dB	: deci Bell

Author details

Dursun Üstündag
Marmara University, Faculty of Science and Letters, Department of Mathematics, Turkey

Mehmet Cevri
Istanbul University, Faculty of Science, Department of Mathematics, Turkey

Acknowledgement

This work is a part of the projects, whose names are "Bayesian Model Selection and Parameter Estimation" with a number FEN - DKR - 200407 - 0082 and "Bayesian Spectrum Estimation of Harmonic Signals" with a number FEN-BGS-060907-0191, supported by the Marmara University, Istanbul, Turkey.

7. References

Aguiare, H., Junior, O., Ingber, L., Petraglia, A., Petraglia, M.R & Machado, M.A.S. (2012). Stochastic Global Optimization and Its Applications with Fuzzy Adaptive Simulated Annealing. Springer Heidelberg New York Dorddrecht London, pp.36-57.

Barbedo, J.G.A. & Lopes, A. (2009). Estimating Frequency, Amplitude and Phase of two Sinusoids with Very Close Frequencies, *International Journal of Signal processing*, Vol.5, No.2, pp. 138-145.

Bernardo, J.M. & Smith, A.F.M. (2000). Bayesian *Theory, Wiley*, New York.

Binder, K. (1986). *Monte Carlo Methods in Statistical Physics*, 2nd Edition, Springer-Verlag.

Bretthorst, G.L. (1988). *Bayesian Spectrum Analysis and Parameter Estimation, Lecture Notes in Statistics*, Springer-Verlag Berlin Heidelberg, New York.

Bretthorst, G. L. (1990). An Introduction to Parameter Estimation Using Bayesian Probability Theory, *Maximum Entropy and Bayesian Methods*, Kluwer Academic Publishers, pp. 53-79.

Brooks S.P. & Morgan B.J.T. (1995). Optimization using simulated annealing, *Journal of Royal Society* (The Statistician) 44, pp. 241-257.

Chan K.W. & So, H. C. (2004). Accurate frequency estimation for real harmonic sinusoids, *IEEE Signal Processing Letters* 11, pp.609-612.

Cooley, J.W. & Tukey, J.W. (1964). An algorithm for the machine calculation of complex Fourier series, Mathematics of Computation *19, pp.* 297-301.

Corana, A., Marchesi, M., Martini, C. and Ridella, S. (1987). Minimizing multimodal functions of continuous variables with the simulated annealing algorithm, *ACM Transactions on Mathematical Software* 13, pp. 262-280.

Edwards, A.W.F. (1972). *Likelihood*, Cambridge University Press.

Gregory, P.C. (2005). Bayesian Logical Data Analysis for the Physical Science, Cambridge University Press.

Goffe, W.L., Ferier, G.D. & Rogers, J. (1994). Global optimization of statistical functions with simulated annealing, Journal of Econometrics *60, pp.* 65-100.

Harney, H.L. (2003). Bayesian Inference: Parameter Estimation and Decisions, Springer-Verlag Berlin Heidelberg.

Hooke, T.R. & Jevees, T.A. (1962). Direct search solution of numerical and statistical problems, *Journal of Association of Computer Machinery* 5, pp. 212-229.

Ingber, L. (1994). Simulated Annealing: Practice versus theory, *Mathematical computer Modeling* 18, pp. 29-57.

Ingber, L. (1996). Adaptive Simulated Annealing ASA: Lessons learned, *Control and Cybernetic 25*, pp. 33-54.

Ireland, J. (2007). Simulated annealing and Bayesian posterior distribution analysis applied to spectral emission line fitting, *Journal of Solar Physics* 243, pp. 237-252.

Jaynes, E. T. (1987). Bayesian Spectrum and Chirp Analysis, in Maximum Entropy and Bayesian Spectral Analysis and Estimation Problems, C. R. Smith and G. J. Erickson (eds.), D. Reidel, Dordrecht, p. 1-37.

Jaynes, E.T. (2003). *Probability Theory: The Logic of Science*, Cambridge University Press.

Jeffreys, H. (1961*). Theory of Probability*, Oxford University Press, London.

Kay, S. M. (1993). *Fundamentals of statistical signal processing: Estimation theory*, Prentice Hall, V.1.

Kirkpatrick, S., Gelatt, C. D., & Vecchi, M.P. (1983). Optimization by Simulated Annealing, *Journal of Science* 220, pp. 671-680.

Laarhoven, P.V. & Aarts, E. (1987). *Simulated annealing: Theory and Applications*, D. Reidel Publishing Company.

Locatelli, M. (2000). Simulated annealing algorithms for Continuous global optimization: Convergence Conditions, *Journal of Optimization Theory and Applications* 106, pp. 121-133.

Lagrange, M., Marchand, S. & Rault, J-Rr. (2005). Improving sinusoidal frequency estimation using a trigonometric approach, *Proceedings of the Digital Audio Effects Conference*, Spain,pp.110-115.

Metropolis, N., Rosenbluth, A., Rosenbluth, M., Teller, A. M. & Teller, E. (1953). Equation of states calculations by fast computing machines, *Journal of Chemical Physics* 21, pp. 1087-1092.

McWhorter T. & Scharf, L. L. (1993). Cramer-Rao Bound for deterministic modal analysis, *IEEE Transactions on Signal Processing* 41, pp. 1847-1866.

Nishiyama, K. (1997). A nonlinear filter for estimating a sinusoidal signal and its parameters in white noise: On the case of a single sinusoid, *IEEE transactions on signal processing* 45, pp. 970-981.

Norton, J.P. (1986). *An introduction to identification,* Academic Press, 1986.

Press, W.H., Flannery, B.P., Teukolshy, S.A. & Vetterling, W.T. (1995). Numerical recipes in C: The art of computing, Cambridge University Press.

Ruanaidh, J.J.K.Q. & Fitzgerald, W.J. (1996). *Numerical Bayesian Methods Applied to Signal Processing*, Springer-Verlag, New York.

Stoica, P. & Nehorai, A.(1989). MUSIC, Maximum Likelihood, and Cramer-Rao Bound, *IEEE Transactions on Acoustic Speech, and Signal Processing* 37, pp.720-741.

Schuster, A. (1905). The Periodogram and its optical analogy, Proceedings of the Royal Society of London, *Series A 77, pp.* 136-140.

Stefankovic, D., Vempala, S. & Vigoda, E. (2009). Adaptive simulated annealing: A near – optimal connection between sampling and counting. *Journal of the ACM* 56, doi:10.1145/1516512.1516520

Tsoulos, G. & Lagaris, I.E. (2006). GenAnneal: Genetically modified Simulated Annealing, *Computer Physics Communications* 174, pp. 846-85.

Üstündag, D., Queen & Bowcock, J.E. (1989). A method for retrieving images from noisy, incomplete data, *Proceeding of the Fifth Alvey Vision Conference*, University of Reading, pp. 239–245.

Üstündag, D., Queen, N.M., Skinner, G.K. & Bowcock, J.E. (1991). Two new methods for retrieving an image from noisy, incomplete data and comparison with the Cambridge MaxEnt package, *10. International Workshop on Maximum Entropy and Bayesian Methods*, MaxEnt 90, pp. 295-301.

Üstündağ, D. & Cevri, M. (2008). Estimating parameters of sinusoids from noisy data using Bayesian inference with simulated annealing, *WSEAS Transactions on Signal Processing* 7, pp. 432–441.

Üstündağ, D. & Cevri, M. (2011). Recovering sinusoids from noisy data using Bayesian inference with simulated annealing, *Mathematical and Computational Applications 16*, pp. 382-391.

Üstündağ, D. (2011). Recovering sinusoids from data using Bayesian inference with RJMCMC, *Proceeding of Seventh International Computation on Natural Computation (ICNN)*, Shanghai, pp. 1850-1854.

Vasan, A. & Raju, K.S. (2009). Comparative analysis of Simulated annealing, Simulated quenching and genetic algorithms for optimal reservoir operation, *Applied Soft Computing* 9, pp.274-281.

Wellin, P., Gayllord, R. & Kamin, S. (2005). An *Introduction to programming with Mathematica*, Cambridge University Press.

Chapter 5

Simulated Annealing: A Novel Application of Image Processing in the Wood Area

Cristhian A. Aguilera, Mario A. Ramos and Angel D. Sappa

Additional information is available at the end of the chapter

http://dx.doi.org/10.5772/50635

1. Introduction

Material's internal structure knowledge is highly relevant to improve quality indexes [1]. For example, the exact information of the internal structure of a wood log or lumber such as density and internal defects is an important economic advantage (see [2]). Internal characteristics of materials have been studied with different non-destructive techniques, including ultrasound [3], microwaves [4-7,8], gamma rays, X-rays, nuclear magnetic resonance and, lately, artificial vision techniques [9]. X-rays have been used to examine the internal characteristics of many materials [10-13].

In the wood industry, some applications have been found that allow the recognition of different types of defects of the raw material [14,15]. Methods for measuring moisture [16-18] have also been developed. Some works have been presented related to knot detection [19,20]. Recently, several approaches have been proposed in the literature along these lines of research; for example in the detection of wood defects [21,22], and more specifically to detect tree-ring knots in wood [23]. However, automatic defect recognition in the manufacture process is ever more necessary and fundamental.

An X-ray computerized tomography reflects variations on the density of an object or body. X-ray sensors reveal the shape of a log below the bark, allowing the detection of macroscopic and microscopic aspects. The latter favors its application in the wood industry, for example Figure 1 shows images of X-ray tomography on wood log. In these illustrations can be clearly seen internal constitutions such as knots and knotty cylinder: a) wood log, b) X-ray computed tomography, c) and d) images of two cuts.

The current work is focused on the visual inspection of wood in order to automatically detect classical defects such as knots and knotty cylinders. The proposed approach is based on the use of simulated annealing in deformable contours and X-ray tomography. The remainder of the manuscript is organized as follow. Firstly, the deformable contour technique, which is used to represent detected defect, is presented. Then, Section 3 introduces the simulated annealing-based optimization that is used to obtain an accurate description of wood's defect. Experimental results and discussions are given in Section 4. Finally, conclusions and future works are provided in Section 5.

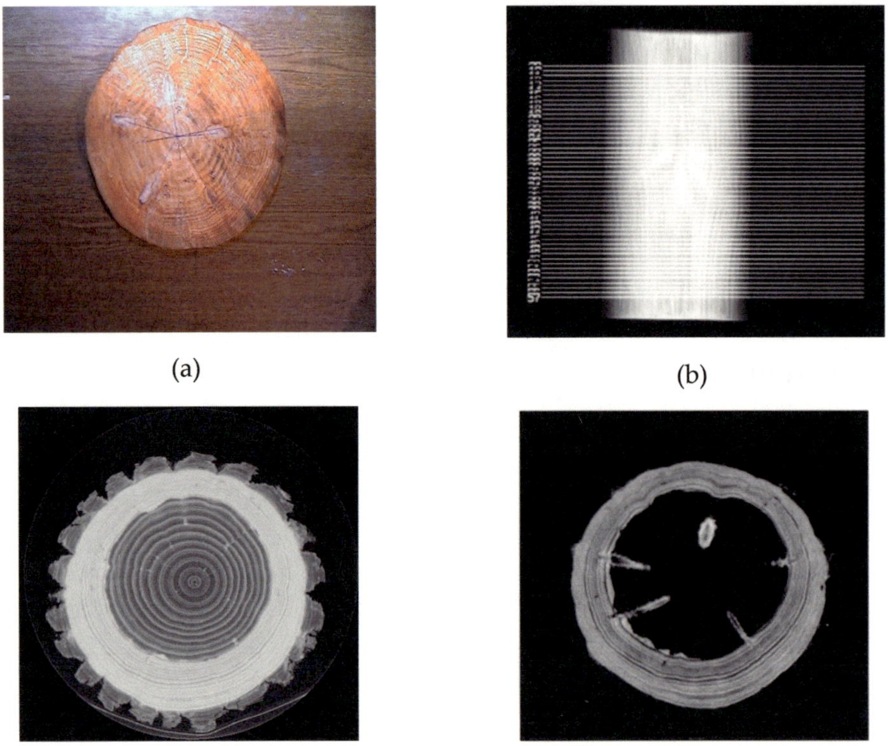

(a) (b) (c) (d)

Figure 1. X-ray tomography process of wood log.

2. The deformable contour

A deformable contour or snake [24] is a parametric curve of the type:

$$\boldsymbol{u}(\boldsymbol{s}) = (\boldsymbol{x}(\boldsymbol{s}), \boldsymbol{y}(\boldsymbol{s})) \tag{1}$$

where *x(s)* and *y(s)* are coordinates along the contour and $s \in [0,1]$. These contours are influenced by internal and external forces and forces typically related to the gradient of the intensity of the image, mathematically:

$$E_{snake} = \int_0^1 [E_{int}(u(s)) + E_{ext}(u(s)) + E_{img}(u(s))]ds \tag{2}$$

where:

$$E_{int}(x) = \underbrace{\alpha(s)\left|x_s(s)\right|^2}_{Tension} + \underbrace{\beta(s)\left|x_{ss}(s)\right|^2}_{Stiffness}$$

$E_{ext}(s)$, = shape energy measurement of external constraints either from higher level shape information or user applied energy.

$$E_{img} = -\alpha\|\nabla I\|^2$$

The internal energy, E_{int} , with alpha and beta parameters, controls the tension and rigidity of the contour; E_{ext} represents measures or external considerations with respect to the shape of the contour; and E_{img} is the force related to the gradient of the image given a Gaussian-type convolution filter. The contour is initially located near the object to be segmented in the image. Then, the attraction and repulsion forces, generated by internal, external image forces, deform the contour until surrounds the object; thereby isolating the object with respect to the overall image. Minimizing the energy equation allows the final solution of the contour (see Fig. 2).

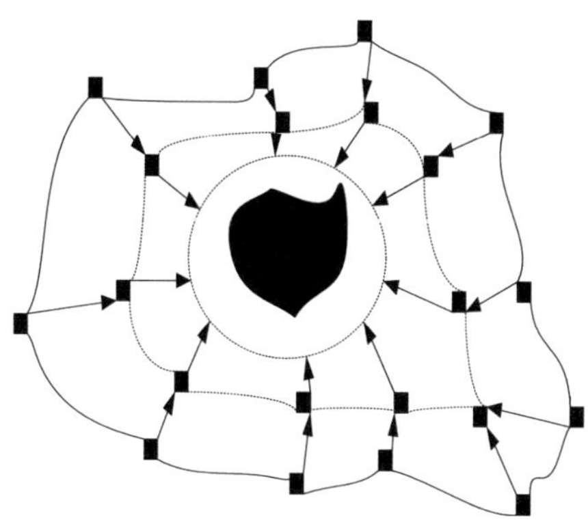

Figure 2. Evolution of a deformable contour.

Figure 3 presents a tomographic image of a log; Fig. 3(a) shows the tomographic image itself, in which two knots, and the appearance of a third knot, can clearly be seen at the bottom of the image; Fig. 3(b) shows the enlarged sector of the image corresponding to a particular knot. Figure 3(c) represents the gradient of the image; the most intense points represent the greatest values of the gradient. These forces guide the contour around the knot in order to completely isolate it, as shown in Fig. 3(d). One of the problems to confront is to identify, given the energy function and initial contour, the location of each new point of the contour. This can be done with different search algorithms.

The deformable contour presented in Fig. 3(d) capture the knot's shape present in the image but also some inaccuracies due to noisy date. Actually, these inaccuracies can be due to noise or distracter points; both points distort the contour by attracting or repulsing it (Fig. 4). This is the case of tomography X-ray images in which other objects or variation in object density (e.g., other knots or water saturation in wood pieces) artificially loses the form of the objects to detect, provoking by this influence, and a large error in segmentation.

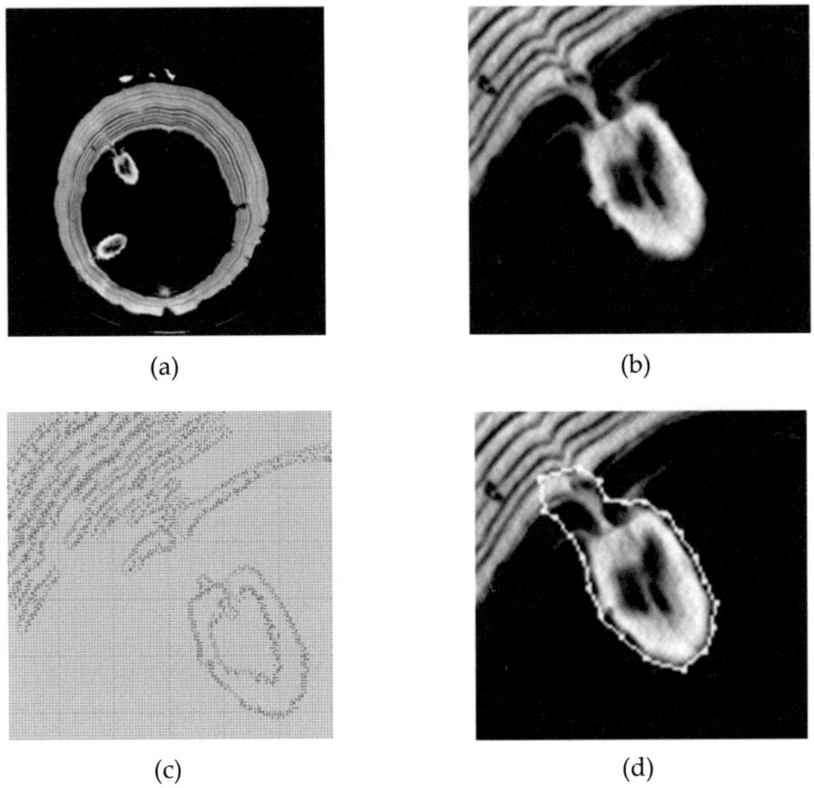

Figure 3. (a) Tomographic image; (b) knot (enlargement); (c) gradient of the image; (d) segmentation through a deformable contour.

For example, a tomography image with distinct singularities or objects can be appreciated in Fig. 5. In a) and b), the tomography with low humidity is presented, involving very clear and easy-to-segment images; however, in c) and d) an image with a very high humidity, can be appreciated. Since it is not homogeneously distributed, the object to be segmented can be deformed.

Several techniques have been employed to improve these aspects, like characteristic extraction techniques and improvements in the energy function. However, when there is a prior-knowledge about the objects to detect, such as in the case of well-defined objects and with known forms and characteristics, this information can be incorporated into the energy function. Some approaches, like Deformable Templates, use templates defined a priori and transform the deformation problem to a template adjustment problem. However, even though this produces good results, it suffers from the rigidity due to the templates.

Besides incorporating more information in the energy function, for the case of images of nodes, can significantly improve the quality of final contours. Hence, the final segmentation can be improved by using methods that allow the contour freely explore neighborhood areas in the stage of evolution. The latter makes the simulated annealing, an ideal candidate for solving this type of problems.

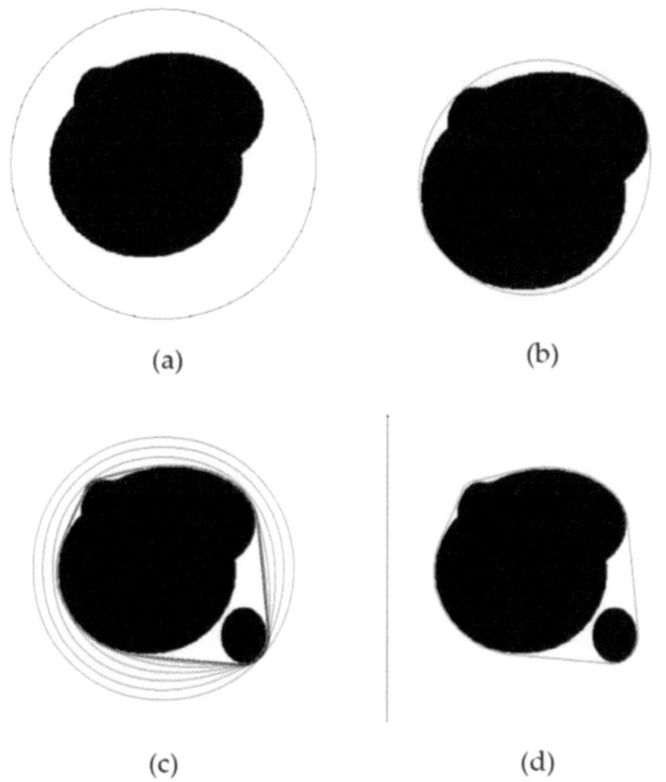

Figure 4. Deformable contour and distracter points; (a) original contour; (b) final contour; (c) original contour with distracter point; and (d) final contour with distracter point.

3. Simulated Annealing

Simulated annealing (SA) is a stochastic optimization technique introduced by Kirkpatrick [25]. This algorithm begins by selecting an initial solution and later generating a new state, randomly generating a new solution in the neighbourhood of the current solution; this is called a neighbour solution. This new state is evaluated and compared with the previous solution. If the solution from the new state is better than the previous one, it is accepted; but if it is not, it is accepted or rejected with some probability. The probability of accepting a new state is given by:

$$e^{-\Delta E/T} > R \tag{3}$$

with: ΔE : Difference between the present and the candidate solutions

T: Temperature

R: Random uniform number between [0,1]

ΔE reflects the change in the objective function and T is the current temperature. The way how the temperature decreases along the algorithm is commonly known as the Cooling Schedule and several types of cooling can be found in the literature as well as stopping criterion of the algorithm.

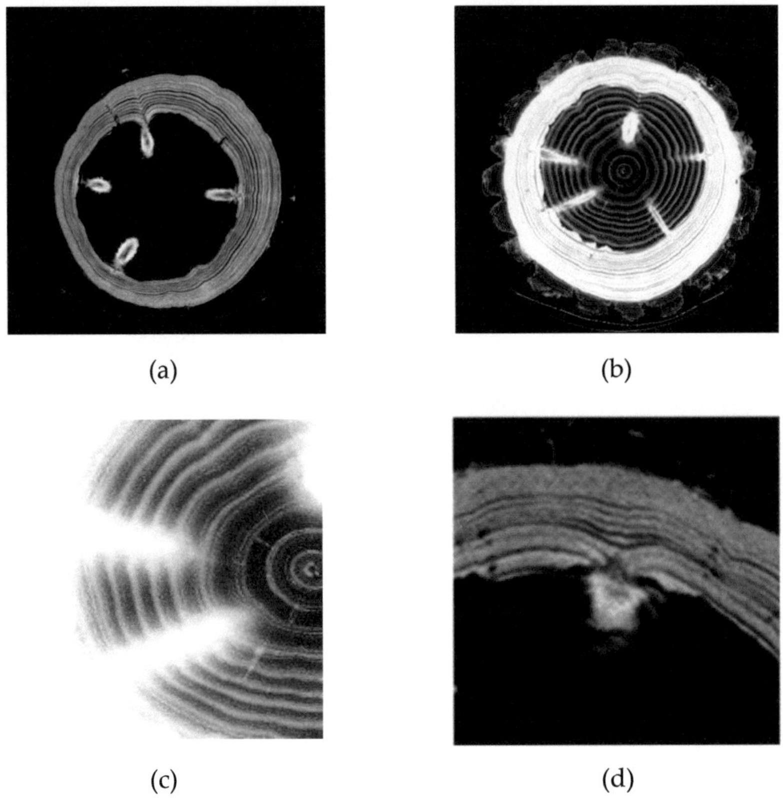

Figure 5. X-ray tomography of de wood log.

Several implementations and variations of SA can be found in the literature: Threshold Accepting (TA), [26]; Boltzmann Annealing (BA), [27]; Simulated Quelching (SQ), [27]; Fast annealing (FA), [27] among others. The latest approaches, SQ and FA, are mainly focused on speeding up the searching for the right solution, which is one of the main drawbacks of BA.

In the current work a classical simulated annealing algorithm is used. A Bolztman distribution (BA) is assumed and consequently, a logarithmic cooling schedule is used: $T(k) = T0/ln\ k$. It should be noticed that since the searching space of our application is not too big, the slowness problem of the method is not relevant. Figure 6 presents the pseudo-code of the algorithm used in the current work.

The ability to accept poor solutions early in the evolution of a deformable contour is a powerful tool to explore complex areas of the image and get better final solutions. Then, SA provides a method of local evolution of each of the points of a contour, allowing the objects in a target to be more accurately represented.

For the case of the images obtained by X-ray computed tomography applied to wood, unlike the steepest descent method that provides a global search method, there are other methods that are based on a local view to find a global minimum. In the current work we propose to use a method based on SA, which is based on a local view, and in this case, local portions of

```
% c: cost function
% j : new solutions
% T(k): Temperature

while (non-stop condition)
    while(all point snake)
          while (k,a conditions)   % Iterations (k) and acceptation (a)
                new j
                if (c(j) – c(i) < 0)
                      i = j
                      a = a + 1
                else
                      new  r (pseudo-random)
                      if (r < exp [(c(i) – c(j))/T])
                            i = j
                            a = a + 1
                      end if
                end if
                      k = k + 1
          end while
    end while
    delete_add_vertex()  % delete or add vertex with dmin and dmax parameters
    T(k)  T0/ln(k)     % Cooling Schedule
end while
```

Figure 6. Simulated Annealing Algorithm used in the current work

the contour, to find a global minimum of the energy function. SA explores the possibilities in a local environment of the contour and it evolves in the direction of minimum energy. The method initially considers the generation of a contour in the 2D image space by using a given number of vertices. A neighborhood patch of *nxn* pixels is created for every pixel in the contour. Then, in that neighborhood a *new candidate vertex* is selected by evaluating it according to the energy function; the acceptance of this new vertex will depend on criteria such as the number of iterations, number of acceptances and the probability of "bad solution" acceptance according to the temperature. Additionally, a criterion for the elimination and creation of new vertices is used. This criterion is based on the distance between the vertices in the contour and allows a homogeneous distribution through the whole contour. Once all the vertices in the contour have been checked for a given temperature *Ti* (Fig. 6) the criterion for elimination and creation is applied. A vertex is eliminated in case the distance to its neighbor is smaller than a minimum distance (*dmin*). Similarly, a vertex is created in case the distance between two vertices is higher than a maximum allowed distance (*dmax*). Figure 7 shows an example where a neighborhood of *3x3* pixels is considered

Figure 8 shows a deformable contour applied to a knot in a X-ray tomography, the contour near to the object to be segmented is attracted and deformed by the force of the image until the contour completely surrounds the object. The forces coming from the gradient of the images guide the contour until it is located around the object's borders. In Fig. 8, we can see the evolution of the contour for the segmentation of knots in *Pinus radiata* wood. We can see that this method can be strongly influenced by other objects in the image, or by points of intensities not related to the object, producing a poor segmentation of the defect.

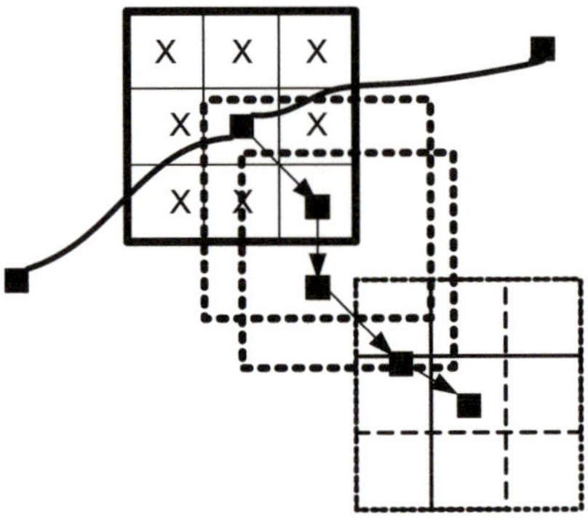

Figure 7. Illustration of a contour evolution for a neighborhood patch of *3x3* pixels. For every contour vertex a new candidate vertex is created and accepted according to a set of user defined parameters.

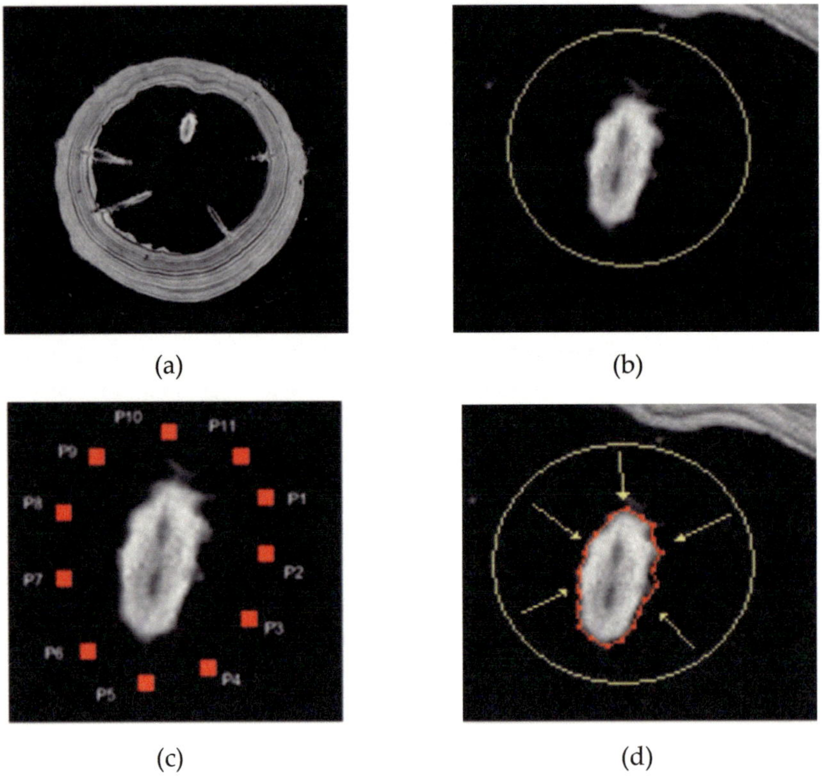

Figure 8. Evolution of a deformable contour: (a) initial image; (b) and (c) initial contour; (d) final contour.

In general, the tomography of logs show the existence of the distortions in the shape of the knots and forces of attraction in other regions of the contour not associated with the knot (in this case, growth rings) and in zones or points that influence the contour, thereby causing a poor final segmentation. Figure 9(a) and (b) present a tomographic image of a wood log; an enlargement of a knot is presented in Fig. 9(c); the corresponding gradient image is depicted in Fig. 9(d). Note that in Fig. 9(c), the most intense points represent the greatest values of the gradient. These forces guide the contour around the knot in order to completely isolate it, but not all are related to force the knot. The latter is one of the problems to be properly solved by the segmentation methods.

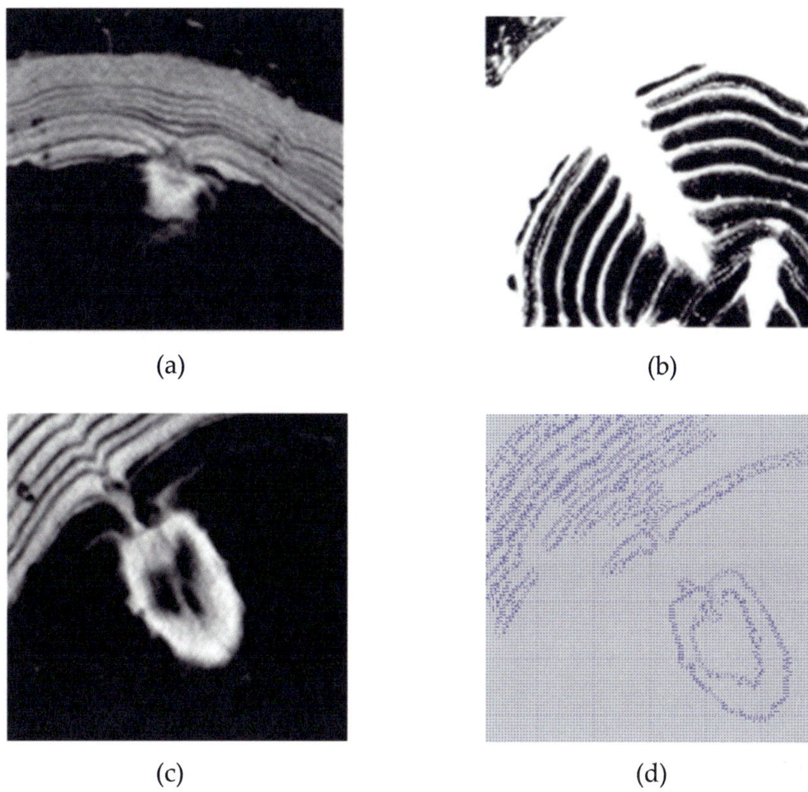

Figure 9. (a) and (b) Tomographic image of a knot, (c) knot (enlargement), (d) gradient of the image.

4. Result and discussion

According to the above description and using the energy functions described in (2), we proceeded to analyse the images from the X-ray computerized topographies of logs with different characteristics. To carry out these experiments, we used discreet versions of the energy functions and a local optimization algorithm based on the greedy algorithm and another based on classic Boltzmann Annealing with T_0=1000. The tests were carried out with topographies of various logs using a medical scanner. In total, 90 images were analysed. The most relevant results are shown in the following figures.

Figure 10 shows a sequence of images revealing the evolution of the knots inside a log. On the left, we can see the images of the complete log and, on the right, an enlargement of the image corresponding to a single knot. Figure 11 shows the situation of a knot that appears to be an isolated object in the image: Fig. 11(a) shows the result of the segmentation using the greedy algorithm; and Fig. 11(b) presents the result using SA. In both cases, the segmentation gives good results; the contour becomes deformed around the knot, isolating it completely. It should be noted here that, in this case, the knot in the image appears to be totally isolated and there are no objects that distort the contour deformation. Figure 12 presents the result of the segmentation in a case in which the knots are confused with the growth rings: Fig. 12(a) shows the result when the greedy algorithm is used; and Fig. 12(b) depicts the result with SA. In this case, a considerable difference can be appreciated between the two methods. The energy gradients generated by the image of the growth rings tend to move the contour away from the knot. However, with SA, due to its random exploration characteristic, in general it tends to better segment the knot. In Fig. 13, the situation is more

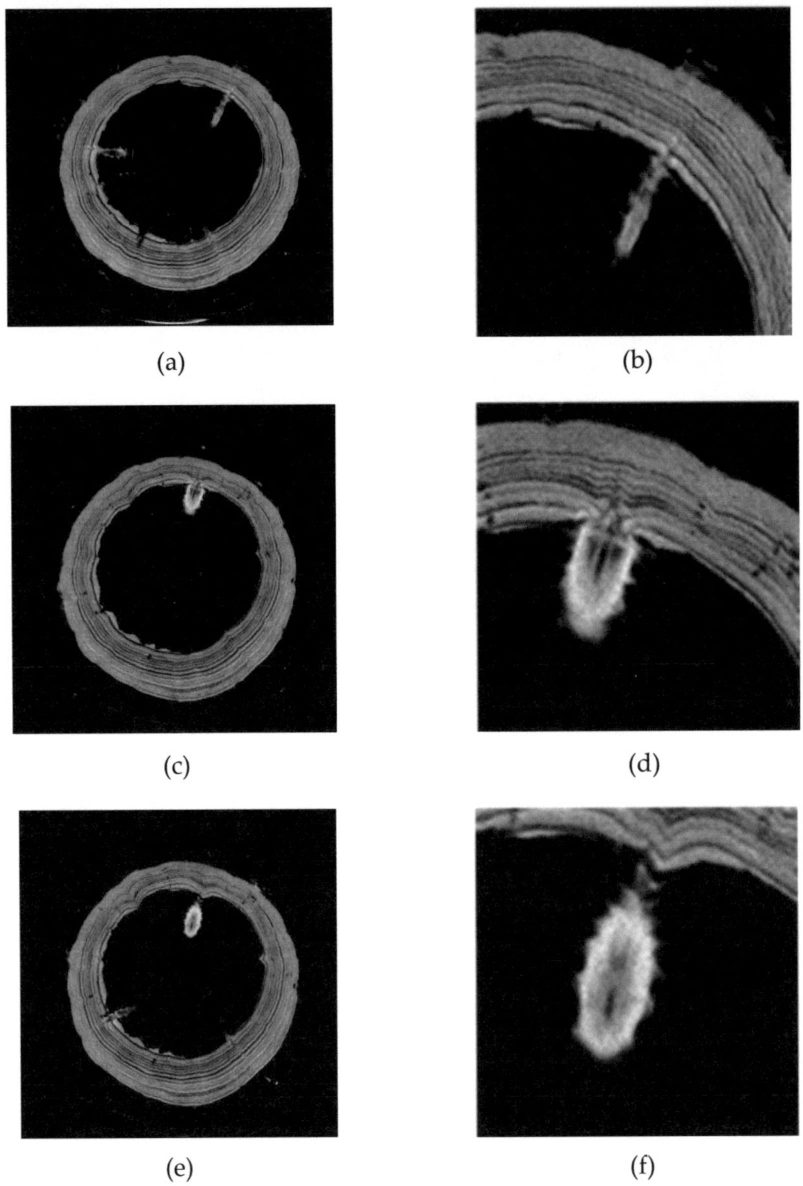

(a) (b) (c) (d) (e) (f)

Figure 10. Tomography of a log; (left) transversal cuts; (right) enlarged image.

critical. In these images the knots are merged with the growth rings; in fact, the knot is only partially visible in the tomographic image, and it appears to be an object with different geometric characteristics than those visible in the previous images. Figure 13(a) shows the result of the evolution of the contour with the greedy algorithm; in this case the algorithm clearly does not surround the knot correctly and the result of the segmentation is extremely poor. Nonetheless, in Fig. 13(b), with SA and thanks to a wider exploration, the general result is perceptibly better.

As we can see in the figures presented above, in some situations the images obtained by X-ray computerized topographies facilitate the segmentation. Nevertheless, in other situations, the segmentation is seriously complicated by confusion with other objects in the image. In particular, growth rings and other objects distract the contours. In such cases, a classical optimization algorithm such as the greedy algorithm does not provide good results, principally due to the tendency of local minima. However, SA provides a more robust method for presenting better results in these situations.

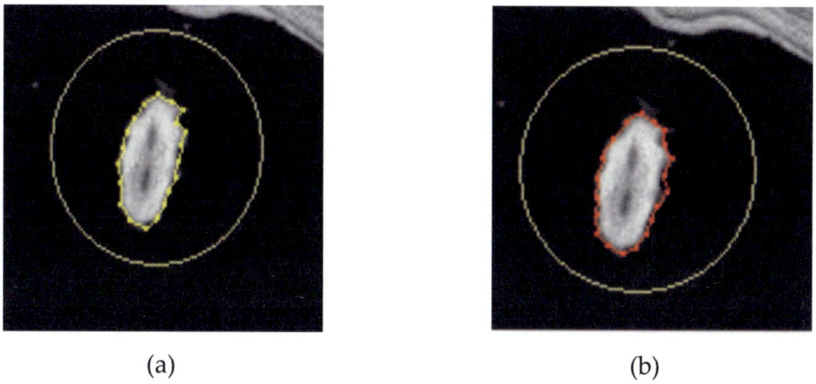

Figure 11. Evolution of an initial contour around an isolated knot; (a) greedy algorithm; (b) simulated annealing algorithm.

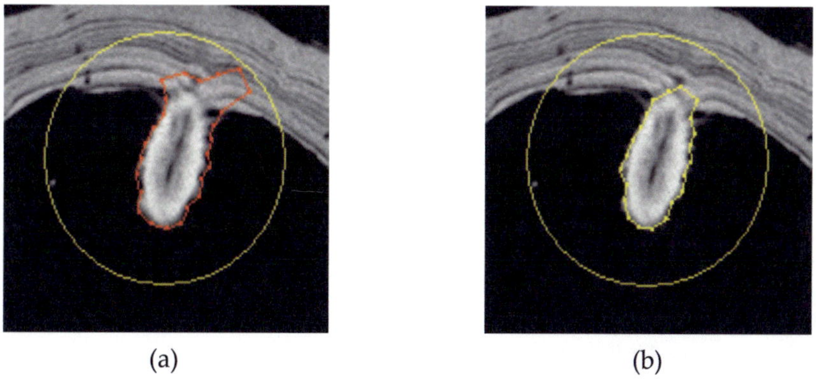

Figure 12. Evolution of an initial contour around a knot with interference from growth rings; (a) greedy algorithm; (b) simulated annealing algorithm.

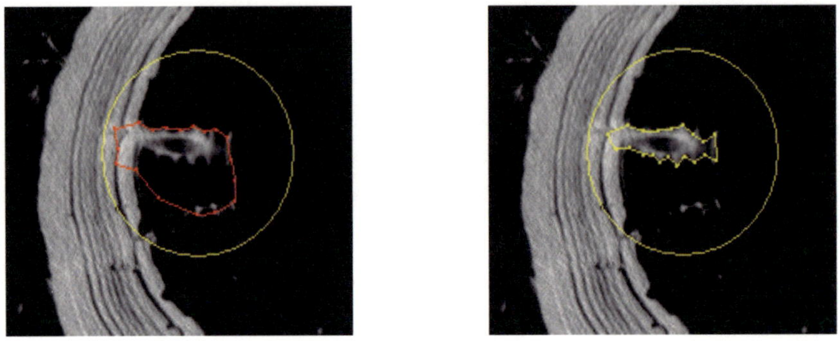

(a) (b)

Figure 13. Evolution of an initial contour around a knot with interference from growth rings; (a) greedy algorithm; (b) simulated annealing algorithm

5. Conclusions

From the experimental results presented above, we can conclude that the proposed method is a very good alternative for knot segmentation. The use of SA allows much more precise segmentation of knots with greater irregularities in the images. Nonetheless, the initial position of the deformable contour is of vital importance for a good identification. We can also use the results obtained to infer that both excessive humidity content in the logs and growth rings present inconveniences. The latter presents a great challenge since the knot shows up in the images obtained with X-rays both in isolation and confused with other objects. In the second case, additional information, for example morphological information on the knots, can be incorporated into the contour's energy function that describes the force of the image, offering a very good alternative. The cases examined correspond to segmentation by X-ray computerized tomography for logs having distortions mainly due to their storage; this causes the density of the material to increase, which then has an effect on the images. This case is of great interest for the wood industry. However, this technique can also be used in other situations such as the study of other defects in logs. The authors are currently developing 3D techniques and a way to reduce the method's computerization costs that was not considered in this first study.

Author details

Cristhian A. Aguilera
Dept. Electrical and Electronic Engineering, University of Bio-Bio, Concepción, Chile

Mario A. Ramos
Dept. Wood Engineering, University of Bio-Bio, Concepción, Chile

Angel D. Sappa
Computer Vision Center, Autonomous University of Barcelona, Barcelona, Spain

Acknowledgement

This work was partially supported from MECESUP2 Postdoctoral program and Regular Project 100710 3/R of the University of Bio-Bio, Chile; and the Spanish Government under Research Program Consolider Ingenio 2010: MIPRCV (CSD2007-00018) and Project TIN2011-25606.

6. References

[1] Benson-Cooper D, Knowles R, Thompsom F, Cown D (1982) Detection of defaults in logs, Forest Research Institute of New Zeeland 8: 9-12.

[2] Harless T, Wagner F, Steele P, Taylor F, Yadama V, McMillin C (1991) Methodology for locating defect within hardwood logs and determining their impact on lumber value yield, Forest Product Journal 41: 25-30.

[3] Sandoz J (1993) Moisture content and temperature effect on ultrasound timber grading, Wood Science and Technology 27: 373-380.

[4] Martin P, Collet P, Barthelemy P, Roussy G (1987) Evaluation of wood characteristics: internal scanning off the material by microwaves, Wood Science and Technology 21: 361-371.

[5] King R (1991) Measurement of basic weight and moisture content of composite board using microwave, VIII Int. Symp. On non-destructive testing osf wood 1: 21-32.

[6] Gapp V, Mallick G, (1995) Capteur intelligent pour la mesure de la humdité dans le matériau par technique micro-ondes, Journées Nationales Micro-ondes 73: 53-57.

[7] Baradit E, Aedo R, Correa J (2006) Knot detection in Wood using microwaves, Wood Science and Technology 40: 118-123.

[8] Lundgren N, Brännström M, Hagman O, Oja J (2007) Predicting the Strength of Norway Spruce by Microwave Scanning: A Comparison with Other Scanning Techniques, Wood and Fiber Science 39: 167-172.

[9] Thomas L, Mili L (2006) Defect Detection on Hardwood Logs Using Laser Scanning, Wood and Fiber Science 38: 243-246.

[10] Funt B, Bryant E (1987) Detection of internal log defects by automatic interpretation of computer tomography images, Forest Product Journal 37(1): 56-62.

[11] Zhu D, Conners R, Schmoldt D, Araman P, (1996) A prototype Vision System for Analyzing CT Imagery of Hardwood Logs. IEEE Transactions on Systems, Man and Cybernetics 26: 522-532.

[12] Rojas G, Hernández R, Condal A, Verret D, Beauregard R (2005): Exploration of the Physical Properties of Internal Characteristics of Sugar Maple Logs and Relationships with CT Images, Wood and Fiber Science 37: 591-604.

[13] Eriksson J, Johansson H, Danvind J (2006) Numerical Determination of Diffusion Coefficients in Wood Using Data From CT-Scanning, Wood and Fiber Science 38: 334-344.

[14] Sarigul E, Abbott L, Schmoldt D (2001) Nondestructive rule-based defect detection and indentification system in CT images of hardwood logs, Review of Progress in Nondestructive Evaluation 20: 1936-1943.

[15] Oja J, Grundberg S, Gronlund A (1998) Measuring the outer shape of pinus sylvestris saw logs with an x-ray log scanner, Scandinavian Journal Forest 13: 340-347.

[16] Sandoz J (1993) Moisture content and temperature effect on ultrasound timber grading, Wood Science and Technology 27: 373-380.

[17] Lindgren L (1991) Medical CAT-Scanning: x-ray CT-numbers and their relation to wood density, Wood Science and Technology 25: 341-349.

[18] Schmoldt D, He J, Abbort A (2000) Automated Labeling of log feature in CT imagery of multiple hardwood species, Wood and Fiber Science 32: 287-300.

[19] Taylor F, Wagner F, McMillin C, Morgan I, Hopkins F (1984) Locating Knots by Industrial Tomography, Forest Product Journal 34: 42-46.

[20] Wagner F, Roder F (1989) Ultrafast CT scanning of an oak log for internal defect, Technical Note, Forest product journal 39: 62-64.

[21] Dawei Q, Zhang P, Xuejing J (2010), Detection of Wood Image, Sixth International Conference on Natural Computation 1459:1463.

[22] Dawei Q, Hongbo M, Mingming Z, Lei Y (2008) Detection of Wood Defects From X-ray Image by ANN, Proceedings of the IEEE International Conference on Automation and Logistics 23:28.

[23] Borianne P, Pernaudat R, Subsol G (2011) Automated Delineation of Tree-Rings in X-Ray Computed Tomography Images of Wood. IEEE International Conference on Image Processing 445-448.

[24] Kass M, Wikins K, Terzopoulos D (1988) Snake: Active Contour Model, International Journal of Computer Vision 4: 321-331.

[25] Kirkpatrick S, Gelatt C, Vecchi M (1983) Optimization by Simulated Annealing, Science 220(4598): 671–680.

[26] Ingber L, (1993) Simulated annealing: Practice versus theory, Mathl. Comput. Modelling, 18(11):29-57.

[27] Dueck G, Scheuer T, (1990) Threshold accepting: A general purpose optimization algorithm appearing superior to simulated annealing, Journal of Computational Physics, 90: 161-175.

Chapter 6

Applications of Simulated Annealing-Based Approaches to Electric Power Systems

Yann-Chang Huang, Huo-Ching Sun and Kun-Yuan Huang

Additional information is available at the end of the chapter

http://dx.doi.org/10.5772/50388

1. Introduction

In the last decade, many heuristic methods have evolved for solving optimization problems that were previously difficult or impossible to solve. These methods include simulated annealing (SA), tabu search (TS), genetic algorithm (GA), differential evolution (DE), evolutionary programming (EP), evolutionary strategy (ES), ant colony optimization (ACO), and particle swarm optimization (PSO). This chapter reviews papers in international journals that present the SA-based methods for electric power system applications.

The numerical simulation of physical phenomena has become a major issue in the design phase or optimization of many systems. The complexity of the phenomena demands that multi-scale and multi-physical aspects are considered. In recent years, many researchers are concerned with the development, study, and implementation of efficient numerical methods that can be used in applications in engineering, natural and other sciences. They also focus on numerical methods for solving highly complex and CPU-time problems using high-performance computing. For the advancement of computing schemes that improve the efficiency and even optimize calculation processes, researchers must be able to generate novel systematic methods for solving multi-physical problems. The applications of SA-based methods for electric power systems are surveyed.

This chapter reviews various SA-based methods (including SA and other meta-heuristics methods) for the planning, operation, and optimization of power systems, including relevant recent and historical developments. Relevant publications in international journals that cover a broad range of applications of SA methods to solving power system problems are reviewed. As is well known among power engineers, many kinds of combinatorial optimization problems arise in the planning and operation of power systems. The generation and transmission expansion planning, generator maintenance scheduling and unit commitment, reactive power planning, load forecasting and economic dispatch, and distribution systems planning and operation are typical problems.

Since finding the global optimal solution to a problem is difficult because of the vast numbers of combinations of tentative solutions, many approximation algorithms have been developed to find acceptable near-optimal solutions over several decades. In the 1960's to 1970's, many mathematical programming methods were proposed and discussed. In the early 1980's, the expert system was introduced to solve such problems. However, it was not very effective in the most complex problems. In the early 1990's, the artificial neural network (ANN), or the simulation of the brains as a living entity, was revived in the field of artificial intelligence by the calculation speeds of contemporary computers. The ANN can solve combinatorial optimization problems and load forecasting problems extremely rapidly although the size of such problems that could be solved is limited. At that time, other methods for solving combinatorial optimization problems attracted the attention of many researchers. These methods skillfully simulate physical phenomena, natural evolution, and mathematical heuristics, and yield good results when applied to the combinatorial optimization problems. These methods are referred to as "modern heuristic approaches" or "meta-heuristic approaches".

Many applications of modern heuristic approaches to power systems have been proposed in the past 20 years, especially in generation and transmission expansion planning, generator maintenance scheduling and unit commitment, reactive power planning, load forecasting and economic dispatch, distribution systems planning and operation, and other applications. In this chapter, articles that have been published on the use of SA, TS, GA, DE, EP, ES, and combinations thereof in relation to power systems are systematically reviewed. These articles were published in international journals and cover a broad range of SA-based methods. A total of 81 journal papers are listed in the references.

This chapter provides a broad and relatively technical treatment of important topics at a level suitable for advanced students and for researchers with a background in electric power systems or in engineering optimization applications. It will review key areas of electric power system planning, operation, and optimization in a citation-rich format that is similar to that used by leading review journals.

The rest of this chapter is organized as follows.

- Formulation of Power System Optimization Problems
- Generation and Transmission Expansion Planning
- Generator Maintenance Scheduling and Unit Commitment
- Reactive Power Planning
- Load Forecasting and Economic Dispatch
- Distribution Systems Planning and Operation
- Other Applications
- Conclusion

2. Formulation of power system optimization problems

Many optimization problems in power system planning, control, and operation can be formulated mathematically as follows:

$$\text{Minimize } f(x) \tag{1}$$

$$\text{subject to } g_i(x) = 0,\ i = 1, 2, \ldots, I$$

$$h_j(x) \geq 0,\ j = 1, 2, \ldots, J$$

where $x^T = [x_1, x_2, \ldots, x_N]$ and the elements of x are the decision variables that are to be determined. The constrained optimization problem that is described by (1) can be transformed to an unconstrained optimization problem using the penalty method. With the equality constraints represented as inequality constraints, the unconstrained optimization problem can be formulated as:

$$\text{Minimize } f(x) + p\ \Sigma\Theta[h_i(x)] \tag{2}$$

where p is the penalty coefficient and Θ is the penalty function. When $f(x)$ is multi-modal, SA can be used to find the global optimal values of the decision variables x that are described in (2). Let x_d be a dependent parameter in a constrained problem that is randomly selected at any stage in the optimization process. The constrained problem can thus be formulated as

$$\text{Minimize } f(x', x_d) \tag{3}$$

$$\text{subject to } h_j(x) \geq 0,\ j = 1, 2, \ldots, J$$

where x' is x excluding x_d. In minimizing the objective function in (3), the values of the non-dependent parameters in x' are firstly determined in each iteration of the optimization process. When the values of the parameters in x' are known, x_d can be determined by applying the relationship between the dependent and non-dependent parameters:

$$x_d = g'(x') \tag{4}$$

This chapter reviews SA-based optimization methods for solving the constrained optimization problem that is formulated as (3) and (4).

3. Generation and transmission expansion planning

Generation expansion planning (GEP) problems are important in planning activities to determine what generating units should be constructed and when such units should come on line over a long-term planning horizon. The major objectives of GEP are to minimize the total investment and operating costs of the generating units, while meeting capacity constraints, energy constraints, operating constraints, and the reliability criteria. The GEP problem is equivalent to finding a set of optimal decision vectors over a planning horizon that minimize the investment and operating costs under various constraints. Therefore, GEP is a highly constrained, nonlinear, discrete optimization problem.

GEP for a power system is challenging because the large-scale, long-term, nonlinear and discrete nature of generation units. Many emerging techniques (including expert systems, fuzzy logic, neural networks, analytic hierarchy process, network flow, decomposition methods, SA and GA) have been used in GEP. These emerging optimization techniques and their potential use in challenging GEP in the future competitive environment of the power industry have been reviewed, and some useful information and resources for future GEP were provided (Zhu & Chow, 1997).

Meta-heuristic techniques, such as GA, DE, EP, ES, ACO, PSO, TS, SA, and hybrid approaches have been used in GEP and compared (Kannan et al., 2005). The original GEP problem has been modified using the virtual mapping procedure and the penalty factor method to improve the efficiency of these meta-heuristic techniques. Further, intelligent initial population generation has been introduced to reduce computational time. The GEP problem considered synthetic test systems for 6-year, 14-year, and 24-year planning horizons and five types of candidate units. The results obtained using these proposed techniques were compared and validated against conventional dynamic programming, and the effectiveness of each of the proposed methods was verified.

A novel GEP model was proposed (Kannan et al., 2007) for developing countries in a partially deregulated environment, in which both utilities and independent power producers (IPPs) participate in the generation market. In this model, the utility purchases electric power from the IPPs and sells it to the consumer. The utility maximizes its profit and ensures profits for all of the participating IPPs. Additionally, the utility checks under/over investment and considers system security, national security (fuel-mix ratio), social welfare and reliability simultaneously. The budget constraints of the utility are considered in the expansion plan. Meta-heuristic methods, such as GA, DE, EP, ES, PSO, TS, SA, and the hybrid approaches have been used to solve the restructured GEP problem, and the performances of each was evaluated and validated against the dynamic programming method for a synthetic test system with five types of candidate plant for the utility and three types of candidate plant for IPPs, with a 6-year planning horizon. The effectiveness of the proposed modifications and techniques has been demonstrated.

Transmission expansion planning (TEP) involves determining when and where new circuits must be installed, as well as the number that must be installed, to ensure that the power system will meet customer's demand with sufficient quality over a long planning horizon, while minimizing investment, operating, and interruption costs. TEP in a deregulated environment is highly complex. The possibility of generation planning and demand-side management as substitutes for transmission expansion must be considered when candidates for transmission expansions are generated. The deregulation of a power system produces new objectives in expansion planning such as the minimization of congestion cost. The optimized TEP considers such primary objectives as network investment cost, congestion cost, reliability, and environmental impact, subject to the relevant technical constraints.

The SA method was proposed for solving long-term TEP problems which are hard, large-scale combinatorial problems (Romero et al., 1996). The proposed approach has been compared with a more conventional optimization technique that is based on mathematical decomposition with an implicit zero-one enumeration procedure. Tests have been performed on three systems. Two smaller systems for which optimal solutions are known have been used to tune the main parameters of the SA process. The SA method has then been applied to a larger example system for which no optimal solutions are known. Therefore, an entire family of interesting solutions have been obtained a cost of approximately 7% less than those of the best solutions for the example system.

A parallel SA (PSA) algorithm for solving the long-term TEP problem was proposed (Gallego et al., 1997). A strategy that does not affect the basic convergence properties of the sequential SA have been implemented and tested. They studied the conditions under which the PSA is most efficient and tested the PSA on three example networks: a small 6-bus network and two complex real-life networks. Excellent results were reported in the test. In addition to reducing computing times, the proposed PSA algorithm greatly improved solution quality when applied to the largest of the test networks.

Three families of non-convex optimization approaches for solving the TEP: SA, GA, and TS, were compared and an integrated view of these methodologies was proposed (Gallego et al., 1998a). Test results obtained using large-scale, real-life networks verified that the presented hybrid approach greatly outperformed those obtained using any individual approach.

An extended GA was proposed to solve the optimal TEP (Gallego et al., 1998b). They improved the GA in two main ways: the initial population was obtained by conventional optimization based methods, and the mutation approach was used in the SA. Test results revealed excellent performance for a difficult large-scale real-life problem, and demonstrated a substantial reduction in investment cost relative to earlier solutions that were obtained by conventional optimization methods and SA.

A parallel TS was proposed to solve the TEP (Gallego et al., 2000). The presented method is a third-generation TS procedure with many advanced features. It exhibits the features of many other approaches, such as heuristic search, SA and GA. In all studied test cases, new generation and load sites can be connected to an existing main network, and these connections may require more than one line and the addition of a transformer, making the problem harder to solve in the sense that more combinations must be considered.

4. Generator maintenance scheduling and unit commitment

The proper generator maintenance scheduling (GMS) in a power system is very important to its economic and reliable operation. To prevent premature aging and the failure of generators in a power system, which would cause unplanned and costly power outages, preventive maintenance must be performed at regular intervals. The generator maintenance in a power system involves scheduling and executing actual maintenance work. The GMS must be solved in the planning of the secure and reliable operation of a power system, primarily because other short- and long-term planning activities, including unit commitment, generation dispatch, import/export of power and GEP are directly influenced by related decisions. Modern power systems have become larger as the demand for electricity has considerably increased, increasing the number of generators, reducing reserve margins, and making the GMS problem even more complex.

The objective of GMS is to plan preventive maintenance scheduling for generators based on load forecasts. The purpose is to minimize the operating cost, maximize the profit or minimize the power shortage annually. Many constraints must be considered, including the system load, the maintenance window of each generator, maintenance resources (including human resources, machinery, equipment, and others), and the estimated power output. Therefore, GMS is a mixed-integer programming problem, in which a combination of 0s and 1s specifies whether a generator unit is undergoing maintenance, subject to some equality and inequality constraints. Before deregulation of the electric power industry, system reliability and loss of load probability were two major objectives of the GMS problem. After deregulation, however, maximizing profit is the driving interest of an IPP. Hence, the profitability of power plants is the objective function for the GMS problem.

The thermal power plant GMS problem has been formulated as a mixed-integer programming problem and solved efficiently using the SA (Satoh & Nara, 1991). The GA, SA and TS methods have been used together to solve the large-scale, long-term GMS problem (Kim et al., 1997). This combined solution algorithm has the advantages of the individual algorithms and supports a reasonable combination of local and global searches. The method considers the maintenance class and many consecutive years scheduling. Several real-scale numerical examples demonstrate the effectiveness of the proposed method.

The application of meta-heuristic approaches, such as GA, SA and their hybrid for GMS in power systems was proposed (Dahal & Chakpitak, 2007). The presented paper mainly focuses on the application of GA/SA and GA/SA/heuristic hybrid approaches. The GA/SA hybrid used the probabilistic acceptance criterion of SA in the GA framework. The GA/SA/heuristic hybrid used heuristic methods with the GA/SA hybrid to seed the initial population. The authors formulated the GMS as an integer programming problem using a reliability-based objective function and typical problem constraints. They discussed the implementation and performance of the meta-heuristic methods and their hybrid in a test case. The obtained results are promising and reveal that the hybrid methods are less sensitive to variations in the parameters of the technique and are effective alternatives to other methods for performing GMS.

The main objective of unit commitment (UC) is how to schedule the on/off status of the generators to minimize the production cost of electricity. A typical UC problem is combinatorial and involves a large set of physical, operating and contractual constraints, making the problem difficult to solve. The SA method was originally proposed to solve the UC problem (Zhuang & Galiana, 1990). It is highly flexible in handling UC constraints, and numerical results on test systems of up to 100 units were reported.

A short-term hydrothermal UC based on the SA method was proposed (Wong & Wong, 1994a). In the algorithm, the power balance constraint, total water discharge constraint, reservoir volume limits and constraints on the operational limits of the hydrothermal generator and the thermal generator are fully considered. The relative operational capacities of the hydroplant and the thermal plant are also considered. A coarse-grained parallel SA algorithm was presented for short-term hydrothermal UC (Wong & Wong, 1994b). The design of the algorithm considers load balancing, processor synchronization reduction, communication overhead reduction and memory contention elimination. The test results were compared with those obtained using a sequential algorithm and the results revealed that the proposed method provides an almost linear reduction in computation time. Two parallel SA concepts, speculative computation and serial subset, were proposed to the UC (Annakkage et al., 1995). A combined scheme in which speculative computation is used in the initial phase and the serial subset is used in the final phase. The test results revealed that the proposed parallel schemes greatly improved the computing performance of SA.

A hybrid GA/SA method was developed to the UC (Wong & Wong, 1995). The proposed method can typically provide feasible schedules in the solution process. The hybrid method can handle the nonconvexity of the UC. The authors subsequently provided a new formulation for short-term UC with a take-or-pay fuel contract (Wong & Wong, 1996) and a used a fuzzy set approach to help to find schedules that yield, as closely as possible, the take-or-pay fuel consumption. They extended the formulation to cover the economic dispatch problem when fuel consumption exceeds the agreed amount in the take-or-pay contract, and the extended formulation was combined with the GA and SA algorithms for determining the UC.

A new formulation for short-term multiple-fuel-constrained UC was presented (Wong & Wong, 1997). In the formulation, the power balance constraint, operating limits of the generators, fuel availability factors of the generators, efficiency factors of the fuels and the supply limits of the fuels are fully considered. They combined the new formulation with GA, SA and hybrid GA/SA methods to establish new algorithms. They demonstrated the new algorithms by using them to determine the most economical generation schedule for 25 generators in a local power system and the schedule of the system for four fuels.

An enhanced SA was adopted to solve the UC by applying mechanisms to ensure that the generated candidate solutions are feasible and satisfy all of the constraints (Wong, 1998). The performance of the enhanced SA was demonstrated and compared with that of conventional methods. The UC was divided into two subproblems: a combinatorial optimization problem and a nonlinear programming problem (Mantawy et al., 1998). They solved the former using the SA and the latter using a quadratic programming routine. Numerical results revealed an improvement in the cost associated with the solutions.

A new algorithm based on integrating GA, TS and SA methods to solve the UC was presented (Mantawy et al., 1999). The core of the proposed algorithm is based on GA. TS is used to generate new population members in the reproduction phase of the GA. The SA is used to accelerate the convergence of the GA by applying the SA test for all the population members. Numerical results showed the superiority of the solutions thus obtained over those obtained using GA, TS and SA methods, and two exact algorithms.

An extended mean field annealing neural network (NN) approach was presented to short-term UC (Liang et al., 2000). The annealing NN provides the high solution quality of the SA with the rapid convergence of the NN. Test results confirmed that their approach was very effective in finding the optimum solution to the UC. An approach combining the feedforward NN and the SA method was presented to solve UC (Nayak & Sharma, 2000). The NN is used to determine the discrete variables that correspond to the state of each unit at various times. The SA is used to generate the continuous variables that correspond to the power output of each unit and the production cost. A set of load profiles as inputs and the corresponding UC schedules as outputs that satisfy the minimum up–down times, spinning reserve and crew constraints were used to train the NN. The experimental results demonstrate that the proposed approach can solve the UC in a reduced computational time.

A combined SA and TS approach was used to solve the UC (Purushothama et al., 2003). In their stochastic extended neighborhood algorithm, SA is the main stochastic algorithm, and TS is used to perform an extended neighborhood search and thus locally improve the solution obtained by SA. The neighborhood search uses local domain-knowledge, resulting in rapid convergence of the SA. The results obtained for many example systems illustrate the potential of the hybrid approach. The SA with local search hybrid algorithm was proposed to solve the UC (Purushothama & Jenkins, 2003). The hybrid algorithm is robust and provides faster convergence than earlier algorithms. The results verified its potential for solving the UC.

A scheduling method for representing the thermal stress of turbine shafts as ramp rate constraints in the UC (Li & Shahidehpour, 2003). In the UC, thermal stress over the elastic limit is used to calculate the ramping cost. Determination of the contribution of the thermal stress to the generation cost requires that a set of solution that includes thermal stress at the end of each time step be calculated; this requirement establishes a complex problem that cannot be solved using an ordinary optimization method. An improved SA was used to determine the optimal trajectory of each generating unit, and they elucidated the economics of frequently ramping up/down of low-cost generating units in relation to the cost of replacing their turbine rotors with a shorter life span. The results demonstrated the effectiveness of the proposed method.

A new SA combined with a dynamic economic dispatch method was designed to solve the short-term UC (Simopoulos et al., 2006a). SA was used to schedule the generating units, while a dynamic economic dispatch method, incorporating the ramp rate constraints, was used to solve the UC. The ramp rates are considered by performing either a backward or a forward sequence of conventional economic dispatches with modified limits on the generating units. The proposed algorithm is relatively fast and provides feasible near-optimal solutions. A new method for the incorporation of the unit unavailability and the uncertainty of the load forecast in the solution of the short-term UC solved by the SA was presented in (Simopoulos et al., 2006b). The required spinning reserve capacity was conducted by imposing reliability constraints, based on the expected unserved energy and the loss of load probability indices. Numerical simulations demonstrated the efficiency of the proposed method.

An absolutely stochastic SA method was proposed to UC (Saber et al., 2007) and fuzzy UC using the absolutely stochastic SA was presented (Saber et al., 2006). In both papers, all of the solutions that involved high and low costs, are associated with acceptance probabilities and an early jump from one local minimum to another, enabling more local minima to be found and compared in a particular time or number of iterations. The number of bits to be flipped is determined by the appropriate control of parameter. Excess units with a system-dependent probability distribution handle constraints efficiently. The sensitivity of the distribution parameters is satisfactory. To reduce the number of required economic load dispatch calculations, a sign bit vector was introduced. Numerical results indicate an improvement in the cost and time required to find a solution relative to those using the proposed algorithms. Besides, An EP based SA was proposed to the short-term UC (Christober Asir Rajan & Mohan, 2007) and the UC using SA embedded EP approach was presented to hydro-thermal UC (Christober Asir Rajan, 2011). Numerical results are used to compare the costs of the solutions and computation times obtained when the proposed approaches and conventional methods are used to determine the optimal UC.

5. Reactive power planning

Reactive power planning, or Var planning seeks to optimize the allocation of reactive power sources in a power system, based on location and size. The objectives of Var planning are to minimize the cost of new reactive power supplies. The many variants of this objective involve the cost of real power losses or the cost of fuel. Furthermore, such technical indices as deviation from a given voltage schedule or the security margin may be used as objectives for optimization. The installation of reactive power sources also releases system capacity and improves the voltage level.

The Var planning greatly influences the secure and economic operation of electric power systems. SA was used to contingency-constrained optimal Var planning in large-scale power systems (Hsiao et al., 1993). Their problem formulation considered practical aspects of Var sources, load constraints, and the operating constraints at different load levels. The proposed SA determines the locations of installed Var sources, the types and sizes of Var sources to be installed, and the settings of Var sources at different loading conditions. Test results confirm that the proposed approach is suitable for large-scale power systems applications.

SA based computer package for multi-objective, Var planning in large-scale power systems was proposed (Hsiao et al., 1994). Optimal Var planning is reformulated as a constrained, multi-objective, nondifferentiable optimization problem. The new formulation considers four objective functions that re related to investment in the system, its operating efficiency, its security and the system service quality. Their new formulation also considered load, operating constraints and contingency constraints. The problem formulation allows both the objective functions and the equality and inequality constraints to be nondifferentiable, making the problem formulation more realistic. The package uses a two-stage solution algorithm that is based on an extended SA and the ε-constraint method. The first-stage of the solution algorithm uses an extended SA to find a global, noninferior solution. The primary objective function and the trade-off tolerances are then used to transform the constrained multi-objective optimization problem into a single-objective optimization problem with more constraints by applying the ε-constraint method. The second-stage uses the SA to obtain the global optimal solution.

A constrained, multi-objective and non-differentiable optimisation for Var planning problem was proposed (Chen & Liu, 1994). The objectives were minimization of active power loss cost, minimization of the cost of investment in Var sources, robustness of the system security margin and minimization of the voltage deviation of the system. The operating constraints, load constraints and expansion constraints of the system were considered. The goal-attainment method, based on SA, for solving general multi-objective optimization problems by assuming that the decision-maker has goals for each of the objective functions was used to solve the problem. The solution methodology involved finding a desirable, global noninferior solution to the problem, even when the objective space is nonconvex.

An interactive satisfying method, a two-level structure, was proposed to solve the multi-objective power system Var planning (Chen & Liu, 1995). The analysis level involved calculation of a possible or set of possible solutions to the multi-objective problem, and the decision level generate noninferior solutions that meet the preferences of the decision makers. In the analysis level, the ε-constraint method that is based on SA is used to find a global noninferior solution. The proposed method guarantees the solution to be a desirable, global noninferior solution for a general multiobjective Var planning, according to the preferences of the decision makers.

Var planning was presented as a multi-objective optimization problem in terms of maximum system security and minimum operation cost (Jwo et al., 1995). An effective algorithm based on hybrid expert system and SA was proposed to obtain the global optimal solution considering both quality and speed. A weak bus-oriented criterion for identifying candidate buses for Var planning was presented in (Chen, 1996). A voltage collapse proximity indicator was first used to identify weak buses. Then appropriate Var planning for those weak buses increased the system security margin to prevent voltage collapse. The goal attainment method, based on the SA, was applied to solving the multi-objective problem by assuming that the decision-maker has goals for each of the objective functions. The presented method both provides a good final solution and reduces the solution space.

An innovative fast global optimization technique, hybrid partial gradient descent/SA (HPGDSA), for optimal Var planning was presented (Liu et al., 1997). The basic concept of the HPGDSA is that partial gradient descent and SA alternate with each other to reduce the CPU time below that of the conventional SA while the ability to find the global optimal of the SA is retained. A hybrid SA/GA approach was proposed to solve the Var planning (Jwo et al., 1999). That approach found the near-global optimal solution in a finite time. Moreover, the solution time was much less than that of the conventional SA. The proposed method yielded promising results, relative to those obtained using SA, GA and HPGDSA.

Three GA/SA/TS combined algorithms for Var planning were proposed (Liu et al., 2002). Trying reasonably to combine local and global search, they adopt the acceptance probability of SA to improve the convergence of the simple GA, and to apply TS to find more accurate solutions. Test results confirmed that the proposed method is effective to find better solutions than those of existing methods within reasonable time. A projection-based two-layer SA was used to solve the multi-objective optimization problems (Chen & Ke, 2004). The SA yielded a desirable, globally efficient solution to such problems, even when the solution space is nonconvex and the objective functions are nondifferentiable.

A new approach was presented to model and solve Var planning under the static voltage stability constraint (Wang et al., 2011). First, the fuzzy clustering method was used to select new candidate Var source locations. Then, modified Gray code was applied to represent a series of non-uniform Var capacity intervals at different candidate buses. Under the new ordering of the Var capacity intervals, a simplified piecewise linear function that relates the total transfer capability with new Var capacity was derived and applied as a static voltage stability constraint in Var planning. Finally, the optimization problem was solved by an enhanced SA using modified Gray code. The proposed SA adopted a modified definition of neighborhood selection and a novel approach to generating new random solutions. Test results demonstrated that the proposed method is a simple and effective approach for voltage stability-constrained Var planning with contingency considered.

A multi-objective SA was proposed to provide decision support in reactive power compensation (Antunes et al., 2011). Their method computed a set of well-distributed and diversified solutions underlying distinct trade-offs, even for a challenging network. The characteristics of the non-dominated front are relevant information that helps planning engineers select satisfactory compromise solutions that improve the operating conditions of the network.

6. Load forecasting and economic dispatch

Accurate forecasting of electricity load has been one of the most important issues in the planning, operation, and control of electric power systems. Recently, following power system privatization and deregulation, the accurate forecasting of electricity load has received increasing attention. An optimal fuzzy inference method for short-term load forecasting was presented (Mori & Kobayashi, 1996). Their proposed method constructs an optimal structure of the simplified fuzzy inference that minimizes model errors and the number of membership functions to capture the nonlinear behavior of short-term loads in the power system. The model was identified by SA and the steepest descent method.

Support vector machines (SVMs) have been successfully used to solve nonlinear regression and time series problems. The SVM with SA approach was proposed to forecast electricity load (Pai & Hong, 2005). They used SA to select the parameters of the SVM model. Empirical results indicate that the proposed model outperforms the autoregressive integrated moving average model and the general regression NN model.

A fuzzy NN combined with a chaos-search GA (CGA) and SA, applied to short-term power-system load forecasting was presented (Liao & Tsao, 2006). They used a fuzzy hyperrectangular composite NN for forecasting the initial load. An integrated CGA and SA method is then used to find the optimal NN parameters. The CGA is effective in global search but ineffective in local search, while the SA is effective in local optimal search. The paper combined the two methods to exploit the advantages of both and to eliminate the known downside of the traditional NN. Their test results demonstrated superior a forecasting accuracy than other commonly used forecasting methods.

Economic dispatch (ED) is an important daily optimization task in the operation of a power system. Most calculus-based industrial algorithms for solving the ED problem, such as the Lagrangian multiplier method, require the incremental cost curves to be monotonically increasing or piece-wise linear. When the generating units have non-monotonically increasing or non-linear incremental cost curves, the conventional procedure either ignores or flattens out the portions of the incremental cost curves that are not continuous or monotonically increasing. Inaccurate dispatch results are thus obtained. In these cases, accurate dispatch results can only be obtained using more general approaches without restrictions on the shape of fuel cost functions.

SA-based ED was presented to obtain a global or near-global optimum dispatch solution (Wong & Fung, 1993). In the algorithm, the load balance constraint and the operating limit constraints of the generators are fully accounted for. Transmission losses were firstly discounted and subsequently incorporated in the algorithm using the B-matrix loss formula. Test results are obtained by the proposed approach more economically than using the dynamic programming method with a zoom feature.

A combination of the incremental GA and SA was presented to obtain the global or near-global optimum solution for the ED and developed to minimize the memory requirement (Wong & Wong 1994c). They proposed a method for solving the problem of discretization in the encoding of generator loadings. The algorithms include a method for ensuring that the dispatch solutions that are generated by the solution process are feasible and valid. The effects of valve-point loading and ramping characteristics of the generators are considered. The developed algorithms were demonstrated by applying them to a power system, and they were shown to be general and are computationally faster than the earlier SA-based method.

A new multi-objective stochastic search technique for ED was proposed (Das & Patvardhan, 1998). Their heuristic combined a real coded GA and SA. The proposed approach provides the values of various parameters that optimize different objectives, and the best compromise between them in a single run. The test results were compared with those obtained using other methods and the proposed heuristic was found to converge rapidly to better solutions. Additionally, perturbation analysis demonstrated that the solutions that were obtained by the proposed algorithm were truly pareto-optimal, meaning that no objective could be further improved without degrading the others. SA approach was used to solve optimal power flow (OPF) problem that involved both the load flow and the ED (Roa-Sepulveda & Pavez-Lazo, 2003). Test results confirmed the effectiveness of the solution technique.

A parallel TS for determining ramp rate constrained ED for generating units with non-monotonically and monotonically increasing incremental cost functions was proposed (Ongsakul et al., 2004). To parallelize TS efficiently, neighborhood decomposition was performed to balance the computing load, and competitive selection was used to update the best solution reached among subneighborhoods. The proposed approach optimizes the compromises between the experimental speedup and the solution quality for the best performance with different subneighborhood sizes. The proposed approach is potentially viable for online ED because of it provides substantial generator fuel cost savings and the high upper bound on speedup.

A novel multiobjective optimization method for economic emission load dispatch of fixed head hydro plants and thermal plants with nonsmooth fuel cost and emission level functions was presented (Basu, 2005). In this problem, economic and emission objectives were competing. Based on the assumption that the decision-maker has goals for each of the objective functions, the multiobjective problem is converted into a single-objective optimization problem by the goal-attainment method, which is then handled by the SA. The solution method yields a global or near-global noninferior solution that will be close to meeting the decision-maker's requirements. Test results confirmed the applicability and validity of the proposed method.

Dynamic ED determines the optimal operation of units with predicted load demands over a certain period with the objective of minimizing total production cost while the system is operating within its ramp rate limits. SA was proposed to obtain the global or near global optimum dynamic ED (Panigrahi et al., 2006). They incorporated load balance constraints, operating limits, valve point loading, ramp constraints, and network losses into the dynamic ED. Numerical results revealed the performance and applicability of the proposed method.

Since generators have quadratic fuel cost functions, classical techniques ignore or flatten out the portions of the incremental fuel cost curves and so may have difficulty in determining the global optimum solution for non-differentiable fuel cost functions. EP based SA approach was presented to ED in a large-scale power system (Christober Asir Rajan, 2010). The proposed techniques can offer global or near-global optimum dispatch solutions. Test results demonstrate that the proposed integrated approach can provide accurate solutions within reasonable times for any fuel cost functions.

7. Distribution systems planning and operation

Network reconfiguration problem was formulated as a constrained, multi-objective and non-differential optimization problem for both loss reduction and load balancing that considers load constraints and operating constraints (Chiang & Jean-Jumeau, 1990a, 1990b). The number of switch-on/switch-off operations that are involved in network reconfiguration was included as a constraint. Then, a two-stage solution method that is based on a modified SA and the ε-constraint method for general multi-objective optimization were presented. The proposed approach allows designers to obtain a desirable, global noninferior solution in a reasonable computation time. Given a target number of switch-on/switch-off operations involved in the network configuration, the solution algorithm can identify the most effective operations. To reduce the required computing time, the researchers studied the idea of approximate calculations and incorporated them into the solution algorithm, in which two efficient load-flow methods were used: one for high temperatures and the other for low temperatures.

A comprehensive approach to strategic planning of Var compensators in a nonsinusoidal distribution system was presented (Chu et al., 1994). The problem was formulated as a nondifferentiable, combinatorial optimization problem to minimize the system costs while meeting various operating constraints and harmonic limits. SA was used to determine the optimal locations, types and sizes, and settings of these Var compensators. Their proposed approach could handle discrete rather than continuous Var compensator values and determine whether the Var compensators were fixed or switchable for different load levels.

A modified SA was presented for network reconfiguration for loss reduction in distribution systems and the switching limitation was considered (Chang & Kuo, 1994). They proposed a set of simplified line flow equations for approximate loss calculation. They then used an efficient perturbation scheme and an initialization procedure to determine a better starting temperature for the SA. The computing time of the SA was greatly reduced without loss of quality of the solution. Additionally, the proposed SA could rapidly provide a global optimal or near-optimal solution to the problem, and numerical results confirmed the effectiveness of the proposed method.

Optimal capacitor placement, replacement and control in large-scale unbalanced, radial or loop distribution networks were formulated in a combinatorial optimization problem with a non-differentiable objective function (Chiang et al., 1995a, 1995b). They solved this problem using SA and the greedy search technique to obtain high-quality of solutions at a high computing speed. The challenge is to determine the optimal locations in which to install (or replace, or remove) capacitors, the types and sizes of the capacitors to be installed (or replaced) and, for each load level, the control schemes for each capacitor in the nodes of a general three-phase unbalanced distribution system, such that a desired objective function is minimized while the load constraints, network constraints and operating constraints at various load levels are satisfied. The objective function incorporated both the cost of energy loss and costs related to capacitor purchase, capacitor installation, capacitor replacement and capacitor removal. Analysis of the computational complexity of the solution algorithm reveals that the algorithm is also effective for large-scale distribution systems as the computational efforts is reasonable.

A new formulation for power system sectionalizing device placement considering outage, maintenance and investments costs was proposed (Billinton & Jonnavithula, 1996). They formulated the problem as a combinatorial constrained optimization problem with a nonlinear, nondifferentiable objective function. The SA was used to determine the number of sectionalizing switches and the locations of the switches. Test results revealed that the proposed approach yielded a global optimal solution to the sectionalizing device placement problem, considering reliability, investment and maintenance costs.

A single comprehensive algorithm for distribution system switch reconfiguration and capacitor control was proposed (Jiang & Baldick, 1996). They used SA to optimize the switch configuration of the distribution system, and a discrete optimization algorithm to find the optimal capacitor control. They evaluated the benefits of the optimal switch configuration and capacitor control, in terms of both reduced loss and decreased voltage bandwidth. A nonlinear constrained, non-differentiable approach for optimal network routing in distribution system planning was presented (Jonnavithula & Billinton, 1996). The main objective was to minimize the total cost, which is the summation of reliability costs, the cost of feeder resistive loss, investment costs and maintenance costs. They used SA to find a global optimum solution to the problem.

SA-based approach for loss minimization by using an automatic switching operation in large-scale distribution systems was presented (Jeon et al., 2002). SA is particularly well suited for a large combinatorial optimization problem since it can avoid local minima by accepting improvements in cost. However, it commonly requires a meaningful cooling schedule and a special strategy, which makes use of the property of distribution systems in finding the optimal solution. This paper expands the cost function by adding the operating conditions of a distribution system, improving the perturbation mechanism with system topology, and using the polynomial-time cooling schedule, which is based on the statistical calculation in the search. Test results validated and confirmed the effectiveness of the proposed approach.

Hybrid SA and TS approach was applied to minimize real power loss in distribution systems (Jeon et al., 2004). SA is very suitable for large combinational optimization problems, but the SA requires excessive computing time. TS attempts to determine a better solution in the manner of a greatest-descent algorithm, but it cannot guarantee convergence. The hybrid SA and TS algorithm was applied to improve the computing time and convergence. Numerical examples validated and established the effectiveness of their hybrid approach.

A method for optimal planning of radial distribution networks was solved by a combination of the steepest descent and the SA (Nahman & Peric, 2008). Their objective was to find the complete network of available routes and the optimization goal was to obtain the routes that provide the minimal total annual cost. The solution with minimum capital cost, obtained using the steepest descent approach, was used as the initial solution in SA to obtain the solution with minimum total cost. The costs associated with capital recovery, energy loss and undelivered energy costs were considered.

8. The other applications

SA for optimal tearing of networks was presented to divide a power system network model into a number of sub-networks to optimize the use of parallel computer systems for network analysis (Irving & Sterling, 1990). Test results were compared with those obtained using the iterative improvement method, and the proposed SA yielded significantly better solutions.

The placement of a minimal set of phasor measurement units (PMUs) to make the system measurement model observable was presented (Baldwin et al., 1993). A PMU at a bus measures the voltage and all of the current phasors at that bus, requiring extension of the topological observability theory. The minimal PMU set is found using a dual search algorithm, which uses both a modified bisecting search and SA. The former fixes the number of PMUs while the latter seeks a placement set that yields an observable network for a fixed number of PMUs. Test results verified the effectiveness of the proposed approach.

A parallel SA was presented to decompose power systems into subsystems that were with equal numbers of nodes and control variables (Mori & Takeda, 1994). The decomposition of a power system is a difficult discrete combinatorial problem. The researchers' numerical results revealed that the parallel SA yielded better solutions than the conventional SA.

SA was applied to multi-partition an observable power system state estimation network into two or more observable sub-networks (Habiballah & Irving, 1995). The proposed SA was theoretically based on combinatorial optimization, rather than a heuristic derivation. Numerical results demonstrated the effectiveness of the proposed SA.

A novel method for designing power system damping controllers was presented (Chen et al., 1996). They used SA to optimize the controller parameters in the nonlinear optimization problem, and considered the all of the design criteria of the controllers simultaneously. Their proposed method can also be used to design controllers that are robust under a specified set of operating conditions.

Design of output feedback controllers for thyristor controlled series compensators in a meshed power system was proposed (Chen et al., 1998). They used SA to optimize the output feedback gains for the controllers. Conflicting design objectives, such as improvement in the damping of the critical modes, any deterioration of the damping of the noncritical modes and the saturation of the controller actuators, were simultaneously considered. Numerical results verified that the SA can be applied to design robust controllers that satisfy the required performance criteria under many operating conditions.

Feeder imbalance describes a situation in which the magnitudes of the voltages of a three-phase voltage source are not equal, or the phase differences between them are not 120 electrical degrees, or both. Phase balancing make the voltages balanced at each load point of the feeder. Phase balancing optimization is currently attracting more attention in the power industry, especially following deregulation. Nonlinear effects, such as voltage drops and energy losses, make the problem difficult to solve. SA was used as an effective method to solve a power distribution phase balancing problem with its nonlinear effects (Zhu et al., 1999). Test results verified the effectiveness of the proposed approach.

Robust design of multi-machine power system stabilizers (PSS) using SA was proposed (Abido, 2000a). The SA can obtain optimal parameter settings of an extensively used conventional fixed-structure lead-lag PSS. The parameters of the proposed SA-based PSS were optimized to shift simultaneously the system electromechanical modes under different loading conditions and system configurations to the left in the s-plane. The incorporation of SA as a derivative-free optimization approach in PSS design greatly reduces the computational burden. Test results demonstrated the effectiveness of the proposed approach under various disturbances and loading conditions for two multimachine power systems.

SA-based approach to PSS and flexible alternating current transmission systems (FACTS) based stabilizer tuning was presented (Abido, 2000b). The problem of designing PSS and FACTS-based stabilizers was formulated as an optimization problem. An eigenvalue-based objective function to increase system damping was proposed, and the SA was used to search for optimal stabilizer parameters. Different control schemes have been proposed and tested on a weakly connected power system under different disturbances, loading conditions, and parameter variations. Nonlinear simulation results indicate the potential usefulness of the SA in the problem of tuning PSS and FACTS-based stabilizer. The effectiveness and robustness of the proposed control schemes under a wide range of loading conditions and system parameter variations have been demonstrated.

A pole placement technique for PSS and thyristor controlled series capacitor (TCSC) based stabilizer using SA was proposed (Abido, 2000c). The design problem is formulated as an optimization problem where SA was used to search for the optimal setting of the design parameters, and considered a pole placement-based objective function to shift the dominant eigenvalues to the left in the s-plane. Eigenvalue analysis and nonlinear simulation results confirmed the effectiveness and the robustness of the proposed stabilizers and their ability to provide efficient damping of low frequency oscillations.

SA was applied to evaluate harmonics and frequency for power system quality analysis and frequency relaying (Soliman et al., 2004). The sum of the squares of errors is the objective function to be minimized for evaluating the amplitude and phase angle of each harmonic component as well as the fundamental frequency of the voltage signal. The proposed algorithm applied digitized samples of the voltage signal where the power quality is to be measured and the frequency relaying is to be implemented. The proposed SA had an adaptive cooling schedule and used a variable discretization to accelerate the convergence of the original SA. Numerical results revealed that the proposed approach can identify the harmonic spectrum in the signal.

Calculation of the optimum installation angle for the fixed solar-cell panels based on GA and SA was presented (Chen et al., 2005). The incident angle of sunlight strongly affects the output power of a solar-cell panel, and its efficiency can be improved if the solar-cell panel is properly installed at the optimum angle. Both GA and SA with climatic data are utilized to calculate the optimum installation angle of the solar-cell panel at various locations in Taiwan. Experimental results reveal that the best monthly installation angles are very close to those determined by the computer simulation results.

Identifying placement sites for PMUs in a power system based on incomplete observability was presented (Nuqui & Phadke, 2005). They introduced the novel concept of depth of unobservability and explained its effect on the number of PMU placements. They extended their model to recognize limitations in the availability of communication facilities around the network and thus formulated the constrained placement problem. The SA was further used to solve the pragmatic phased installation of PMUs. Results demonstrated that the proposed SA provides utilities with systematic approach for incrementally placing PMUs, to help manage the impact of their cost.

SA was applied to optimize size of a PV/wind integrated hybrid energy system with battery storage (Ekren & Ekren, 2010). Their SA used a stochastic gradient search to find the global optimization solution that minimized the total cost of the hybrid energy system. The decision variables included PV size, area swept by the wind turbine rotor, and battery capacity. Test results confirmed that SA yielded better results than those obtained from response surface method.

9. Conclusions

This chapter reviewed journal papers that have presented the applications of SA-based approaches to electric power systems, especially in generation and transmission expansion planning, generator maintenance scheduling and unit commitment, reactive power planning, load forecasting and economic dispatch, distribution systems planning and operation, and the other applications. The SA-based approaches have the following advantages. They may find a global optimum; they can produce a number of alternative solutions; no mathematical restrictions on the problem formulation exist, and they are relatively easy to program and numerically robust. The purpose of the review of papers and example applications in this chapter is to illustrate the potential application of the SA-based approaches in the optimization of electric power systems, and the advantages of such methods. Recently, these new heuristic tools have been combined with each other and with knowledge to solve extremely challenging problems. Hybrid approaches typically seem both to combine complementary strengths and to overcome the drawbacks of single methods by embedding in them one or more steps that involve different techniques. Developing solutions using such tools provides two major advantages: development time is much shorter than when more traditional approaches are used, and the solutions are very robust.

Author details

Yann-Chang Huang, Huo-Ching Sun, and Kun-Yuan Huang
Department of Electrical Engineering, Cheng Shiu University, Kaohsiung, Taiwan

10. References

Abido, M. A. (2000a). Robust design of multimachine power system stabilizers using simulated annealing. *IEEE Transactions on Energy Conversion*, Vol. 15, No. 3, (September 2000), pp. 297-304

Abido, M. A. (2000b). Simulated annealing based approach to PSS and FACTS based stabilizer tuning. *International Journal of Electrical Power & Energy Systems*, Vol. 22, No. 4, (May 2000), pp. 247-258

Abido, M. A. (2000c). Pole placement technique for PSS and TCSC-based stabilizer design using simulated annealing. *International Journal of Electrical Power & Energy Systems*, Vol. 22, No. 8, (November 2000), pp. 543-554

Annakkage, U. D.; Numnonda, T. & Pahalawaththa, N. C. (1995). Unit commitment by parallel simulated annealing. *IEE Proceedings-Generation, Transmission and Distribution*, Vol. 142, No. 6, (November 1995), pp. 595-600

Antunes, C. H.; Lima, P.; Oliveira, E. & Pires, D. F. (2011). A multi-objective simulated annealing approach to reactive power compensation. *Engineering Optimization*, Vol. 43, No. 10, (October 2011), pp. 1063-1077

Baldwin, T. L.; Mili, L.; Boisen, M. B., Jr. & Adapa, R. (1993). Power system observability with minimal phasor measurement placement. *IEEE Transactions on Power Systems*, Vol. 8, No. 2, (May 1993), pp. 707-715

Basu, M. (2005). A simulated annealing-based goal-attainment method for economic emission load dispatch of fixed head hydrothermal power systems. *International Journal of Electrical Power & Energy Systems*, Vol. 27, No. 2, (February 2005), pp. 147-153

Billinton, R. & Jonnavithula, S. (1996). Optimal switching device placement in radial distribution systems. *IEEE Transactions on Power Delivery*, Vol. 11, No. 3, (July 1996), pp. 1646-1651

Chang, H. C. & Kuo, C. C. (1994). Network reconfiguration in distribution systems using simulated annealing. *Electric Power Systems Research*, Vol. 29, No. 3, (May 1994), pp. 227-238

Chen, X. R.; Pahalawaththa, N. C.; Annakkage, U. D. & Kumble, C. S. (1996). Simulated annealing for the design of power system damping controllers. *Electric Power Systems Research*, Vol. 39, No. 1, (October 1996), pp. 67-72

Chen, X. R.; Pahalawaththa, N. C.; Annakkage, U. D. & Kumble, C. S. (1998). Design of decentralised output feedback TCSC damping controllers by using simulated annealing. *IEE Proceedings-Generation, Transmission and Distribution*, Vol. 145, No. 5, (September 1998), pp. 553-558

Chen, Y. L. & Liu, C. C. (1994). Multiobjective Var planning using the goal-attainment method. *IEE Proceedings-Generation, Transmission and Distribution*, Vol. 141, No. 3, (May 1994), pp. 227-232

Chen, Y. L. & Liu, C. C. (1995). Optimal multi-objective Var planning using an interactive satisfying method. *IEEE Transactions on Power Systems*, Vol. 10, No. 2, (May 1995), pp. 664-670

Chen, Y. L. (1996). Weak bus-oriented optimal multi-objective Var planning. *IEEE Transactions on Power Systems*, Vol. 11, No. 4, (November 1996), pp. 1885-1890

Chen, Y. M.; Lee, C. H. & Wu, H. C. (2005). Calculation of the optimum installation angle for fixed solar-cell panels based on the genetic algorithm and the simulated-annealing method. *IEEE Transactions on Energy Conversion*, Vol. 20, No. 2, (June 2005), pp. 467- 473

Chen, Y. L. & Ke, Y. L. (2004). Multi-objective Var planning for large-scale power systems using projection-based two-layer simulated annealing algorithms. *IEE Proceedings-Generation, Transmission and Distribution*, Vol. 151, No. 4, (July 2004), pp. 555-560

Chiang, H. D. & Jean-Jumeau, R. (1990a). Optimal network reconfigurations in distribution systems: part I: a new formulation and a solution methodology. *IEEE Transactions on Power Delivery*, Vol. 5, No. 4, (October 1990), pp. 1902-1909

Chiang, H. D. & Jean-Jumeau, R. (1990b). Optimal network reconfigurations in distribution systems: part II: solution algorithms and numerical results. *IEEE Transactions on Power Delivery*, Vol. 5, No. 3, (July 1990), pp. 1568-1574

Chiang, H. D.; Wang, J. C.; Tong, J. & Darling, G. (1995a). Optimal capacitor placement, replacement and control in large-scale unbalanced distribution systems: modeling and a new formulation. *IEEE Transactions on Power Systems*, Vol. 10, No. 1, (February 1995), pp. 356-362

Chiang, H. D.; Wang, J. C.; Tong, J. & Darling, G. (1995b). Optimal capacitor placement, replacement and control in large-scale unbalanced distribution systems: system solution algorithms and numerical studies. *IEEE Transactions on Power Systems*, Vol. 10, No. 1, (February 1995), pp. 363-369

Christober Asir Rajan, C. & Mohan, M. R. (2007). An evolutionary programming based simulated annealing method for solving the unit commitment problem. *International Journal of Electrical Power & Energy Systems*, Vol. 29, No. 7, (September 2007), pp. 540-550

Christober Asir Rajan, C. (2010). A solution to the economic dispatch using EP based SA algorithm on large scale power system. *International Journal of Electrical Power & Energy Systems*, Vol. 32, No. 6, (July 2010), pp. 583-591

Christober Asir Rajan, C. (2011). Hydro-thermal unit commitment problem using simulated annealing embedded evolutionary programming approach. *International Journal of Electrical Power & Energy Systems*, Vol. 33, No. 4, (May 2011), pp. 939-946

Chu, R. F.; Wang, J. C. & Chiang H. D. (1994). Strategic planning of LC compensators in nonsinusoidal distribution systems. *IEEE Transactions on Power Delivery*, Vol. 9, No. 3, (July 1994), pp. 1558-1563

Dahal, K. P. & Chakpitak, N. (2007). Generator maintenance scheduling in power systems using metaheuristic-based hybrid approaches. *Electric Power Systems Research*, Vol. 77, No. 7, (May 2007), pp. 771-779

Das, D. B. & Patvardhan, C.; (1998). New multi-objective stochastic search technique for economic load dispatch. *IEE Proceedings-Generation, Transmission and Distribution*, Vol. 145, No. 6, (November 1998), pp. 747-752

Ekren, O. & Ekren, B. Y. (2010). Size optimization of a PV/wind hybrid energy conversion system with battery storage using simulated annealing. *Applied Energy*, Vol. 87, No. 2, (February 2010), pp. 592-598

Gallego, R. A.; Romero, R. & Monticelli, A. J. (2000). Tabu search algorithm for network synthesis. *IEEE Transactions on Power Systems*, Vol. 15, No. 2, (May 2000), pp. 490-495

Gallego, R. A.; Alves, A. B.; Monticelli, A. & Romero, R. (1997). Parallel simulated annealing applied to long term transmission network expansion planning. *IEEE Transactions on Power Systems*, Vol. 12, No. 1, (February 1997), pp. 181-188

Gallego, R. A.; Monticelli, A. & Romero, R. (1998a). Comparative studies on nonconvex optimization methods for transmission network expansion planning. *IEEE Transactions on Power Systems*, Vol. 13, No. 3, (August 1998), pp. 822-828

Gallego, R. A.; Monticelli, A. & Romero, R. (1998b). Transmission system expansion planning by an extended genetic algorithm. *IEE Proceedings-Generation, Transmission and Distribution*, Vol. 145, No. 3, (May 1998), pp. 329-335

Habiballah, I. O. & Irving, M. R. (1995). Multipartitioning of power system state estimation networks using simulated annealing. *Electric Power Systems Research*, Vol. 34, No. 2, (August 1995), pp. 117-120

Hsiao, Y. T.; Chiang, H. D.; Liu, C. C. & Chen, Y. L. (1994). A computer package for optimal multi-objective Var planning in large scale power systems. *IEEE Transactions on Power Systems*, Vol. 9, No. 2, (May 1994), pp. 668-676

Hsiao, Y. T.; Liu, C. C.; Chiang, H. D. & Chen, Y. L. (1993). A new approach for optimal Var sources planning in large scale electric power systems. *IEEE Transactions on Power Systems*, Vol. 8, No. 3, (August 1993), pp. 988-996

Irving, M. R. & Sterling, M. J. H. (1990). Optimal network tearing using simulated annealing. *IEE Proceedings-Generation, Transmission and Distribution*, Vol. 137, No. 1, (January 1990), pp. 69-72

Jeon, Y. J.; Kim, J. C.; Kim, J. O; Shin, J. R. & Lee, K. Y. (2002). An efficient simulated annealing algorithm for network reconfiguration in large-scale distribution systems. *IEEE Transactions on Power Delivery*, Vol. 17, No. 4, (October 2002), pp. 1070-1078

Jeon,Y. J. & Kim, J. C. (2004). Application of simulated annealing and tabu search for loss minimization in distribution systems. *International Journal of Electrical Power & Energy Systems*, Vol. 26, No. 1, (January 2004), pp. 9-18

Jiang, D. & Baldick, R. (1996). Optimal electric distribution system switch reconfiguration and capacitor control. *IEEE Transactions on Power Systems*, Vol. 11, No. 2, (May 1996), pp. 890-897

Jonnavithula, S. & Billinton, R. (1996). Minimum cost analysis of feeder routing in distribution system planning. *IEEE Transactions on Power Delivery*, Vol. 11, No. 4, (October 1996), pp. 1935-1940

Jwo, W. S.; Liu, C. W. & Liu, C. C. (1999). Large-scale optimal Var planning by hybrid simulated annealing/genetic algorithm. *International Journal of Electrical Power & Energy Systems*, Vol. 21, No. 1, (January 1999), pp. 39-44

Jwo, W. S.; Liu, C. W.; Liu, C. C. & Hsiao, Y. T. (1995). Hybrid expert system and simulated annealing approach to optimal reactive power planning. *IEE Proceedings-Generation, Transmission and Distribution*, Vol. 142, No. 4, (July 1995), pp. 381-385

Kannan, S.; Slochanal, S. M. R. & Padhy, N. P. (2005). Application and comparison of metaheuristic techniques to generation expansion planning problem. *IEEE Transactions on Power Systems*, Vol. 20, No. 1, (February 2005), pp. 466-475

Kannan, S.; Slochanal, S. M. R.; Baskar, S.; Murugan, P. (2007). Application and comparison of metaheuristic techniques to generation expansion planning in the partially deregulated environment. *IET Generation, Transmission & Distribution*, Vol. 1, No. 1, (January 2007), pp. 111-118

Kim, H.; Hayashi, Y. & Nara, K. (1997). An algorithm for thermal unit maintenance scheduling through combined use of GA, SA and TS. *IEEE Transactions on Power Systems*, Vol. 12, No. 1, (February 1997), pp. 329-335

Li, Z. & Shahidehpour, M. (2003). Generation scheduling with thermal stress constraints. *IEEE Transactions on Power Systems*, Vol. 18, No. 4, (November 2003), pp. 1402-1409

Liang, R. H. & Kang, F. C. (2000). Thermal generating unit commitment using an extended mean field annealing neural network. *IEE Proceedings-Generation, Transmission and Distribution*, Vol. 147, No. 3, (May 2000), pp. 164-170

Liao, G. C. & Tsao, T. P. (2006). Application of a fuzzy neural network combined with a chaos genetic algorithm and simulated annealing to short-term load forecasting. *IEEE Transactions on Evolutionary Computation*, Vol. 10, No. 3, (June 2006), pp. 330-340

Liu, C. W.; Jwo, W. S.; Liu, C. C. & Hsiao, Y. T. (1997). A fast global optimization approach to Var planning for the large scale electric power systems. *IEEE Transactions on Power Systems*, Vol. 12, No. 1, (February 1997), pp. 437-443

Liu, Y.; Ma, L. & Zhang, J. (2002). Reactive power optimization by GA/SA/TS combined algorithms. *International Journal of Electrical Power & Energy Systems*, Vol. 24, No. 9, (November 2002), pp. 765-769

Mantawy, A. H.; Abdel-Magid, Y. L. & Selim, S. Z. (1998). A simulated annealing algorithm for unit commitment. *IEEE Transactions on Power Systems*, Vol. 13, No. 1, (February 1998), pp. 197-204

Mantawy, A. H.; Abdel-Magid, Y. L. & Selim, S. Z. (1999). Integrating genetic algorithms, tabu search, and simulated annealing for the unit commitment problem. *IEEE Transactions on Power Systems*, Vol. 14, No. 3, (August 1999), pp.829-836

Mori, H. & Kobayashi, H. (1996). Optimal fuzzy inference for short-term load forecasting. *IEEE Transactions on Power Systems*, Vol. 11, No. 1, (February 1996), pp. 390-396

Mori, H. & Takeda, K. (1994). Parallel simulated annealing for power system decomposition. *IEEE Transactions on Power Systems*, Vol. 9, No. 2, (May 1994), pp. 789-795

Nahman, J. M. & Peric, D. M. (2008). Optimal planning of radial distribution networks by simulated annealing technique. *IEEE Transactions on Power Systems*, Vol. 23, No. 2, (May 2008), pp. 790-795

Nayak, R. & Sharma, J. D. (2000). A hybrid neural network and simulated annealing approach to the unit commitment problem. *Computers & Electrical Engineering*, Vol. 26, No. 6, (August 2000), pp. 461-477

Nuqui, R. F. & Phadke, A. G. (2005). Phasor measurement unit placement techniques for complete and incomplete observability. *IEEE Transactions on Power Delivery*, Vol. 20, No. 4, (October. 2005), pp. 2381-2388

Ongsakul, W.; Dechanupaprittha, S. & Ngamroo, I. (2004). Parallel tabu search algorithm for constrained economic dispatch. *IEE Proceedings-Generation, Transmission and Distribution*, Vol. 151, No. 2, (March 2004), pp. 157-166

Pai, P. F. & Hong, W. C. (2005). Support vector machines with simulated annealing algorithms in electricity load forecasting. *Energy Conversion and Management*, Vol. 46, No. 17, (October 2005), pp. 2669-2688

Panigrahi, C. K.; Chattopadhyay, P. K.; Chakrabarti, R. N. & Basu, M. (2006). Simulated annealing technique for dynamic economic dispatch. *Electric Power Components and Systems*, Vol. 34, No. 5, (May 2006), pp. 577-586

Purushothama, G. K.; Jenkins, L. (2003). Simulated annealing with local search-a hybrid algorithm for unit commitment. *IEEE Transactions on Power Systems*, Vol. 18, No. 1, (February 2003), pp. 273-278

Purushothama, G. K.; Narendranath, U. A.; Jenkins, L. (2003). Unit commitment using a stochastic extended neighbourhood search. *IEE Proceedings-Generation, Transmission and Distribution*, Vol. 150, No. 1, (January 2003), pp. 67-72

Roa-Sepulveda, C. A. & Pavez-Lazo, B. J. (2003). A solution to the optimal power flow using simulated annealing. *International Journal of Electrical Power & Energy Systems*, Vol. 25, No. 1, (January 2003), pp. 47-57

Romero, R.; Gallego, R. A.; Monticelli, A. (1996). Transmission system expansion planning by simulated annealing. *IEEE Transactions on Power Systems*, Vol. 11, No. 1, (February 1996), pp. 364-369

Saber, A. Y.; Senjyu, T.; Miyagi, T.; Urasaki, N. & Funabashi, T. (2007). Unit commitment by heuristics and absolutely stochastic simulated annealing. *IET Generation, Transmission & Distribution*, Vol. 1, No. 2, (March 2007), pp. 234-243

Saber, A. Y.; Senjyu, T.; Miyagi, T.; Urasaki, N. & Funabashi, T. (2006). Fuzzy unit commitment scheduling using absolutely stochastic simulated annealing. *IEEE Transactions on Power Systems*, Vol. 21, No. 2, (May 2006), pp. 955-964

Satoh, T. & Nara, K. (1991). Maintenance scheduling by using simulated annealing method. *IEEE Transactions on Power Systems*, Vol. 6, No. 2, (May 1991), pp. 850-857

Simopoulos, D. N.; Kavatza, S. D. & Vournas, C. D. (2006b). Reliability constrained unit commitment using simulated annealing. *IEEE Transactions on Power Systems*, Vol. 21, No. 4, (November 2006), pp. 1699-1706

Simopoulos, D. N.; Kavatza, S. D. & Vournas, C. D. (2006a). Unit commitment by an enhanced simulated annealing algorithm. *IEEE Transactions on Power Systems*, Vol. 21, No. 1, (February 2006), pp. 68-76

Soliman, S. A.; Mantaway, A. H. & El-Hawary, M. E. (2004). Simulated annealing optimization algorithm for power systems quality analysis. International Journal of Electrical Power & Energy Systems, Vol. 26, No. 1, (January 2004), pp. 31-36

Wang, Y.; Li, F.; Wan, Q. & Chen, H. (2011). Reactive power planning based on fuzzy clustering, gray code, and simulated annealing. *IEEE Transactions on Power Systems*, Vol. 26, No. 4, (November 2011), pp. 2246-2255

Wong, K. P. & Fung, C. C. (1993). Simulated annealing based economic dispatch algorithm. *IEE Proceedings-Generation, Transmission and Distribution*, Vol. 140, No. 6, (November 1993), pp. 509-515

Wong, K. P. & Wong, Y. W. (1994a). Short-term hydrothermal scheduling. Part I: simulated annealing approach. *IEE Proceedings-Generation, Transmission and Distribution*, Vol. 141, No. 5, (September 1994), pp. 497-501

Wong, K. P. & Wong, Y. W. (1994b). Short-term hydrothermal scheduling. Part II: parallel simulated annealing approach. *IEE Proceedings-Generation, Transmission and Distribution*, Vol. 141, No. 5, (September 1994), pp. 502-506

Wong, K. P. & Wong, Y. W. (1994c). Genetic and genetic/simulated-annealing approaches to economic dispatch. *IEE Proceedings-Generation, Transmission and Distribution*, Vol. 141, No. 5, (September 1994), pp. 507-513

Wong, K. P. & Wong, Y. W. (1995). Thermal generator scheduling using hybrid genetic/simulated-annealing approach. *IEE Proceedings-Generation, Transmission and Distribution*, Vol. 142, No. 4, (July 1995), pp. 372-380

Wong, K. P. & Wong, Y. W. (1996). Combined genetic algorithm/simulated annealing/fuzzy set approach to short-term generation scheduling with take-or-pay fuel contract. *IEEE Transactions on Power Systems*, Vol. 11, No. 1, (February 1996), pp. 128-136

Wong, K. P. & Wong, Y. W. (1997). Hybrid genetic/simulated annealing approach to short-term multiple-fuel-constrained generation scheduling. *IEEE Transactions on Power Systems*, Vol. 12, No. 2, (May 1997), pp. 776-784

Wong, Y. W. (1998). An enhanced simulated annealing approach to unit commitment. *International Journal of Electrical Power & Energy Systems*, Vol. 20, No. 5, (June 1998), pp. 359-368

Zhu, J. & Chow, M. Y. (1997). A review of emerging techniques on generation expansion planning. *IEEE Transactions on Power Systems*, Vol. 12, No. 4, (November 1997), pp. 1722-1728

Zhu, J.; Bilbro, G. & Chow, M. Y. (1999). Phase balancing using simulated annealing. *IEEE Transactions on Power Systems*, Vol. 14, No. 4, (November 1999), pp. 1508-1513

Zhuang, F.; Galiana, F. D. (1990). Unit commitment by simulated annealing. *IEEE Transactions on Power Systems*, Vol. 5, No. 1, (February 1990), pp. 311-318

Chapter 7

Improvements in Simulated Quenching Method for Vehicle Routing Problem with Time Windows by Using Search History and Devising Means for Reducing the Number of Vehicles

Hisafumi Kokubugata, Yuji Shimazaki, Shuichi Matsumoto, Hironao Kawashima and Tatsuru Daimon

Additional information is available at the end of the chapter

http://dx.doi.org/10.5772/47854

1. Introduction

In many countries, rationalization of freight transportation is recognized to be an important problem. For example in Japan surrounded by sea, about 60% of freight transportation is carried out by road traffic. The proportion of freight road transportation to total road transportation is close to half. Although development of information technology accelerates electronic communication, physical distribution of goods is left behind. On the contrary, because electronic commerce has enhanced door-to-door delivery services, delivery distribution of goods has increased in urban areas. The demands for high-quality delivery services such as small-amount high frequency deliveries with time windows have been made by many clients (including companies and individuals).

From the aspect of freight carrier, decease of fuel consumption makes big profit, since the proportion of fuel to total cost is large. The rationalization in terms of increasing the loading rate and decreasing the total travel time is aimed not only for reducing operational costs in each freight carrier but also for relieving traffic congestion, saving energy and reducing exhaust gas. Effective distribution of goods should be realized by sophisticated delivery planning.

A typical delivery problem is modelled mathematically in Vehicle Routing Problem (VRP). In VRP, scattered clients are serviced only once by exactly one of plural vehicles with load capacity which depart from a depot and return to it after touring the assigned clients. As mentioned above, clients often impose the earliest delivery time and the latest delivery time. A variation of VRP in which delivery time windows are included is called Vehicle Routing Problem with Time Windows (VRPTW). VRPTW is also applied to pick up operations such as cargo collection and garbage collection.

At the beginning of this chapter, VRP and VRPTW are introduced and followed by the explanation of precedent solution methods for VRPTW. And then, a practical solution method is proposed. It is composed by a data model, transformation rules of a solution on the data model and an overall search algorithm based on the refined Simulated Quenching (SQ) for VRPTW. The refined SQ procedures are derived from incorporating information of good solutions found in search history into basic SQ scheme. In the last section, the evaluation of the proposed method is conducted by comparisons on computational experiments with basic SQ.

2. Vehicle Routing Problem with Time Windows

Typical routing problems are abstracted from actual logistics operations in urban areas and formalized as mathematical programming problems. They are categorized as the combinatorial optimization problems.

2.1. Vehicle Routing Problem (VRP)

The Vehicle Routing Problem (VRP) is the most popular problem in routing problems. It involves the design of a set of minimum cost vehicle trips, originating and ending at a depot, for a fleet of vehicles with loading capacity that services a set of client spots with required demands. The problems studied in this chapter can be described in the style used by Crescenzi & Kann in [1] for their compendium of *NP* optimization problems. Although VRP is not listed in the compendium, it is given by Prins & Bouchenoua in [2] as follows.

- INSTANCE: Complete undirected graph $G = (V,E)$, initial node $s \in V$, vehicle capacity $W \in N$, length $c(e) \in N$ for each $e \in E$, demand $q(i) \in N$ for each $i \in V$, where N is the set of natural numbers.
- SOLUTION: A set of cycles (trips), each containing the initial node 0, that collectively traverses every node at least once. A node must be serviced by one single trip and the total demand processed by any trip cannot exceed W.
- MEASURE: The total cost of the trips, to be minimized. The cost of a trip is the sum of its traversed edges.

Although the VRP in a narrow sense is defined above, the VRP in a broader sense includes the more comprehensive class of routing problems related to various conditions in which demands are located on nodes. It includes VRP with time windows imposed by clients, VRP with multiple depots, periodic VRP and etc. In this case, the simplest VRP defined above is called capacitated VRP (CVRP).

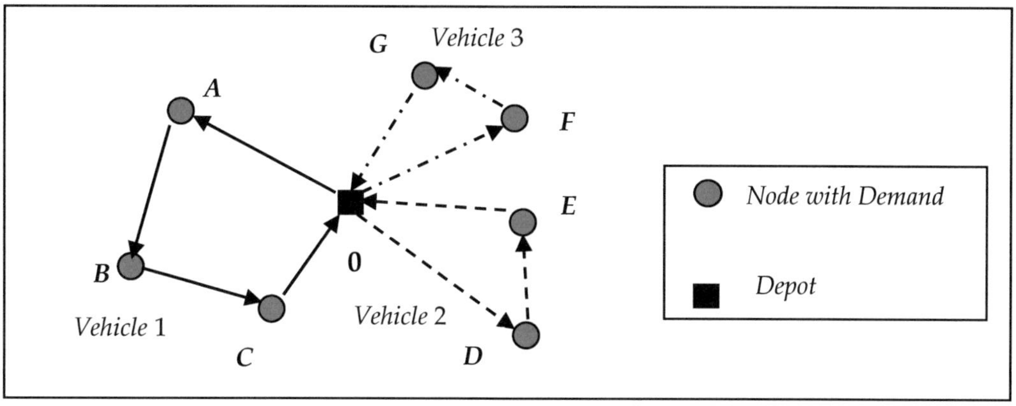

Figure 1. Vehicle Routing Problem (VRP).

2.2. Vehicle Routing Problem with Time Windows (VRPTW))

In actual delivery operations, delivery time windows are often imposed by clients. Time window at node i is described as $[e_i, l_i]$, where e_i is the earliest service starting time, l_i is the latest service starting time at node i. Vehicle routing problem taking account of time windows is called Vehicle Routing Problem with Time Windows (VRPTW). The first characteristic of VRPTW is the strictness of restriction on solutions. It often imposes increasing the number of operating vehicles. The second characteristic is the existence of waiting time. If the vehicle arrives before e_i, it must wait until e_i and then starts unloading service. Because VRP belongs to *NP*-hard problems, VRPTW belongs to them, too. Moreover, time windows make sequential delivery order restrictive. Hence, although both of VRP and VRPTW belong to same *NP*-hard problems in computational complexity theory, from a point of view with making practical algorithms, VRPTW is more difficult than VRP because of its tight constraint.

3. Precedent studies on heuristics for VRPTW

Because VRPTW belongs to *NP*-hard problems, exact methods are not fit for large problems. Therefore, heuristics have been important in the application of the VRPTW. Before the proposed method is explained, precedent studies on heuristics for VRPTW are introduced briefly. The heuristics for solving routing problems are classified into two major classes. The one is the family of traditional heuristics and the other is the family of metaheuristics including Simulated Annealing.

3.1. Traditional heuristics approaches for VRPTW

Comprehensive survey on traditional heuristics for VRPTW is presented in [3] by Bräysy & Gendreau. In this section, an outline of it is sketched. The traditional heuristics have been specially invented for solving VRPTW. They utilize the proper characteristics of VRPTW. They are further classified into two types.

The first one is the type of constructive heuristics that produce vehicle routes by merging existing routes or inserting nodes into existing routes. Ioannou et al. proposed an efficient constructive heuristic in [4]. They use the generic sequential insertion framework proposed in [5] by Solomon.

The second one is the type of improvement heuristics which make changes in one vehicle route or between several vehicle routes. Bräysy proposed several efficient local search heuristics in [6] using a three-phase approach. In the first phase, several initial solutions are created using the route construction heuristics with different combinations of parameter values. In the second phase, an effort is put to reduce the number of routes. In the third phase, classical Or-opt exchanges, which replace three edges in the original tour by three new ones without modifying the orientation of the route, are used to minimize total travelled distance.

3.2. Metaheuristics for VRPTW

Metaheuristics have been introduced into the solutions for VRPTW in the last two decades. Because metaheuristics are generally recognized to fit combinatorial optimizations, Simulated Annealing (SA), Tabu Search (TS), Genetic Algorithm (GA) and Ant Colony Optimization (ACO) have been tried to apply to VRPTW. Traditional heuristics explained in Sec. 3.1 are often embedded into these metaheuristics.

Cordeau et al. presented an efficient TS heuristic in [7]. Among the methods incorporating GA, the methods proposed by Homberger & Gehring in [8] and Berger et al. in [9] are reported to get good results. With respect to ACO, although not so many works on VRPTW are appeared in the literature, Gambardella et al. use an ACO approach with a hierarchy of two cooperative artificial ant colonies in [10]. Chiang & Russell developed a SA approach for VRPTW in [11]. They combined the SA process with the parallel construction approach that incorporates improvement procedures during the construction process.

In a comprehensive survey on metaheuristics for VRPTW given by Bräysy & Gendreau in [12], it is described that some hybrid methods are very effective and competitive with two good GA algorithms listed above. They are briefly introduced as follows. Bent & Van Hentenryck present a two-stage hybrid metaheuristic in [13], where in the first stage is a basic SA used to minimize the number of routes, and the second stage focuses on distance minimization using the large neighbourhood search. Bräysy presents a four-phase deterministic metaheuristic algorithm in [14] which is based on a modification of the variable neighbourhood search. Ibaraki et al. propose three metaheuristics in [15] to improve randomly generated initial solutions.

4. Data model and method of generating neighbours in searching process of simulated quenching for VRPTW

Although some precedent methods based on metaheuristics mentioned above show good performance, their procedures are considerably complex. In particular, the local search procedures incorporated into them are rather complicated. In practical application of VRPTW algorithms to real-world problems, ease of implementation and flexibility are very important as well as quality of solution, running time and robustness. Hence, the authors of this chapter have proposed a simpler data model and a one-phase algorithm to solve VRPTW in [16] which is not the two-phase algorithm composed by construction and improvement.

4.1. Data modelling for VRPTW

The model to express a state of solution of VRPTW is realized as a sequence of integers, i.e., a string. In the string, the position of an integer, which is a symbol of the node with demand, implies not only which vehicle tours the node but also the routing order of it. An example of the string model is illustrated in Figure 2. The special number '0' should be interpreted not only as the depot but also as the delimiter which partitions the trips. If the number of vehicles is denoted by m, $(m-1)$ '0's are provided in the string. If there is no integer between '0' and '0', the relevant vehicle is not in use.

This data model is coincidentally similar to that invented for the solution based on a kind of GA. It was introduced by Gendreau et al. in [17] as the original idea was given by Van Breedam in [18]. However, the proposed transformation rules in this chapter based on the data model are quite different from those of precedent methods. They will be described in the following section.

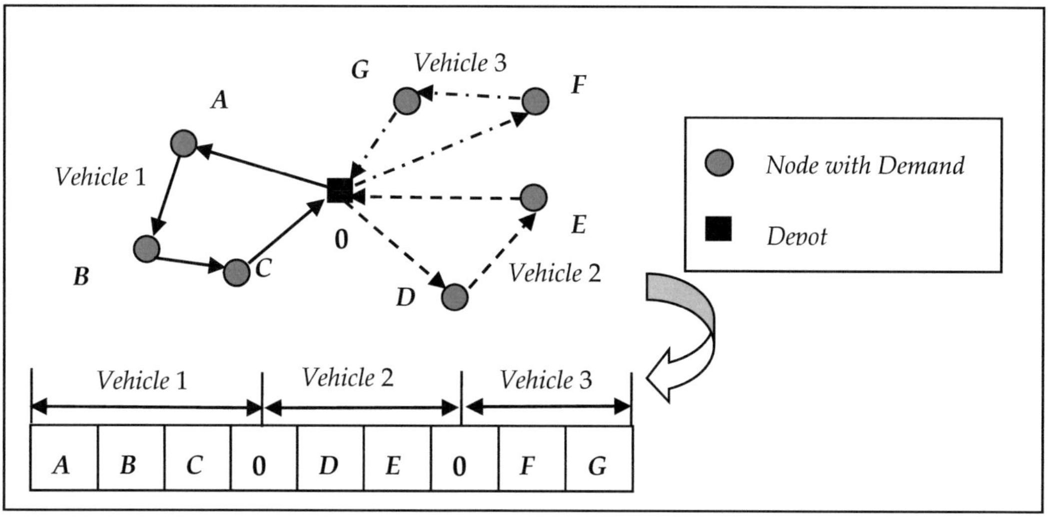

Figure 2. Proposed data model for VRPTW.

4.2. Transformation rules for generating neighbours

In a repetition in the proposed procedure, a new state of solution is generated from the present state by one of the following three types of transformation rules for generating neighbours. The first rule is to exchange an integer with another one in the string. The second rule is to delete an arbitrary integer and then insert it to another position in the string. The third rule is that after a substring is taken out temporally, the direction of the substring is reversed, and then embedded in the place where the substring is taken out. These three transformation rules are illustrated in Figure 3.

Note that the rules are also applied to the special number '0' in the string illustrated in Figure 2. In other words, '0' is treated impartially with other integers. If 'one-to-one exchange' is executed within a substring partitioned by two '0's, only a route is changed. An example of the case is illustrated in Figure 4. If 'one-to-one exchange' is executed between two non-zero integers striding over '0', two nodes are exchanged between two routes. An example of this case is illustrated in Figure 5. If 'one-to-one exchange' is executed between a non-zero integer and '0', two routes are merged, while another route is divided into two routes. An example is illustrated in Figure 6.

When the second transformations rule 'delete and insert' is applied, several different cases also arise. If a non-zero integer is deleted and inserted at '0', a node is moved to another vehicle route. An example is illustrated in Figure 7.

When the third transformations rule 'partial reversal' is applied, several different cases also arise. If a substring including '0' is reversed, the relevant plural routes are changed. An example is illustrated in Figure 8. These three transformation rules were originally invented for VRP in [19] by the authors of this chapter.

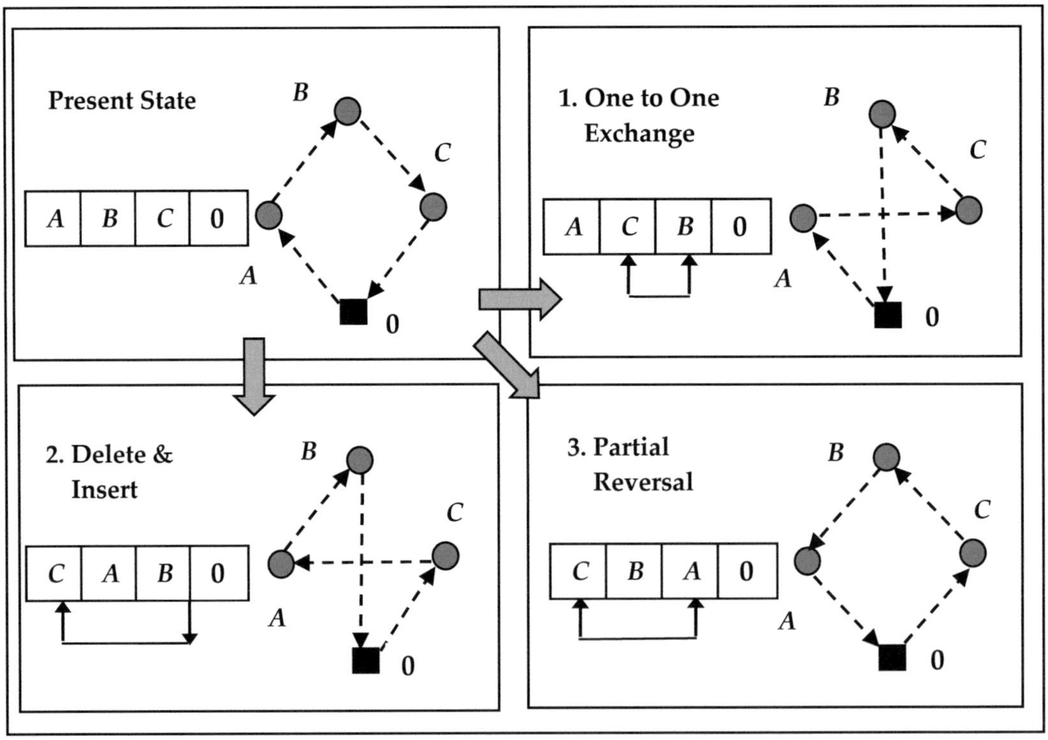

Figure 3. Transformation rules for generating neighbours.

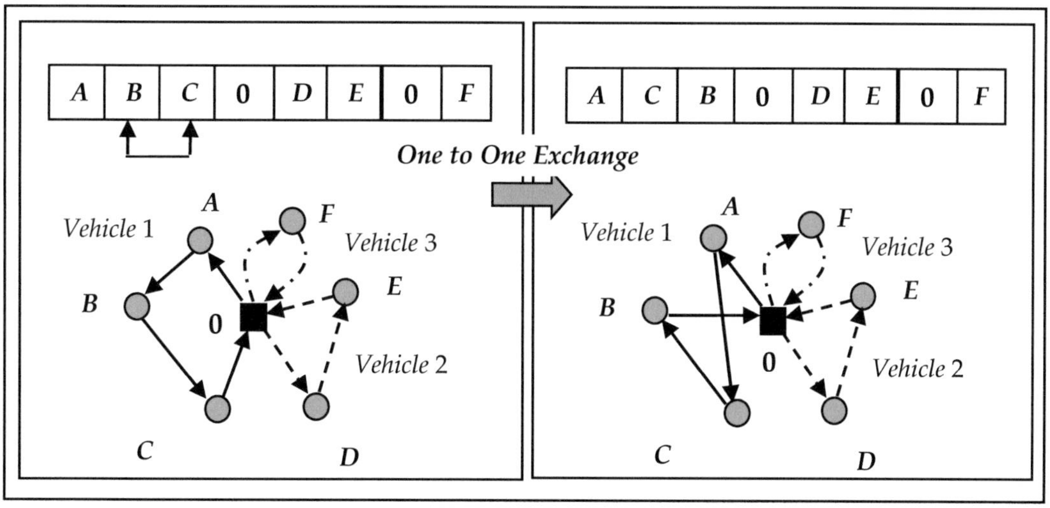

Figure 4. A result of 'one-to-one exchange' within a route.

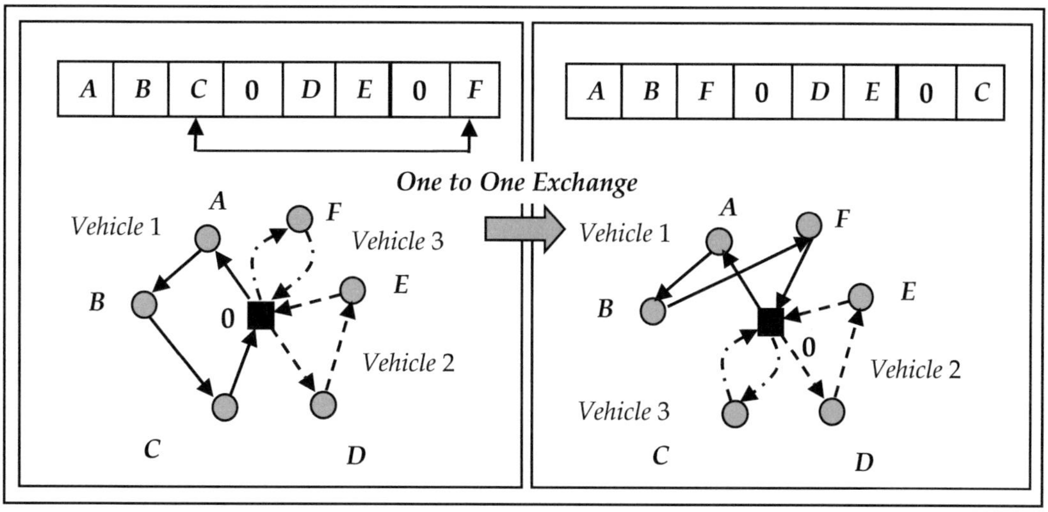

Figure 5. A result of 'one-to-one exchange' between two non-zero integers striding over '0'.

4.3. Objective function

The objective of the VRPTW is the minimization of total cost which is subject to constraints including the loading capacity of each vehicle and the time windows imposed by clients. The objective function of the VRPTW is formulated as follows.

$$E(s) = \sum_{i=1}^{n} c_{s_i} + \sum_{i=0}^{n} d_{s_i, s_{i+1}} \tag{1}$$

where $s = (s_1, s_2, \cdots, s_n)$ is a string that consists of the nodes with demands and a depot; s_0 and s_{n+1} are the implicit expressions of the depot omitted in the string s; c_k is the servicing

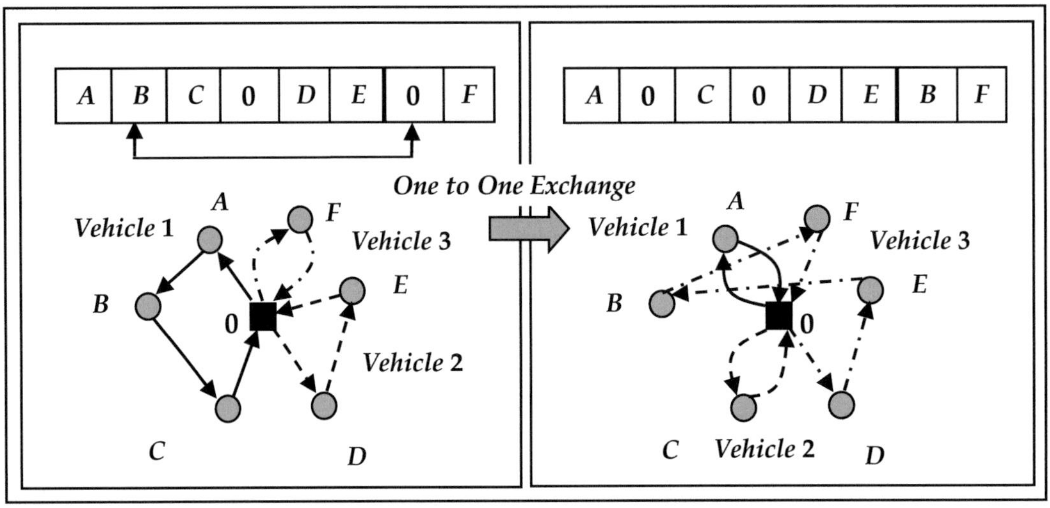

Figure 6. A result of 'one-to-one exchange' between non-zero integer and '0'.

cost at the node k (if k = 0,then c_k = 0); $d_{k,l}$ is the minimal traversing cost from the node k to the node l. Each value of $d_{k,l}$ might be given by input data; or calculated as the Euclidean distances between a pair of coordinates of nodes; or calculated by the shortest path search algorithm (Warshall-Floyd's algorithm) when road network is given and vehicles must follow the roads in the network.

In order to impose solutions of VRPTW to satisfy time window constraints and load capacities and to reduce the number of vehicles in use, three penalty terms are added to the objective function (1) as follows:

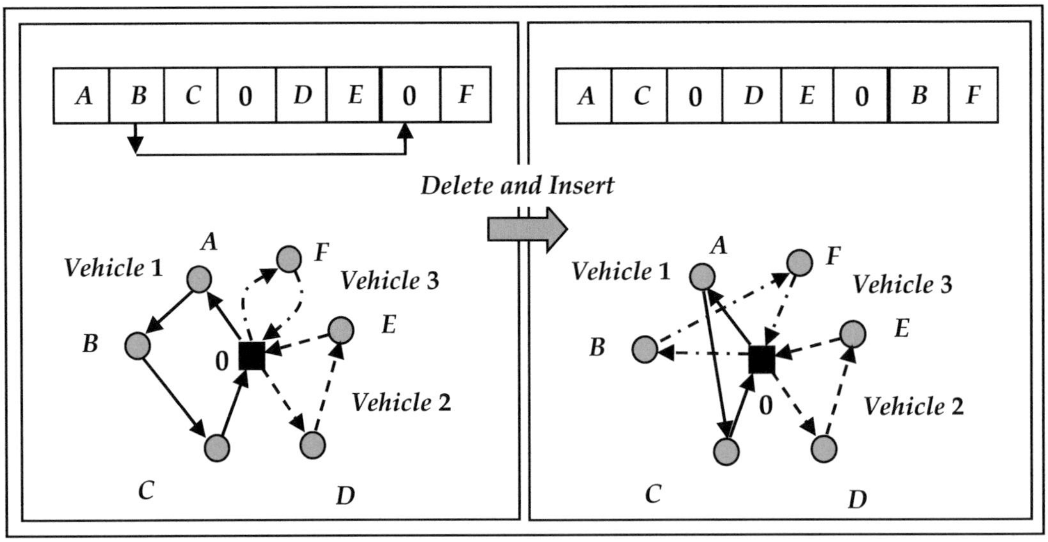

Figure 7. A result of deleting non-zero and inserting it at '0'.

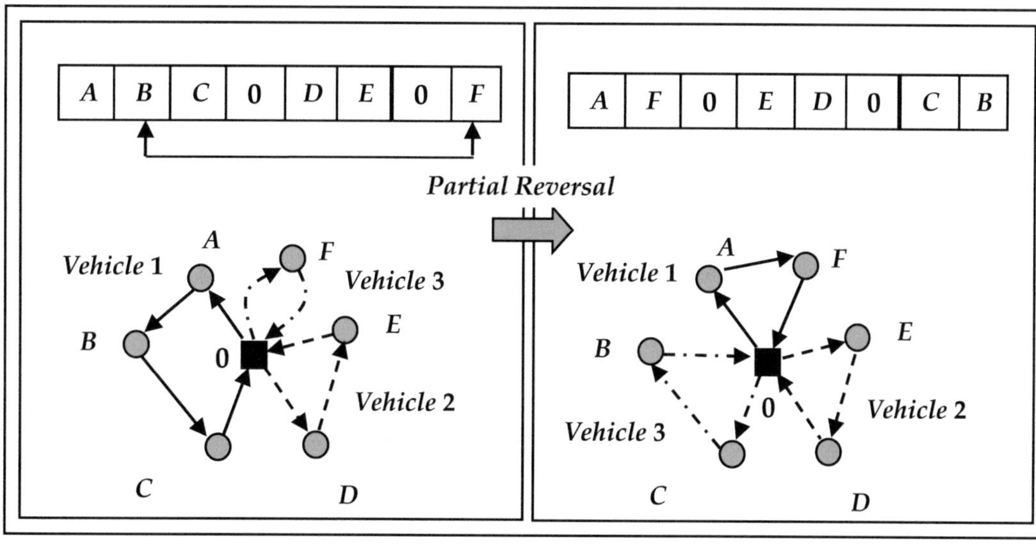

Figure 8. A result of 'partial reversal' striding over '0'.

$$E'(s)=\left(\sum_{i=1}^{n}c_{s_i}+\sum_{i=0}^{n}d_{s_i,s_{i+1}}\right)+\alpha\left(\sum_{i=1}^{n+1}\max\left(0,a_{s_i}-l_{s_i}\right)\right)+\beta\left(\sum_{k=1}^{m}\left(\sum_{i=z_{k-1}+1}^{z_k}w_{s_i}-W_k\right)\right)+\gamma m \quad (2)$$

where a_{s_i} is arriving time at node s_i; l_{s_i} is the latest service starting time at node s_i; m is the number of vehicles in use; w_{s_i} is the amount of demand at node s_i; z_k is the position of kth '0' in the string $s = (s_1, s_2, \cdots, s_n)$ (provided that $z_0 = 0$; $z_m = n+1$) and W_k is the loading capacity of vehicle k. α, β and γ are weight parameters.

4.4. Optimization algorithm using Simulated Quenching

Simulated Quenching is adopted as the optimization technique for the proposed method since it is characterized by simple stochastic procedures and by global searching scope. In the original Simulated Annealing (SA), starting with a random initial state, it is expected to approach an equilibrium point. In order to obtain global optimum, cooling schedule should be logarithmic. However, it spends too much time to implement it. Hence, in practical applications, exponential cooling schedule (3) is often adopted.

$$T(t+1) := kT(t) \ (0 < k < 1) \quad (3)$$

In the proposed method, it is adopted too. According to the strict theory of Simulated Annealing, the optimization technique using exponential cooling schedule (3) belongs to Simulated Qeunching (SQ) as described in [20]. SQ is considerd to be a practical variant of SA.

In the proposed method, the three transformation rules described in Sec. 4.2 are applied randomly to the string model. The entire algorithm for the VRP is described as follows.

{I. *Preparation*}
Read input data;
If the link cost are not given from the input data, calculate the minimum path cost $d_{i,j}$ *between all pair of clients i, j including the depot* 0;
{II. *Initialization*} *Generate an initial feasible solution* s_0 *by assigning nodes to vehicles in ascending order of the specified earliest arriving time;* $s := s_0$; $s^* := s$;
T := INITTEMP; Set N as the averaged neighbourhood size;
{III. *Optimization by SQ*} *Minimize E by repetition of applying randomly one of the three transformation rules to the string model corresponding to x in the framework of SQ;*
{IV. *Output*} *Output the best solution* s^*. (4)

Step III, that is the main part of this algorithm, is detailed as follows.

Repeat
trials := 0; *changes* := 0;
Repeat
trials := *trials* + 1;
Generate a new state s' *from the current state* s *by applying randomly one of the three transformation rules to the string model of* s;
Calculate $\Delta E' = E'(s') - E'(s)$;
If $\Delta E' < 0$ *Then* (5)
s' *is accepted as a new state;*
If If ($E'(s') < E'(s^*)$ *and* s' *is feasible*) *Then* $s^* := s'$
Else s' *is accepted with probability* $\exp(-\Delta E'/T)$
If s' *is accepted Then changes* := *changes* + 1; $s := s'$
Until trials ≥ *SIZEFACTOR* · *N or changes* ≥ *CUTOFF* · *N;*
T := *T* · *TEMPFACTOR*
Until $T \leq$ *INITTEMP / FINDIVISOR*

The words noted by capital letters are parameters used in SA and SQ and values of them are specified in Sec 6.2. As descibed in Sec. 4.2, the transformation procedure of a solution of the proposed method is carried out randomly to all over the string data model. Hence, the transformation might derive changes in a vehicle route on one occation, it might derive changes over several vehicle routes on other occation. This method is applied to VRP in [19], VRP with backhaouls (VRPB) in [21], Pick up and Delivery Problem (PDP) and VRPTW in [16] by the autors of this chaper. It is also applied to other types of routing problems including Capacitated Arc Routing Problem (CARP) in [22] and a general routing problem with nodes, edges, and arcs (NEARP) in [23]. A precise analysis of this method is presented in [24].

5. Improvement of optimization algorithm based on SQ by adaptation of devices inspired by ACO

Most of metaheuristics belong to stochastic local search (SLS) which starts at some position in search space and iteratively moves to neighbour, using randomised choices in generating and modifying candidates. In application of metaheuristics, both intensification and diversification are important. For sufficient convergence of solution, intensification of search scope is necessary. On the other hand, in order to avoid stagnation in local but not global minimum area, diversification of search scope is also necessary. As compared to 'stupid fly', search process in pure SA and SQ is completely random. Making use of history records during search processes might be possible to improvement of optimization algorithm based on SQ.

5.1. Ant Colony Optimization

ACO was introduced by Dorigo et al. in [25]. It was inspired by foraging behaviour of ants. Ants often communicate via chemicals known as pheromones, which are deposited on the ground in the form of trails. With time, paths leading to the more interesting food sources become more frequented and are marked with larger amounts of pheromone. Pheromone trails provide the basis for stochastic trail-following behaviour underlying, for example, the collective ability to find shortest paths between a food source and the nest. ACO is described as follows.

Initialise pheromone trails;
While termination criterion is not satisfied
Generate population sp of candidate solutions
using subsidiary randomised constructive search
Perform subsidiary local search on sp;
Update pheromone trails based on sp (6)

In applying ACO to TSP (Travelling Salesman Problem) which is single vehicle version of VRP, details are specified as follows.

1. Pheromone trail τ_{ij} is associated with each edge (i, j) in G, while heuristic values $\eta_{ij} = 1 / d_{ij}$ is used, where d_{ij} is traversing cost of edge (i, j).
2. In the beginning, all weights are initialised to small value τ_0.
3. In constructive search, each artificial ant starts with randomly chosen node and iteratively extends partial round trip φ by selecting node not contained in φ with probability:

$$\frac{[\tau_{ij}]^a \cdot [\eta_{ij}]^b}{\sum_{l \in N'(i)} [\tau_{il}]^a \cdot [\eta_{il}]^b}, \tag{7}$$

   where $N'(i)$ is the feasible neighbourhood of node i, that is, the set of all neighbours of i that are not contained in φ. a and b are parameters which control the relative impact of the weights *vs.* the heuristic values.

4. After the constructive search, subsidiary local search which is iterative improvement based on standard 2-exchange neighbourhood is operated until local minimum is reached.

5. In the end of loop, pheromone trail is updated according to

$$\tau_{ij}(t+1) := (1-\rho)\cdot\tau_{ij}(t) + \sum_{k=1}^{m}\Delta\tau_{ij}^{k}, \tag{8}$$

where $0 < \rho \leq 1$ is a parameter regarding vaporization of pheromone,

$$\Delta\tau_{ij}^{k} = \begin{cases} Q/E(s^{k}) & if\ (i,j)\in s^{k} \\ 0 & else \end{cases}. \tag{9}$$

$E(s^k)$ is total cost of the k th ant's cycle s^k, m is the number of ants (= size of *sp*) and Q is a constant. Criterion for weight increase is based on intuition that edges contained in short round trips should be preferably used in subsequent constructions.

As mentioned in Sec. 3.2, not so many methods based on ACO for VRPTW are appeared in the literature.

5.2. Application of information obtained in search history to SQ

One of main drawbacks of SA and SQ which are pointed out by users of other metaheuristics is lack of learning in search history, that is, blind random search often compared to 'stupid fly'. Although the complete ACO belongs population-based SLS methods in which genetic algorithm is also contained, a predecessor of ACO is Ant System which is a single ant version of ACO. It was also proposed by Dorigo et al. in [25] and it is recognized as a member of Adaptive Iterated Construction Search (AICS) methods. In the Ant System single artificial ant works and uses information obtained in its own preceding searches. Utilization of some information about good solutions obtained in the preceding search processes is able to be incorporated into SQ procedures. It would be possible to overcome the blind random searches in SQ. Because traversed arcs in good solutions of VRPTW are recognized as characteristics of them, such arcs are expected to be not drastically changed in the succeeding search processes.

In order to embody the idea described above, artificial pheromone trail τ_{ij} is associated with each edge (i, j) and τ_{ij} is updated at the end of the loop of temperature T in SQ procedures. The 'characteristic of good solution' is embodied in increase of probability of selecting better candidate in random search process in SQ. In end of loop, weight is updated according to

$$\tau_{ij}(t+1) := (1-\rho)\cdot\tau_{ij}(t) + \Delta\tau_{ij} \tag{10}$$

, where $0 < \rho \leq 1$ is a parameter regarding vaporization of pheromone,

$$\Delta\tau_{ij} = \begin{cases} Q/E(s^{*}) & if\ (i,j)\in s^{*} \\ 0 & else \end{cases}. \tag{11}$$

However, in order to avoid extreme effect of pheromone, lower bound and upper bound of τ_{ij} is set as $0.2 \leq \tau_{ij} \leq 1.4$. $E(s^{*})$ is the total cost of the best found solution at the present s^*, Q is a constant. Let $r_i = d_{ai} / \tau_{ai} + d_{ib} / \tau_{ib}$, where d_{ij} is traversing cost of edge (*i*, *j*). When edges (*a*, *i*) and (*i*, *b*) which are connected with node *i* are frequently contained in the best routes and costs of the edges are small, the value of r_i becomes small. Because such a situation of node *i* is agreeable to good solutions, it should not be drastically changed in the succeeding search processes. In order to embody the idea stated before, r_i is used for assigning node *i* the biased small probability with which node *i* is selected for change, instead of obeying uniform distribution as described in Sec. 4.2. That is to say, node *i* is selected for transformation with probability:

$$p_i = \frac{r_i}{\sum_i r_i} \tag{12}$$

The core part of the basic SQ algorithm (5) is replaced by the revised algorithm which is called SQ_{ph} described as follows.

$\tau_{ij} := 0.2$ *for all* (i,j) ;
$p_i := 1 / \sum_i i$ *for all* i ;
Repeat
    trials := 0; *changes* := 0;
    Repeat
        trials := *trials* + 1;
        Generate a new state s′ from the current state s by applying randomly one of the three transformation rules to the string model of s, where posiition i to be changed in the string is selected with the probability p_i ;
        Calculate $\Delta E = E(s') - E(s)$;
        If $\Delta E < 0$ *Then* (13)
            s′ is accepted as a new state;
            If ($E(s') < E(s^*)$ *and s′ is feasible*) *Then* $s^* := s'$
        Else s′ is accepted with probability $\exp(-\Delta E/T)$
    If s′ is accepted Then changes := changes + 1; s := s′
    Until trials ≥ *SIZEFACTOR* · *N or changes* ≥ *CUTOFF* · *N;*
    T := *T* · *TEMPFACTOR;*

$\tau_{ij} := (1-\rho)\cdot\tau_{ij} + \Delta\tau_{ij}$; *if* $\tau_{ij} > 1.4$ *then* $\tau_{ij} := 1.4$; *elseif* $\tau_{ij} < 0.2$ *then* $\tau_{ij} := 0.2$ *for all* (i,j);
$r_i := d_{ai} / \tau_{ai} + d_{ib} / \tau_{ib}$, $p_i := r_i / \sum_i r_i$ *for all* i ;
Until T ≤ *INITTEMP / FINDIVISOR*

5.3. A device for decreasing the number of vehicles in use

When performance of plural solutions for VRPTW is compared, the first measure is the number of vehicles in use, which is denoted by *m*, while the second is total cost *E*. Although SQ_{ph} is expected to utilize characteristics of good solutions already found and to reduce *E*, it could not reduce *m* directly. In order to reduce it, another device should be included in SQ procedures.

In the string model described in Sec. 4.1, successive substring of '0's is interpreted as there is no tours between two '0's, that is to say that there is a vehicle not in use. For example shown in Figure 9, when '0' is replaced by another symbol in the string, one vehicle will become not in use. In order to urge to carry out this kind of transformation, artificial pheromone trail associated with edges (i, 0) or (0, i) should be decreased.

$$\tau_{i0}(t+1) := \{(1-\rho)\tau_{i0}(t) + \Delta\tau_{i0}(t)\} \times \delta, \quad \tau_{0i}(t+1) := \{(1-\rho)\tau_{0i}(t) + \Delta\tau_{i0}(t)\} \times \delta,$$

$$0 \le \delta \le 1 \tag{14}$$

The effect of the device (14) could make the probability p_0 large according to mechanism described in Sec.5.2. This further revised algorithm in which the device (14) is incorporated with pheromone update (10) is called SQ_{ph}^{*} in this chapter.

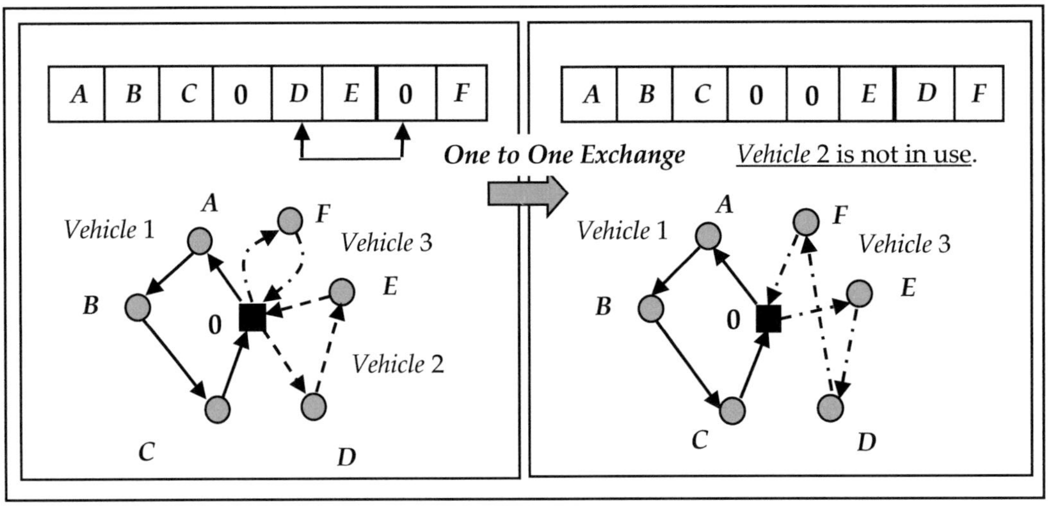

Figure 9. Reduction of m as a result of one to one exchange including '0'.

6. Computational experiments on the proposed method

Computational experiments have been attempted for testing the performance of the proposed method compared with basic SQ method. They have been tried on typical instances for VRPTW.

6.1. Solomon's benchmark problems and extended problems for VRPTW

Solomon's benchmark problem sets are produced by Solomon in [5] and provided from Solomon's own website in [26]. They are extremely popular VRPTW instances, and have been used for testing performance of methods by many researchers. Although in some of instances, optimum solutions have been already found by using exact methods, in others, they have not found yet. In both cases, the best solutions found by heuristics have been presented in the literature. Instances including 25, 50, and 100 clients have been provided from Solomon's problem sets. In the instances, each position of clients is given as x-coordinate and y-coordinate. Link cost between client i and client j is calculated with the Euclidian distance. Service time c_i is also given to each client i, in addition to the earliest arriving time e_i and the latest arriving time l_i. In this benchmark problems, link cost is directly considered as traversing time of (i, j). Arriving time a_i of each node i is calculated by summing up link cost of traversing edges and service time of traversing nodes. Concerning 100 clients problem sets, the geographical data are clustered in C-series 23 instances, randomly generated in R-series 17 instances, and a mix of random and clustered structures in RC-series 16 instances.

Gehring & Homberger extended Solomon's benchmark problems to larger scale problem sets including 200, 400, 600, 800 and 1000 clients in [27]. They are provided from their website [28]. Concerning 200 clients problem sets, the geographical data are clustered in C-series 20instances, randomly generated in R-series 19 instances, and a mix of random and clustered structures in RC-series 20 instances.

In this chapter, all of 56 instances with 100 clients from Solomon's problems and all of 59 instances with 200 clients from Gehring & Homberger's problems are chosen for computational experiments.

6.2. Values of parameters used in the algorithms and specs of the computer in use

In the computations, the values of the parameters with respect to SQ that appear in the basic SQ, SQ_{ph} and SQ_{ph}^* are set commonly as follows according to the preliminary experiments and the reference to the recommended values by Johnson et al. in [29-30].

$N = 2L^2$ (L : length of string, that is $L = n + vn - 1$, where n is the number of clients, vn is the number of vehicles superfluously allocated)
$SIZEFACTOR = 8$
$CUTOFF = 0.2$ (Repeat iterations in the same temperature T, until ($trials \geq SIZEFACTOR \cdot N$ or $changes \geq CUTOFF \cdot N$))
$INITTEMP = 10$ (Initial temperature)

$$TEMPFACTOR = 0.95 \left(T_{n+1} = 0.95T_n\right) \tag{15}$$

$FINDIVISOR = 20$ (If $T \leq INITTEMP / FINDIVISOR$, terminate the whole of the iterations.)
Values of parameters appeared in energy function (2)

$$\alpha = 25, \beta = 1, \gamma = 500 ; \tag{16}$$

and values of parameters used in the proposed method

$$\rho = 0.5 \text{ in } (10) \text{ and } (14); Q = 1000 \text{ in } (11) \tag{17}$$

are set according to the preliminary execution.

The computational experiments are executed on Windows 7, with Core i7 960, 3.2GHzCPU.

6.3. Comparison between basic SQ and SQ_{ph}

Ratio of application of artificial pheromone trail τ_{ij} to all transformations is set to 3 cases regarding 50%, 75% and 100%. In other words, the ratio of random transformation is 50%, 25% and 0% in each case. Computational experiments are performed ten times on all benchmark problems with 100 clients and 200 clients.

Concerning the number of vehicles in use m, significant difference is not detected. Regarding total traversing cost E, improvement ratio of SQ_{ph} to SQ is illustrated in Figure 10. Values corresponding to the best cost solutions and the worst cost solutions of SQ in ten executions are shown by two bars. Computing time consumed by the methods using SQ and SQ_{ph} is illustrated in Figure 11. According to these results, larger ratios of application of artificial pheromone trail bring better total traversing cost and longer computing time.

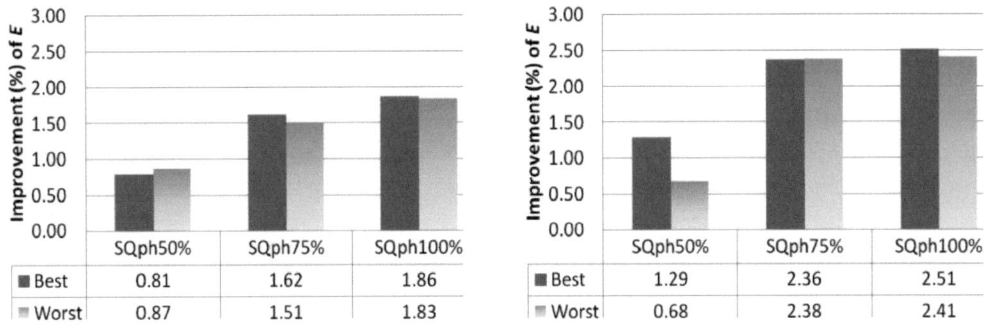

(a) 100 clients (b) 200 clients

Figure 10. Improvement ratio of total traversing cost E using SQ_{ph} to that using SQ.

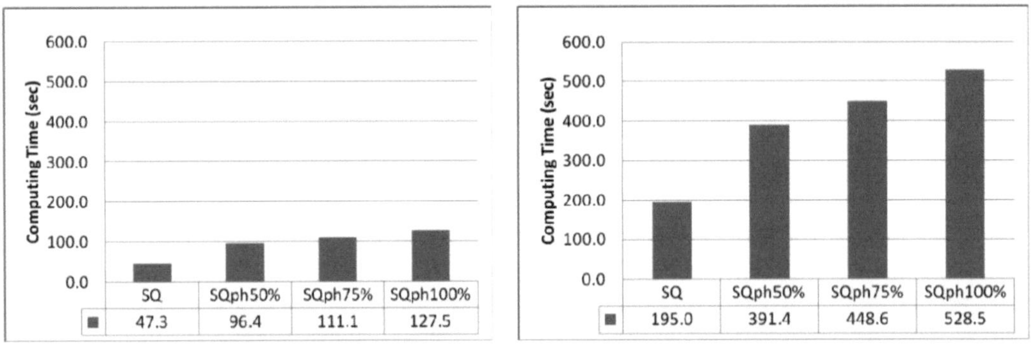

(a) 100 clients (b) 200 clients

Figure 11. Computing time consumed by the methods using SQ and SQ_{ph}.

6.4. Comparison between SQ, SQ_{ph} and SQ_{ph}^*

In order to evaluate effect of device for reducing the number of vehicles in use m, results of experiments using four types of SQ_{ph}^* are compared. Processes in SQ are divided into two parts. The first half processes correspond to higher temperature, the latter processes to lower temperature. δ (coefficient of reducing pheromone on the edges connecting depot and other clients) is set to 0.25, 0.5 and 1(= not reducing). Four types of SQ_{ph}^* (SQ_{ph}^*1, SQ_{ph}^*2, SQ_{ph}^*3, SQ_{ph}^*4) are defined in Table 1.

		In the latter half processes in SQ		
		$\delta = 0.25$	$\delta = 0.5$	$\delta = 1$
In the first half processes in SQ	$\delta = 0.25$	SQ_{ph}^*3	---	SQ_{ph}^*4
	$\delta = 0.5$	---	SQ_{ph}^*1	SQ_{ph}^*2

Table 1. Four types of SQ_{ph}^* for experiments.

6.4.1. *Comparison of the number of vehicles m*

Computational experiments are performed ten times on all benchmark problems with 100 clients and 200 clients. Regarding the number of vehicles in use m, improvement ratio of SQ_{ph}^* to $SQ_{ph}100\%$ (also to SQ) is illustrated in Figure 12.

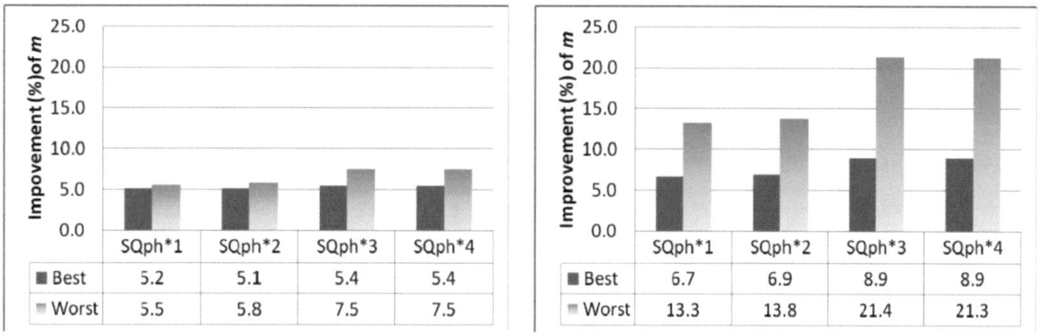

(a) 100 clients (b) 200 clients

Figure 12. Improvement ratio (%) of the number of vehicles in use m using SQ_{ph}^* to that using SQ_{ph}.

According to this result, severer reducing coefficient δ (corresponding to SQ_{ph}^*3 and SQ_{ph}^*4) brings smaller number of vehicles in use. Improvement in worst cases is larger than in best cases. This situation might be caused by the fact that the value of m is so large in worst case using $SQ_{ph}100\%$ that there is potential to be greatly improved by using SQ_{ph}^*.

6.4.2. *Comparison of traversing cost E*

Comparison of traversing cost E is significant only between situations based on same number of vehicles m. There are 15 instances with 100 clients and 23 benchmark instances with 200 clients in which optimal m is already obtained by basic SQ. Because in these instances further reduction of m cannot be brought by SQ_{ph}^*, comparison of E is attempted in these instances. Regarding E, improvement ratio of SQ_{ph}^* to SQ is illustrated in Figure 13.

Values of E by using SQ_{ph}^*2 and SQ_{ph}^*4 are better than those by $SQ_{ph}100\%$, SQ_{ph}^*1 and SQ_{ph}^*3. As defined in Table 1, SQ_{ph}^*2 and SQ_{ph}^*4 accompany reduction of pheromone on the edges connecting depot conducted only in the first processes in SQ_{ph}, while SQ_{ph}^*1 and SQ_{ph}^*3

accompany reduction of the pheromone in all processes in SQ_{ph}. These results are interpreted to mean that most of reduction of the number of vehicles in use is likely attained in the first processes in SQ_{ph}^*, while convergence of E mainly depends on the latter processes in SQ_{ph}. Computing time consumed by the methods using SQ, SQ_{ph} and SQ_{ph}^* is compared in Figure 14.

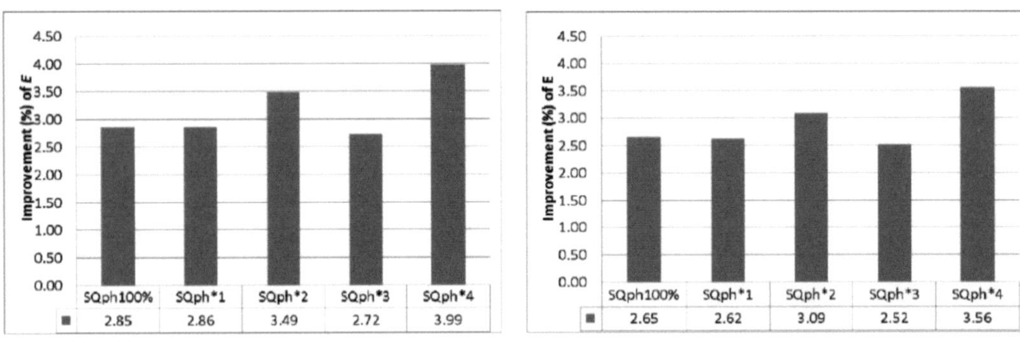

(a) 100 clients (b) 200 clients

Figure 13. Improvement ratio of total traversing cost E using SQ_{ph} and SQ_{ph}^* to that using SQ.

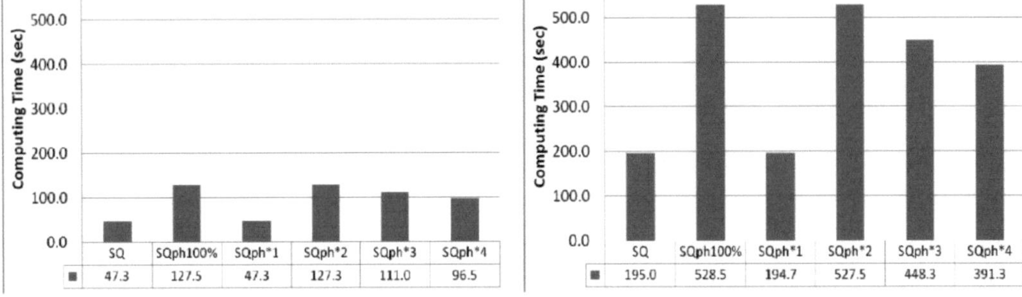

(a) 100 clients (b) 200 clients

Figure 14. Computing time consumed by the methods using SQ, SQ_{ph} and SQ_{ph}^*.

To summarize these experiments, $SQ_{ph}^{*}4$ is the most well-balanced method among methods discussed in this chapter by taking account of both of reduction of the number of vehicles in use m and reduction of total traversing cost E. Moreover, regarding computing time, $SQ_{ph}^{*}4$ is moderate. Although computing time consumed by $SQ_{ph}^{*}4$ is about two times longer than that by basic SQ, it takes about only 1.6 min for solving 100 clients problems, and it takes about 6.5 min even in 200 clients problems. It is applicable to make actual vehicle routing plans in freight carriers.

7. Conclusion

In this chapter, in order to relieve blind searches in Simulated Quenching (SQ), that is are a practical variant of Simulated Annealing (SA), utilization good characteristics of history records during SQ search processes is attempted. Two new devices which are inspired by the effect of pheromone in ant colony optimization (ACO) are adjusted and incorporated into SQ procedures to solve VRPTW. The one is a device to reduce total traversing cost E and the other is a device to reduce the number of vehicles in use m. By computational experiments on all of 56 benchmark instances with 100 clients and all of 59 benchmark instances with 200 clients, it is shown that both of two devices are effective. However, there is a trade-off between effects for reducing E and for reducing m. Taking account of putting the right device in the right place, $SQ_{ph}^{*}4$ in which the device for reducing m is set in the first half processes and the device for reducing E is set in all processes in SQ seems to be the best method. Moreover, it is moderate in computing time consumed. Reducing m in the first half processes is corresponding to diversification, while reducing E in all processes is corresponding to intensification of search. Hence, it is considered that this method improves both diversification and intensification in SQ procedures.

As mentioned before, ease of implementation and flexibility are very important as well as quality of solution, running time and robustness in practical application of VRPTW algorithms to real-world problems. The proposed method is composed by a simple data model and straightforward one-phase algorithm to solve VRPTW. Therefore, the proposed method has comparative ease of implementation and much flexibility.

Two devices incorporated in SQ procedures in this chapter are able to be incorporated into SQ procedures in other routing problems which are embodied in the string model. As introduced in Sec.4.4, VRP with backhaouls (VRPB), Pickup and Delivery Problem (PDP), Capacitated Arc Routing Problem (CARP) and a general routing problem with nodes, edges, and arcs (NEARP) have aleady been embodied in string model and solved by SQ method. Application of two devices to these problems are left for future study.

Author details

Hisafumi Kokubugata*, Yuji Shimazaki, Shuichi Matsumoto,
Hironao Kawashima and Tatsuru Daimon
Department of Administration Engineering, Keio University, Japan

* Corresponding Author

8. References

[1] Crescenzi, P. & Kann, V. (2000). A Compendium of NP Optimization Problem, Web site: http://www.nada.kth.se/~viggo/wwwcompendium/node103.html

[2] Prins, C. & Bouchenoua, S. (2004). A Memetic Algorithm Solving the VRP, the CARP and more General Routing Problems with Nodes, Edges and Arcs, In: *Recent Advances in Memetic Algorithms, Studies in Fuzziness and Soft Computing 166*, Hart W.; Kranogor N. & Smith J. (Eds.), pp. 65-85, Springer, Berlin.

[3] Bräysy, O. & Gendreau, M. (2005). Vehicle Routing Problem with Time Windows Part I: Route construction and local search algorithms, *Trans. Sci.*, Vol.39, No.1, pp.104–118.

[4] Ioannou, G.; Kritikos M. & Prastacos G. (2001). A Greedy Look-ahead Heuristic for the Vehicle Routing Problem with Time Windows, *J. Oper. Res. Soc.* Vol.52, pp.523–537.

[5] Solomon, M. (1987). Algorithms for the Vehicle Routing and Scheduling Problems with Time Window Constraints, *Operations Research*, Vol. 35, No. 2, pp. 254-265.

[6] Bräysy, O. (2003). Fast Local Searches for the Vehicle Routing Problem with Time Windows. *Inform. Systems Oper. Res.* Vol.41, pp.179–194.

[7] Cordeau, J.-F.; Laporte G., Mercier. A. (2001). A unified tabu search heuristic for vehicle routing problems with time windows, *J. Oper. Res. Soc.* Vol.52, pp.928–936.

[8] Homberger, J. & Gehring, H. (2005). A two-phase hybrid metaheuristic for the vehicle routing problem with time windows. *Eur. J. Oper. Res.* Vol.162, pp.220–238.

[9] Berger, J.; Barkaoui, M. & Bräysy, O. (2003). A Route-Directed Hybrid Genetic Approach for the Vehicle Routing Problem with Time Windows, *Inform. Systems Oper. Res.*, Vol.41, pp.179–194.

[10] Gambardella, L. M.; Taillard, E. & Agazzi, G. (1999). MACS-VRPTW: A Multiple Ant Colony System for Vehicle Routing Problems with Time Windows. In: New Ideas in Optimization, Corne, D., Dorigo, M. & Glover, F. (Ed). pp. 63–76, McGraw-Hill, London, UK.

[11] Chiang, W. C. & Russell, R. A. (1996). Simulated Annealing Metaheuristics for the Vehicle Routing Problem with Time Windows. *Ann. Oper. Res.* Vol. 63 pp.3–27.

[12] Bräysy, O. & Gendreau, M. (2005). Vehicle Routing Problem with Time Windows Part II: Metaheuristics, *Trans. Sci.*, No.39, No.1, pp.119–139.

[13] Bent, R. & P. Van Hentenryck. (2004). A two-stage hybrid local search for the vehicle routing problem with time windows, *Transportation Sci.*, Vol.38, No.4, pp.515–530.

[14] Bräysy, O. (2003b). A Reactive Variable Neighborhood Search for the Vehicle Routing Problem with Time Windows, *INFORMS J. Comput.*, Vol.15, pp.347–368.

[15] Ibaraki, T.; Imahori, S., Kubo, M., Masuda, T., Uno, T. & Yagiura, M. (2005). Effective Local Search Algorithms for Routing and Scheduling Problems with General Time Window Constraints, Transportation Science, Vol. 39, No. 2, pp.206-232.

[16] Hasama, T.; Kokubugata, H. & Kawashima H. (1999). A Heuristic Approach Based on the String Model to Solve Vehicle Routing Problem with Various Conditions, *Preprint for World Congress on Intelligent Transport Systems*, No.3027, Toronto, Canada, Nov. 1999.

[17] Gendreau, M.; Laporte, G. & Potvin, J.-Y. (2002). Metaheuristics for the Capacitated VRP, In: *The Vehicle Routing Problem*, Toth P. & Vigo, D. (Ed), pp. 129–154, SIAM, Philadelphia, USA.

[18] Van Breedam, A. (1996). An Analysis of the Effect of Local Improvement Operators in GA and SA for the Vehicle Routing Problem, *RUCA working paper 96/14*, University of Antwerp, Belgium.

[19] Kokubugata, H.; Itoyama, H. & Kawashima, H. (1997). Vehicle Routing Methods for City Logistics Operations, *Preprint for 8th IFAC Symposium on Transportation Systems*, pp.727-732, Hania, Greece, June 1997.

[20] Ingber, A. L. (1993). Simulated annealing: Practice versus theory, *J Mathl. Comput. Modelling*, Vol.18, No.11, pp.29-57.

[21] Hasama, T.; Kokubugata H. & Kawashima H. (1998). A Heuristic Approach Based on the String Model to Solve Vehicle Routing Problem with Backhauls, *Preprint for 5th Annual World Congress on Intelligent Transport Systems*, No. 3025, Seoul, Korea, Oct. 1998.

[22] Kokubugata, H.; Hirashima, K. & Kawashima, H. (2006). A Practical Solution of Capacitated Arc Routing for City Logistics, *Proceeding of 11th IFAC Symposium on Control in Transportation Systems*, No.222.

[23] Kokubugata, H.; Moriyama, A. & Kawashima, H. (2007). A Practical Solution Using Simulated Annealing for General Routing Problems with Nodes, Edges, and Arcs, *Lecture Notes in Computer Science*, Vol. 4638 (SLS2007), pp. 136-149, Springer, Berlin.

[24] Kokubugata, H. & Kawashima, H. (2008). Application of Simulated Annealing to Routing Problems in City Logistics, in *Simulated Annealing*, Cher Ming Tan (Ed.), pp.131-154, I-Tech Education and Publishing, Vienna.

[25] Dorigo, M.; Maniezzo, V. & Colorni, A. (1996).Ant System: Optimization by a Colony of Cooperating Agents. *IEEE Transactions on Systems, Man, and Cybernetics - Part B*, Vol. 26 No.1, pp.29-41.

[26] Solomon, M., (2005). Web site: http://w.cba.neu.edu/~msolomon/home.htm

[27] Gehring, H. & Homberger J. (2001). A Parallel Two-phase Metaheuristic for Routing Problems with Time Windows, *Asia-Pacific Journal of Operational Research*, 18, 35-47.

[28] Gehring, H. & Homberger J. (2001). Web site: http://www.fernuni-hagen.de/WINF/touren/inhalte/probinst.htm

[29] Johnson, D.S.; Aragon, C.R.; MacGeoch, L.A. & Schevon, C. (1989). Optimization by Simulated Annealing : An Experimental Evaluation, Part I, Graph Partitioning, *Operations research*, Vol. 37, pp. 865-892.

[30] Johnson, D.S.; Aragon, C.R.; MacGeoch, L.A. & Schevon, C. (1991). Optimization by Simulated Annealing : An Experimental Evaluation, Part II, Graph Colouring and Number Partitioning, *Operations research*, Vol. 39, pp. 378-406.

Chapter 8

Lot Sizing and Scheduling in Parallel Uniform Machines – A Case Study

F. Charrua Santos, Francisco Brojo and Pedro M. Vilarinho

Additional information is available at the end of the chapter

http://dx.doi.org/10.5772/50975

1. Introduction

This chapter presents the problem of scheduling tasks in uniform parallel machines with sequence-dependent setup times. The problem under analysis has a particular set of constraints, including equipment capacity, precedence tasks, lot sizing and task delivery plan. This kind of problem should be considered an operational planning problem and is found in different types of industries, namely, in the textile industry.

The main goal of this work is to minimise the total production time, including processing, setup and tardiness time.

The complexity of the studied model does not allow to locate the optimal solution. To solve this, the authors used a meta-heuristic known as the simulated annealing algorithm to obtain "nearly optimal" solutions for the problem.

The results obtained through the application of this particular heuristic show the importance of using a structured approach in production scheduling when compared with the results obtained from a reference plant where mainly "ad hoc" actions were normally taken.

The classical parallel machine scheduling problem considers p tasks with a fixed processing time that must be processed in k machines. This type of problem involves two kinds of decisions: i) which tasks are assigned to a specific machine and ii) tasks processing order (Mokotoff, 2001).

This problem becomes more complex as the number and characteristics of the available machines increases. The following levels of complexity can be considered: i) identical parallel machines; ii) uniform parallel machines and iii) unrelated parallel machines.

In the case of uniform parallel machines, each machine or group of machines has a different processing velocity. A processing time Tp for each task p in each machine k, described by $Tppk$, can be considered.

The work described in this paper takes into consideration lot sizing because it is an important aspect of these kinds of problems. Scheduling and lot sizing are indivisible problems in a significant number of real-world operations, where the setup times are sequence dependent. The analysis of lot sizing usually results in a decrease in the setup numbers as a result of tasks grouping in batches (Potts and Wassenhove, 1992; Potts and Kovalyov, 2000). Often, as a consequence of their size and complexity, problems are not solvable by exact methods. This paper introduces a heuristic approach to solving the problem (Drexl and Kimms 1997; Allahverdi, Gupta and Aldowaisan, 1999).

In general, the problem can have different formulations that are dependent on the objective function and the algorithm used to solve the problem.

In next section of this chapter (Section 2) the bibliographic review is done, in (Section 3), the research problem characterisation is presented, followed by the problem statement and formulation (Section 4). In the next (Section 5), the simulated annealing algorithm is presented. The developed heuristic to solve the problem is shown in Section 6. In Section 7, is presented a set of computational results obtained with the developed heuristic. Finally, the conclusions and future areas of research are presented (Section 8).

2. Bibliographic review

The classical problem of scheduling parallel machines considered "n" tasks or assignments and "m" machines. Each work must be performed on a machine with a fixed processing time. Accordingly, the aim is to find the sequence of jobs to optimize determined performance indicators (Mokotoff (2001)). This type of problem involves two decisions: assigning work to a particular machine and its sequencing in the queue of this machine. The complexity of the problem increases exponentially with the number of machines considered.

The problems of parallel machines may also consider three cases (Gupta and Magnusson (2005) (Mokotoff (2001)):

- Identical parallel machines;
- Uniform parallel machines;
- Unrelated parallel machines.

In the case of parallel identical machines, the processing time of a task is independent of the machine where it is processed.

For uniform machines, each machine has a different processing speed. In practice we have to consider the processing time of job i on machine j (tij). This time is equal to the processing requirements of the job i divided by the speed of the machine j (sj) (tij = ti / sj). This feature leads to an increased complexity in solving the problem.

In unrelated machines sequencing, there is no particular relationship between the processing times in the different machines, i.e., there is no proportionality between the processing time of a task on a given machine and the time of the same task on another machine. It is necessary to define the processing time of each task in each machine.

In these cases the problem of sequencing seeks the best distribution of the tasks in the machines and the best result in each machine, with the aim of optimizing certain performance criteria.

Among all possible performance criteria, the most widely used is the production flow total time, since it provides the best balance between all factors considered. The problems in real environments are usually of this type. The solution of a problem may be different depending on the criteria considered and when the criteria do not conflict with each other, it is possible to find an optimal overall solution. Such problems were reviewed in depth in an article published by Nagar et al. (1995).

In general, the scheduling problems in parallel machines assume that:

- Each task or work is comprised of a single operation cannot be processed in more than one machine simultaneously;
- When a task is started, it must be completed before another task is started on the same machine;
- Interruption is not allowed, so each job once started must be completed;
- Processing time is independent of the sequence;
- Is permitted accumulation of work, i.e., the work can wait for a free machine;
- Is allowed downtime of machines;
- No machine can process more than one task at a time;
- During the sequenced period, stops due to equipment damage are not considered;
- The preparation time is zero for all jobs;
- The number of jobs is known and fixed;
- The number of machines is known and fixed;
- Processing times are known and fixed;
- All other specifications to define a particular problem must be known and fixed.

The characteristics presented herein refer to the classical approach of the problem, but there are other ways such as the proposed by Allahverdi et al. (1999), analysing the problems of scheduling parallel machines considering independent setup times and sequence dependent.

Also the optimization problems on parallel machines are identical, mostly, "NP-Hard." As is the case with the problems of single machine, there are two ways to deal with this problem:

- exact methods;
- heuristic approach.

Mokotoff (2001) published a review article on production scheduling in parallel machines. In this paper is made a thorough literature review on the work in different environments: identical parallel, uniform and unrelated machines.

The problem of sequencing with preparation time dependent of sequence in parallel identical machines, in order to minimize the time of preparation was made by Sumichrast and Baker (1987). Subsequently, França et al. (1996), again look into this matter.

Over the past three years there are many publications on identical parallel machines scheduling with the objective of minimizing production time. These publications present some variations on the objective that are:

- Minimization of production times with independent setup times (Tang and Luo (2006) and Haouari et al. (2006)). With preparation time dependent sequence refers to an article by Chen et al. (2006) who also consider batch processing and resizing. Gheun-shie et al. (2004) analyze the sequencing batch using genetic algorithms. Allahverdi et al. (2008) presented a review paper on exact methods and heuristic approach to solving problems with positive time or cost of preparation in many industrial environments.
- Minimization of production times and meeting delivery dates. Shakhlevich and Strusevich (2005) analyze the sequencing of work while Jaehwan and Marc (2005) study the sequencing batch resorting to using several heuristics, no one considers the dependence of the sequence. In turn, using a tabu search algorithm, Shin and Leon (2004) analyze the production batch with setup times sequence dependent.
- Sequencing with replanning of work such as the work presented by Curry and Peters (2005).

There are also several works that address the sequencing problem in unrelated parallel machines environments. The objective focuses again around the minimization of the production time. Also in parallel study of unrelated machines are papers analyzing the sequence of separate operations such as the ones of Chen et al. (2006), while Rabadi et al. (2006) focus on the study of dependent sequences. Minimizing delays is also examined in environments of unrelated parallel machines. Chen et al. (2006) analyze the sequencing of jobs dependent on using for this purpose an algorithm of simulated recrystallization, while Kim et al. (2006) compared the use of simulated cooling algorithm with the genetic algorithm in the sequencing batch of time dependent preparation sequences. Meanwhile, Shim and Kim (2007) use techniques "Branch and Bound" to find the best sequence of jobs on machines not connected with preparation times independent of the sequence.

The problem of parallel machines may also appear associated with problems in continuous environments. Gupta et al. (1997) present a continuous production model consisting of two steps in which the first stage is comprised of multiple identical parallel machines. In this case the aim is to minimize the maximum processing time. Such problems are very common in practice, especially in the process industry, where in each step of the process there are multiple parallel machines, which may be identical or not.

Also the concept of the machine can be broadened if we consider parallel assembly lines. Considering the line as a black box, the sequence of batches in these lines can be considered as a problem of parallel sequencing machines. The problem of sequencing on automatic parallel production lines and considering the problem of scale batch was analyzed by Meyr (2002).

Table 1 presents a summary of works published in the literature that were considered relevant when compared with the work performed by the authors.

Ref.	Year	Parallel Machine Problem	Objective Function	Algorithm
Sivrikaya-Serifoglu, and Ulusoy	1999	Identical	Minimise Earliness (ME) Tardiness (MT) Penalties (MP)	Genetic algorithm(GA)
Allahverdi, Gupta and Aldowaisan	1999	Generalised	Review	
Radhakrishnan and Ventura	2000	Identical	ME/MT/MP	Simulated Annealing (SA)
Mokotoff	2001	Generalised	Review	
Anagnostopoulos and Rabadi	2002	Unrelated	Minimise Makespan(MM)	SA
Eom, shin, Kwun, Shim, and Kim	2002	Generalised	MT	Three-phase Heuristic, Tabu Search (TS)
Ref.	Year	Parallel Machine Problem	Objective Function	Algorithm
Meyr	2002	Generalised	Simultaneous Lot Sizing and Scheduling	
Liao and Lin	2003	Uniform	MM	Optimal Lexicographic Search
Jaehwan and Marc	2005	Generalised	Rescheduling to Increasing	Branch-and-Price
Tang and Luo	2006	Generalised	MM	Iterated Local Search (ILS)
Wen-Chiunge, Chin-Chia and Chen	2006	Identical	MM	SA
Chen	2006	Unrelated	(MT)	Tabu Lists
Haouari, Gharbi and Jemmali	2006	Identical	MM	Lower Bounding Strategies
Kim, Choi and Lee	2006	Identical	MT	TB, and SA
Lee, Wu and Chen	2006	Identical	MM	SA
Nessah, Yalaoui and Chu	2006	Identical	MM	Branch and Bound (BB)
Rabadi, Moraga and Al-Salem	2006	Single machine	E/T	Heuristic Tailored and SA
Shim and Kim	2007	Unrelated	MT	BB
Anghinolfi and. Paolucci	2007	Generalised	MT	Hybrid Metaheuristic Approach
Logendran, McDonell and Smucker	2007	Unrelated	MT	Six TS
Oh and Kim	2008	Identical	MT	BB

Ref.	Year	Parallel Machine Problem	Objective Function	Algorithm
Allahverdi, Cheng, Mikhail and Kovalyov	2008	Generalised	Review	
Raja, Selladurai, Saravanan and. Arumugam	2008	uniform	E/T	SA and Fuzzy Logic
Rocha, Ravetti and Mateus	2008	Unrelated	Due Dates and Weighted Jobs	Exact Methods (EM)
Su	2009	Identical	MM	
Eren, and Guner	2009	Identical	MM	Bi-criteria Scheduling with a Learning Effect
Mellouli, Sadfi, and Chu	2009	Identical	MM	Three EM
Chaudhry and Drake	2009	Identical	MT	GA
Wang and Cheng	2009	Identical	Minimise the Weighted Sum of the Last Arrival Time of the Jobs	Heuristics
Chen, and Chen	2009	Unrelated	MT	Hybrid Metaheuristics
Chen	2009	Unrelated	MT	SA
Kang and Shin	2010	Identical	MT and Reworks	Dispatching Algorithm
Ref.	Year	Parallel Machine Problem	Objective Function	Algorithm
Huo and Leung	2010	Identical	MM	Algorithm With a Worst-case Bound
Huang, Cai, and Zhang	2010	Identical	MM	Hybrid GA
Chuang, Liao, and Chao	2010	Unrelated	MM	BB
Unlu and Mason	2010	Identical	MM	Different Mixed Integer Programming
Gacias, Artigues and Lopez,	2010	Identical	MM	BB and Climbing Discrepancy Search
Behnamian; Zandieh and Ghomi,	2010	Identical	MM/MT/ME	GA
Fanjul-Peyro and Ruiz,	2010	Unrelated	MM	ILS

Ref.	Year	Parallel Machine Problem	Objective Function	Algorithm
Okolowski and Gawiejnowicz	2010	Identical	MM	EM
Moradi and Zandieh	2010	Identical	MM and System Unavailability for the Maintenance	GA
Lin, Lin and Zhou	2010	Identical	MM	Approximation Algorithm
Lin, Pfund and Fowler	2011	Unrelated	MM/MT	Review
Jouglet and Savourey	2011	Unrelated	MT	Dominance Rules and Filtering Methods
Edis and Ozkarahan	2011	Identical	MM	Three Optimisation Models
Hsu, Kuo Yang	2011	Unrelated	MM	Polynomial Time Solution

Table 1. Problems referred to in the literature.

The analysed problem of production lots sizing and scheduling is intended to minimize, for a given planning horizon, the setup and processing time of a set of orders with due dates set and the delays incurred in the due date of these orders. The problem occurs in an environment where the equipment for the production of such packages are configured as uniform parallel machines and setup times that are dependent on the sequence of orders generated by manufacturing orders. The equipment used induces width restrictions.

The problems of minimizing the production time for a sequence of jobs minimizing delays and work with common due dates have been solved as a problem of route salesman (TSP) by Baker (1974) and Picard and Queyranne (1978). These problems were further extended to include time dependent setup sequence between batches, and some of the preparation times are void (Monma and Potts 1989).

As it turned out, since the review work carried out by Mokotoff (2001) has been extensive published literature on the sequencing of parallel machines. However, the literature on uniform parallel machines is scarce ((Blazewicz et al. (2000), Liao and Lin (2003), Jiang et al. (2005) and Raja et al. (2008)).

Given the research done, the problem of sequencing and lot sizing in uniform parallel machines, with penalties for failure to meet delivery deadlines, setup times depend on the sequence and width restrictions and size of batch, identified during the study carried out in an industrial environment, has never been analyzed, it is therefore important to carry out the formulation and development of a method for its resolution.

In the objective function we choose a model that reflects the performance indicators in use in the company that accompanied the study, including the objective function to minimize the processing time, preparation (Setup) and delays in deliveries.

The high complexity of the proposed model, to the problem under analysis, prevents its resolution by use of the optimizing methods, even in cases where problems are of average size. Given this difficulty, we developed a heuristic for finding sub-optimal solutions to the problem, which is based on simulated annealing algorithm. Heuristics based on this algorithm have been used to solve different combinatorial optimization problems whose resolution by use of optimizing methods is not possible.

Simulated Annealing is a technique of non-deterministic local search, which uses an analogy between the minimum energy state of a physical system with minimal cost in a combinatorial optimization problem (Metropolis et al. (1953), Kirkpatrick et al. (1983), Radhakrishnan and Ventura (2000) and Wen-Chiunge et al. (2006)).

At liquid state, a given material has a random atomic ordering, it can be considered chaotic. As it cools, converges to a state of minimum energy structure with a rigidly defined and ordered. If cooling is done quickly, in most materials is produced a deformation of the microstructure which can lead to cracking. The different states which a material during the controlled cooling goes through correspond by analogy to the various possible solutions of an optimization problem, with the system energy corresponding to the function to be optimized.

The simulated annealing has been widely used in recent years in all kinds of problems. A search for the last 5 years reporting over 500 scientific articles concerning the use of simulated annealing algorithm optimization problems.

3. Problem characterisation

The analysis made to the production process of three of the largest producers of fabrics of Beira Interior, one of them the largest in Europe in the production of wool fabrics, shows that the planning methodology does not differ greatly from company to company. It is possible to observe significant differences at computerization level, but the planning philosophy is the same.

The companies analyzed have the following common features:

- The flow between sections is typical Batch Shop. Within each section you can find environments challenging to frame in a specific classification, being the most frequent a combination of a continuous environment with steps and parallel machines;
- The work is done mainly to order, having only, during some phases of the production process, work made for stock and that just to reduce order response times;
- Each section is a supplier or customer of the section that follows or precedes, respectively. In this perspective, planning and production control are highly complex;
- The production is characterized by strong seasonality and the fashion factor. These facts, coupled with the growing trend of product differentiation, but with higher added value, translates into unpredictability of sales increase;
- It is clear a decrease in the size of orders, with a visible impact on the size of the lots;
- The problem of sequencing and lot sizing is placed in all sections of the companies.

In Beira Interior (Portugal), the more credited tissue producing companies, wager on verticality. This option brings greater difficulties in terms of planning.

In the analyzed companies and in line with the literature, there are three identifiable levels of planning:

- first, of administration responsibility, typically includes strategic decisions;
- second, usually designated central, is responsibility of the company senior management, including tactical decisions, namely, the determination of supplies requirements and management of abilities. Sometimes, at this level of planning and at some sections of the company, are also made decisions regarding lot size and production scheduling, actions that are normally field of operational planning;
- third planning level, arises at the sequencing level and is based on some simple heuristics rules and occurs in certain sections, being the responsibility of the section responsible. At this level, the planning may also include some decisions issued by central planning, related to the distribution of production orders by the equipment.

The problem of production sequencing cannot be analyzed in an integrated way to all sections of the companies. Such problem would be highly complex and impossible to solve. The question that arises is: which of the sequencing problems that arise in the different sections may be more relevant to a greater efficiency of the company as a whole? To try to answer this question, the planning process will be analyzed in some detail.

In practice, it is observed that the orders breakdown and advances are not made from the orders backlog, but from the programming of these at weaving. If, in terms of quantities to produce, the difference is not significant, in terms of delivery dates of the sections that are upstream is important because the reference will be the weaving needs and not the delivery date to the client. This reasoning is always implicit in the decisions that are taken, i.e., underlying the planning process of the textile industry there is always the question of maximizing the weaving production.

The problem of weaving lot sizing lays on the assumption: the larger the lot the greater the efficiency of the company. The reasons for this assumption are the following:

- setup times associated with the change of batch. For larger batches are fewer setups and the weaving efficiency will be higher;
- since the components needs (strings) are generated from the weaving programming, the larger the batch, the greater the size of the lots in sections upstream (dyeing and spinning).

Production planning in the analyzed companies is a highly centralized planning system. This fact is understood by the need to integrate information in a complex environment dependent of demand.

In spinning and dyeing, the sequencing is responsibility of the sections responsible, with only a delivery date associated with each manufacturing order that the person responsible has to respect.

In weaving, the interference of the central planning in scheduling operations is significant and justified by the disaggregation model presented above.

On the other hand, there are other reasons why weaving deserves special attention:

- the productive capacity of a company is given by the ability to weave, even when including outsourcing;
- marketed product (fabric), even not finished, is physically produced in this section, which induces two clearly distinct types of control. Upstream weaving we have a pull process, where the weaving facing its needs "pulls" the production, downstream of weaving, because the relation order/product is very visible, we have a push system, where the delivery pressure is exerted by both the commercial and production sectors.

Given the foregoing, it is concluded that the problem of scaling and sequencing of production batches arises in all sections of the companies. However, by the observations made, the problem that seems to have a greater impact on productive efficiency is the weaving batch sizing and sequencing.

It was concluded that a set of parts are a lot, and that within the same lot, the batch preparation time between parts is zero. However, there is a limit to the number of parts that can form a batch. This limit is given by the technical limitations of product and equipment.

Once grouped into batches, the pieces can lead to two types of setup, depending on their technical characteristics. So, considering the sequencing of parts, one can consider the following setup times:

- zero, when the parts are equal and belong to the same lot,
- 8 hours, when the parts are equal but are from different batches,
- 24 hours, when the parts are different.

For the analyzed problem, it was found that the demand is known in advance, being this way deterministic and stable for a planning horizon of 10 weeks. In each week, there is only a delivery date, corresponding to the last day of the week. Thus, we are facing a problem where the planning horizon is finite, with multiple periods of time. The periods are constant, which means that the difference between two consecutive delivery dates are equal.

In order to characterize the demand, was made a survey of confirmed orders for a period of 10 weeks, in one of the companies, concluding that there were 87 different products totaling 5828 pieces to schedule in 48 machines.

The processing time per piece is known.

From data analysis of the problem and characteristics of the industry producing tissues, it was concluded that a factor of competitiveness for the sector is the fulfillment of deadlines, so this feature should be included in the formulation of the problem. In order to be compared with current practice in the company, was chosen an objective function that reflected the heuristics applied in the company, which took into account the production time, the setup time and a tardiness penalty.

Beyond the characteristics presented above, it was verified that there are still a number of assumptions that, in terms of problem formulation, is important to consider, including:

- a machine can only process one piece at a time;
- one piece can be processed in only one machine;
- the production of one piece on a machine cannot be interrupted;
- a piece can move from the period;
- initial characteristics of the machines are known.

As previously stated, the problem under research was inspired by the textile industry, weaving in particular, where n jobs are sequenced on k machines. The machines are grouped into G groups according to their processing velocity, turning this into a uniform parallel machine problem. Not all tasks can be processed on all machines because the plant has looms with different widths. This technical aspect confers to the problem an increased level of complexity, and it is a distinguishing factor when compared with the other situations reported in the literature.

The tasks have sequence-dependent setup times. Same type of tasks sequenced in the same machine are joined in batches, so the setup time between same type of tasks (i.e., for tasks belonging to the same batch) is zero. Nevertheless, when the amount of same type of tasks reaches an established lot size, a setup should then take place. The lot size varies according to the type of tasks. This is a major achievement compared with what was found in the literature. For same type of tasks with a continuous sequence on the same machine, two setup times take place: a) the setup time will be zero if the limit lot size has not been reached; and b) it will have a certain value (the setup time) if another situation occurs.

The demand is deterministic for the planning horizon with a due date associated with each task. On the planning horizon, we can have n due dates.

4. Problem statement and formulation

The generalised version of the problem can be described as the problem of scheduling tasks in uniform parallel machines, with due dates and sequence-dependent setup times.

The problem under analysis has the following characteristics:

i. The throughput is stable during the period of production;
ii. The demand is known in the planning horizon;
iii. The production can be adequately represented by a limited set of tasks (p=1,…, P);
iv. The production is performed in a set of machines joined together according to their characteristics in G groups (g=1,…, G), each one having k machines (k=1,…, K);
v. For each task p, a due date dp exists;
vi. The task can be delivered after the due date but incurs tardiness. This tardiness is penalised by a factor ρ
vii. Each task is processed in a single machine of a defined group;
viii. Each task p can be processed in machines belonging to any group, whereas the required characteristics need verifying, considering $Tppg$, the processing time of task p on one of the machines belonging to group g (p=1, …, P and g=1, …, G);

ix. The setup time for each machine that processes task *j*, after having processed task *i*, is known and represented by sij and is independent of the machine in which the task has been processed but always considers the characteristics of the sequence of the tasks;
x. The setup time of the machine that produces task *i* in the beginning of the job sequence on a machine is known and represented by s0i;
xi. *L*max represents the maximum number of tasks than can be joined and processed together continuously without requesting setup time;
xii. There is no need to consider a setup time for tasks with the same technical characteristics. The amount of different assembly combinations is *A*, whereas Na (*a*=1,…,*A*) is the set of tasks that have the same characteristics, assembly and colour, with a as the reference.
xiii. *M* is a large positive number;
xiv. $\beta_{ig} = \begin{cases} 1 \text{ if task i can be processed in group g} \\ 0 \text{ otherwise} \end{cases}$

Before providing a mathematical formulation, the following variables are defined:

*C*p date of conclusion of task *p*
*t*p tardiness of task *p*
*s'*ij setup time for task *j* when this occurs after task *i*

$$x_{ijkg} = \begin{cases} 1 & \text{if task i is followed by j in machine k of group g} \\ 0 & \text{otherwise} \end{cases}$$

$$y_{ikg} = \begin{cases} 1 & \text{if i is assigned to a machine in group g} \\ 0 & \text{otherwise} \end{cases}$$

$$z_{ij} = \begin{cases} 1 & \text{the tasks i and j are processed in the same machine k of group g} \\ 0 & \text{otherwise} \end{cases}$$

The problem can then be modelled as one of mixed integer programming by the following expressions, which are explained next.

$$\min \sum_i \sum_k \sum_g y_{ikg} \cdot \rho_{ig} + \sum_i \sum_j \sum_k \sum_g x_{ijkg} + \sum_{i,j \in Na; a=1}^{A} \sum_k \sum_g S_{ijg} \cdot R_{akg} + \rho \cdot \sum_i t_i \tag{1}$$

$$\sum_j x_{0jkg} = 1, \text{k}=1,\ldots,\text{M}_g;\ g = 1,\ldots,G \tag{2}$$

$$c_j + M.\left(1 - x_{0jkg}\right) \geq s_{0jg} + p_{jg},\ j = 1,\ldots,P; k = 1,\ldots,M_g; g = 1,\ldots,G \tag{3}$$

$$\sum_{i;i \neq j} x_{ijkg} = y_{jkg},\ j = 1,\ldots,P;\ k = 1,\ldots,M_g; g = 1,\ldots,G \tag{4}$$

$$\sum_{j;i \neq j} x_{ijkg} = y_{jkg},\ i = 1,\ldots,P;\ k = 1,\ldots,M_g; g = 1,\ldots,G \tag{5}$$

$$c_j + M.\left(1 - x_{ijkg}\right) \geq c_i + p_{jg}, i, j = 4,\ldots,P; k = 1,\ldots,M_g; g = 1,\ldots,G \tag{6}$$

$$c_i - d_i \leq t_i,\ \ i = 1,\ldots,P \tag{7}$$

$$\sum_k \sum_g y_{ikg} * \beta_{ikg} = 1,\ i = 1,\ldots,P \tag{8}$$

$$y_{ikg} + y_{jkg} - 1 \leq z_{ij},\ \ i, j = 1,\ldots,P;\ k = 1,\ldots,M;\ \ g = 1,\ldots,G \tag{9}$$

$$s'_{ij} = s_{ij} \cdot z_{ij}, \forall_{i,j} \in N_a;\ a = 1,\ldots,A \tag{10}$$

$$\left\lfloor \frac{\sum_{\forall(i,j) \in N_a} x_{ijkg}}{L_{a\max}} \right\rfloor \leq R_{akg}, \forall_{i,j} \in N_a; a = 1,\ldots,A \tag{11}$$

In the objective function,

(1) is the total production time and is the sum of the following four items: i) tardiness time, ii) setup time among different tasks, iii) setup time among same type of tasks and iv) processing time.

- The tardiness time is given by the sum of all the delays that occurred during the planning horizon, multiplied by a penalty factor.
- The setup time is given by the sum of all the setup times during the planning horizon.
- The processing time is given by the sum of all of the task processing times included in the planning analysis.

The model constraints can be interpreted as follows:

2) Constraints ensuring that only one initial setup can take place in each machine k belonging to group g;

3) Constraints ensuring that the conclusion time of task j, when processed at the beginning of the job sequence on a machine k, belongs to group g and is equal to the initial setup time of task j plus the processing time of task j in machine g.

4) Constraints ensuring that if the setup time for task j takes place in machine k of group g, then j is either preceded by another task i or is the first task in the queue and will be processed in machine k of group g.

5) Constraints ensuring that if the setup time for task *j* takes place in machine *k* of group *g*, then *j* is either preceded by another task *i* or is the last task in the queue and will be processed in machine *k* of group *g*.

6) Constraints ensuring that the processing of each task once started cannot be interrupted;

7) Constraints ensuring that the tardiness time of task *i* is given by the difference between the conclusion date and the due date. From this problem, it can be concluded that *t*i is a positive number and tardiness time only occurs when *c*i is larger than *di*;

8) Constraints ensuring that each task is manufactured only once and in a compatible machine;

9) Constraints ensuring that if one task *i* is processed in machine *k* of group *g*, and another task *j* is processed in machine *k* of the same group *g*, then both tasks are processed in the same machine.

10) Constraints ensuring that the setup time taken by changing from task *i* to task *j* in group *g* is zero if *i* and *j* have the same technical characteristics, assembly and colour. Otherwise, it assumes a known value *S*ij;

11) Constraints ensuring that when the number of same type of tasks, continuously sequenced, exceeds the maximum lot size, then the resulting setup time is taken into account.

Restrictions 4) and 5) ensure that tasks *p* will be processed by the *k* machines of *g* groups, which means that all the tasks will be processed. Simultaneously, these conditions ensure that each machine only processes a single task at a time, considering that the number of setups plus the initial setup gives the total number of processed tasks.

5. Simulated annealing algorithm

According Radhakrishna and Ventura (2000) the simulated annealing (SA) algorithm is a stochastic approach that endeavours to overcome local optimality by accepting bad solutions with a definite probability. It has been used to solve many combinatorial problems. SA is a technique of searching for good solutions in a configuration space which draws an analogy between minimum cost configuration in a combinatorial optimization problem (Metropolis et al (1953); Kirkpatrick et al (1983) Liao and Lin (2003); Lee, Wu and Chen(2006); Logendran et al (2007), and Kimet et al, (2006)).

The motivation for the SA algorithm comes from an analogy between the physical annealing of solids and combinatorial optimization problems. Physical annealing refers to the process of finding low-energy states of a solid by melting the substance initially and then lowering the temperature slowly, to temperatures close to the freezing point. An example would be to produce a crystal from the molten substance. In a liquid, the particles are arranged randomly, but the ground state of the solid, which corresponds to the minimum energy configuration, will have a particular structure, such as seen in a crystal. If the cooling is not performed slowly, the resulting solid will not attain the ground state but will be frozen into a meta-stable, locally optimal structure, e.g. glass or crystal with several defects in the structure. In the analogy, the different states of the substance correspond to the different feasible solutions of the combinatorial optimization problem, and the energy of the system corresponds to the function to be minimized. (Radhakrishna and Ventura, 2000).

The basic idea of the utilized algorithm is as follows. An initial feasible solution to the problem is generated using Heuristic, for which, a starting value of the temperature parameter, (T0) is estimated. A neighbouring solution is then randomly generated, which is accepted or rejected using the SA methodology, e.g. if the energy for the new solution is smaller, the solution is accepted, if the energy for the new solution is higher, then Metropolis criterion is employed for acceptance. This process is repeated until the algorithm converges to the final temperature value (Tf). SA ensures that the solution found increase the possibility to find the global minimum.

For the present case, the energy is substituted by the cost. A specific cost, Zx, matches the initial solution. From this initial solution, a neighbouring solution is randomly reached, with a cost represented by Zy. The difference between Zy and Zx is represented by ΔZyx. If the cost decreases (i.e., if ΔZyx = Zy - Zx < 0), then the neighbouring solution Zy will be accepted. If the opposite occurs, solution Zx will be retained if respecting the Metropolis criterion. This procedure will be executed as long as a local minimum is not reached. The local search algorithms are very easy to apply, but they may converge to a unique local minimum. As a result, a considerable deviation from the optimal solution can be introduced.

The SA algorithm improves the overall performance of this type of method because it makes it possible to overtake one local minimum with a given probability, i.e., there is a probability of accepting a worse solution, expressed by ΔZyx = Zy - Zx > 0.

The pseudocode developed by Madhavan (1993), is shown below:

Begin

Select an initial solution (where π is the solution space)

Select an initial temperature, T=$T0$;

Set temperature counter, u=0

Repeat

    Set interaction counter, n=0

    Repeat

Generate solution, $y \in \Pi$, a neighbour of x;

        Calculate ΔZyx = Zy - Zx

        If ΔZyx<0 then x=y

Else

if > random number from $U(0,1)$, then $x=y$;

$n=n+1$

Until ($n=R$);

$u=u+1$;

$T=Tu$;

Until (stopping criterion ($Tu<=Tf$) is satisfied);

End

6. The developed heuristic

To solve the optimisation problem using the SA algorithm, we must establish the parameters, namely:

- the initial and final values of the temperatures given in Eq. (12) and (13), which are the control parameters.

$$T0 = \Delta_{Zmin} + \left(\Delta_{Zmax} - \Delta_{Zmin}\right)/10 \tag{12}$$

$$T_f = \Delta_{Zmin} \tag{13}$$

- the dimension of each temperature stage, i.e., the number of iterations executed for each value of the temperature. In the case being analysed, the number of iterations in each stage was considered proportional to the problem dimension and is given by $2N$, where N represents the number of batches to be scheduled.
- the cooling rate:

$$T_{u+1} = \frac{T_u}{1+\beta T_u}, \tag{14}$$

where

$$\beta = \frac{\left(T0 + Tf\right)}{p\,T0\,Tf} \tag{15}$$

and

$$p = \frac{50\delta N\left(N-1\right))}{2}, \tag{16}$$

where δ controls the cooling rate.

The utilisation of the SA algorithm still requires the determination of the following:

- the initial solution;
- neighbouring solutions;
- the cost function.

6.1. Initial solution heuristic

The restrictions on the problem must be observed by the heuristic used to find its initial solution, namely, the compatibility between the width of the task to be performed and the required equipment. The task with a larger width is assigned to the less busy compatible equipment. The initial solution will have the following characteristics:

i. All the tasks with the same characteristics and due date are joined in the same batch.
ii. Each machine has a queue in which the number of batches and the processing time are known.
iii. The setup time required by the batch depends on its sequence inside the queue.
iv. Cost Z is expressed in time units. This value is obtained by adding the processing time, the setup times and the time penalty resulting from tardiness.
v. The batch is considered the carry unit.

From the application of this heuristic, a possible solution is obtained and is used to generate the neighbouring solutions.

6.2. Heuristic for the determination of neighbouring solutions

Neighbour solutions are obtained through random batch transferences and swaps.

A neighbour solution can be generated by two different methods:

- Random transference in which the width is a restriction

This method consists of changing the batch position inside the queue (see Figure 1) or transferring it to another queue (see Figure 2). This task only takes into account the width restriction.

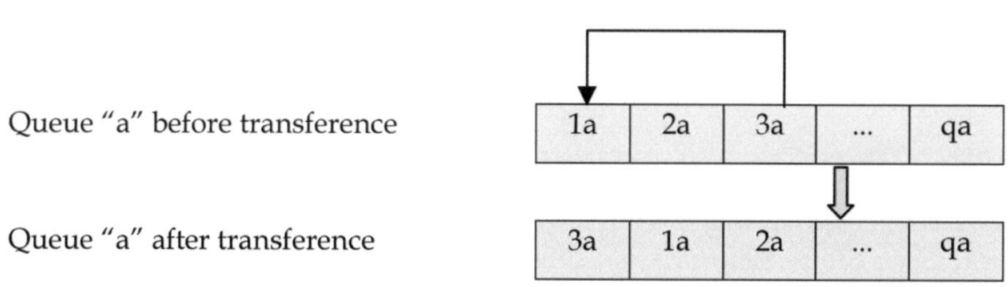

Figure 1. Transference of batches in the same queue

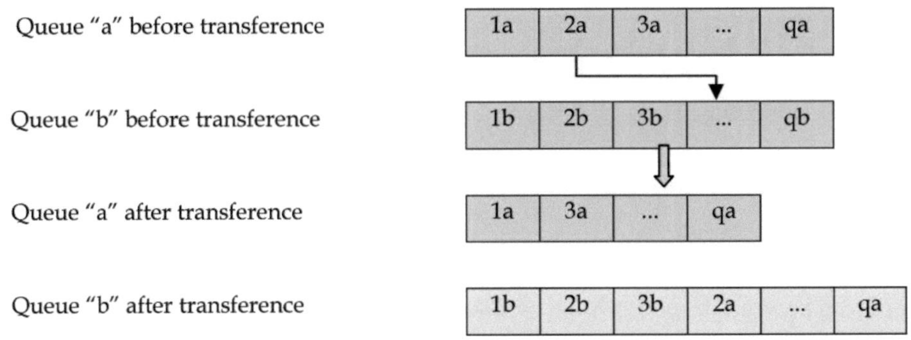

Figure 2. Batch transference between queues of compatibles machines.

- Random swaps in which the width is a restriction

In this situation, in the same queue, two batches are allowed to swap positions, as represented in figure 3. If the two batches are from different queues, they can also be exchanged, as shown in figure 4.

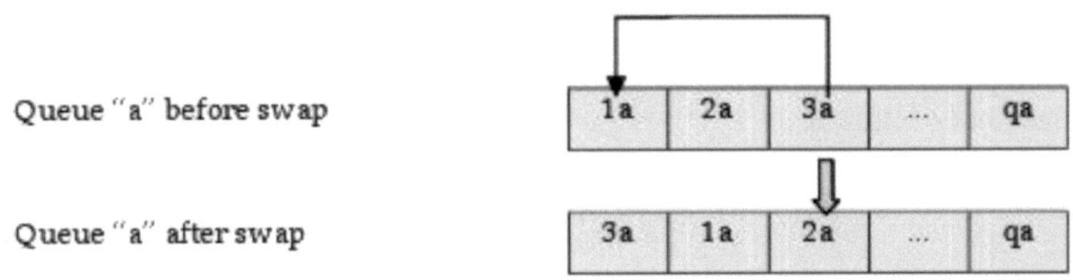

Figure 3. Swapping batches in the same queue

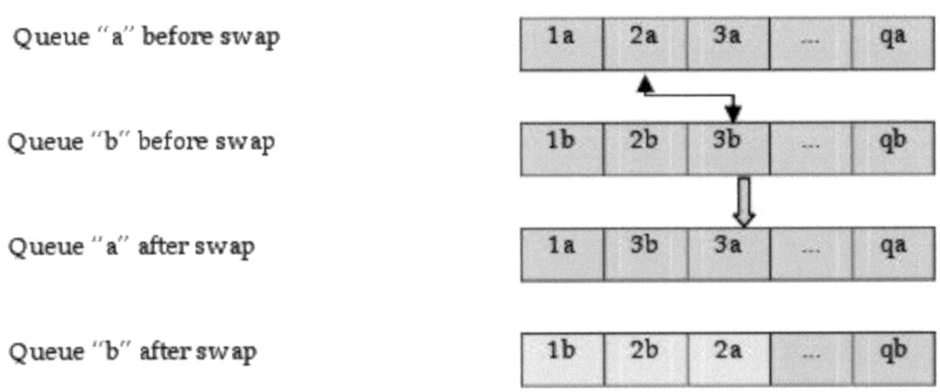

Figure 4. Swapping batches between queues of compatibles machines.

The heuristic for the determination of neighbouring solutions, as presented in this paper, which is based only on random transfers or swapping, leads to a wider search space.

Although the proposed method takes more time to converge to a minimum local, it has the advantage of reaching improved solutions. This option has proved to be a good choice for the particular problem under analysis, where the quality of the solution was considered much more important than the computational time.

As a result of the swap or transference described in this paragraph results a new possible solution to the problem. This new solution is acceptable if the cost z is less than or greater in some cases. The cost Z referred to in paragraph 6.3 reflects the objective function. This is, the solution resulting from the swap is evaluated according to the setup and processing time and delays it generates. Thus a relation between the objective function and the generation process of surrounding solutions is established.

As shown in the figure 5 is generated the initial solution, applying the objective function to this solution results in a certain cost Z. From the initial solution is generated a neighbouring solution as described in section 6.2. The objective function is applied to the new solution and tested in the simulated annealing algorithm and thus accepted or rejected. The solution obtained will give rise to another solution neighbour and so on until the process is interrupted according to the criteria described.

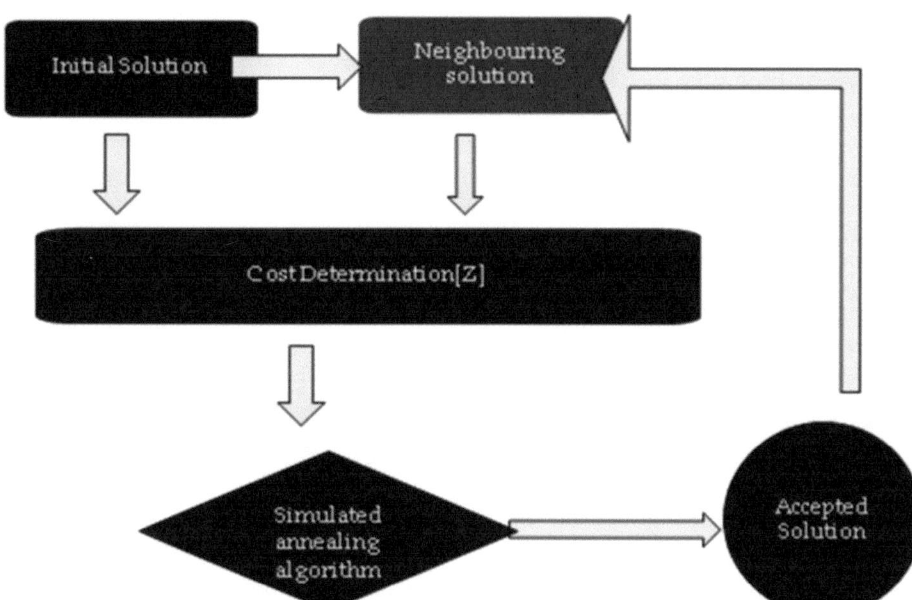

Figure 5. Schematic representation of performed algorithm

6.3. Description of the heuristic for evaluating the "Z" cost

As previously explained, the initial solution is a sequence of batches ordered by machine. Therefore, it is possible to find the time required to process the batches. In this case, the algorithm, queue by queue, or, in other words, machine by machine, will evaluate the sum of the processing times associated with each batch.

After task exchange, or transference, the heuristic will compute queue by queue the number and type of setups and then proceed to the total calculation for all of the setups.

Of all of the production times, the one that presents more calculation difficulties is the tardiness time, period by period. The system should take into account the possibility of an advance or delay for a machine that lasts for a certain period of time and should transfer the work to the following period. Such evaluation requires an additional computational effort, especially when large dimension problems are faced.

Cost "Z" is then the result of adding up processing and setup times plus the penalties that arise from failing to meet the due dates.

7. Computational results

To validate the solutions, a heuristic for the determination of a lower bound has been used.

7.1. Determination of "lower bound"

A good solution is the one resulting from the ratio of the production capacity versus the processing and setup requirements.

If the processing time plus the setup time requirements exceed the capacity, the lower limit should take into account the penalty time for the tardiness.

The penalty factor can take the form of a parameter.

Considering the magnitude of the problem that we are dealing with, it is not an easy job to find an algorithm for computing the lower limit that can successfully address every situation.

An algorithm that demonstrated good performance in environments with many groups of machines, many machines per group and many periods of time was chosen; the authors think that it is the best adjusted algorithm for the practical case that they were studying.

7.2. Description of the tests that were conducted

Tests were designed, keeping in mind the structure of the identified problem. Ten scenarios with increasing complexity were created, with the last one corresponding to the practical case being analysed. For each of the scenarios, a significant number of tests have been done using different parameters.

The characteristics of the performed tests are shown in Table 2.

In Table 3, the information contained in each column of Table 2 is described.

From the analysis of the results, the authors concluded that the algorithm performance was strongly related to the structure and the dimensions of the problem. This was expected once the evaluation criteria (convergence to the lower bound) were developed because of the previous experience of the skilled planner with large dimension problems.

In fact, the lower bound reflects the value that would be considered best by the planner for a problem of such size. It should be emphasized that in the analysed industrial environments, the methodology used for lot sizing and scheduling is very close to the heuristic applied to generate the initial solution in the presented work.

Scenario	Machine Group	Machine Number	Period	Assembly	Colour	Task Number	Batch Number
1	1	1	3	1	3	50	3
2	1	1	3	2	3	105	9
3	1	1	5	2	5	292	20
4	2	2	3	2	6	189	6
5	2	2	3	2	6	235	12
6	2	2	4	4	10	242	32
7	2	4	2	4	7	295	13
8	4	4	2	4	7	295	13
9	4	8	4	6	14	1148	50
10	4	48	10	39	87	5828	232

Table 2. Performed tests

Column	Description.
Scenario	Identifies the constructed scenario
Machine Group	Indicates the number of groups considered in the scenario.
Machine Number	Indicates the total number of machines considered in the scenario.
Period	Indicates the number of periods—the number of different due dates in the planning horizon. Within the same period, all the tasks have the same due date.
Assembly	Number of different assemblies considered in the scenario.
Colour	Number of different colours considered in the scenario.
Task Number	Indicates the number of tasks.
Batch Number	Indicates the number of batches generated by the initial solution.

Table 3. Description of Table 2

For the smaller problems, scenario 1 and 4 of Table 4, the best solution was obtained by the initial solution. In these cases, the solution equals the value found for the lower limit because this is the optimal solution. In other scenarios, there is a significant improvement in the value found through the heuristic when compared with the value of the initial solution.

Scenario	Zm	Z0	LB	((Zm-LB)/LB)*100	((Z0-Zm)/Zm)*100
1	12,790	12,790	12,790	0.00	0,00
2	37,910	49,995	28,040	35.20	31.88
3	79,027	164,908	65,819	20.07	108.67
4	50,739	50,739	50,739	0.00	0,00
5	62,160	66,910	57,640	7.84	7.64
6	55,345	74,071	46,495	19.03	33.84
7	61,550	73,585	58,025	6.07	19.55
8	70,445	80,925	61,950	13.71	14.88
9	248,494	318,428	231,411	7.38	28.14
10	2,053,181	2,273,021	1,946,458	5.48	10.71

Table 4. Obtained results

The results obtained for the real scenario (shown in table 4 and figure 6) show that there was a reduction of 10.71% when the best solution obtained was compared with the value of the initial solution.

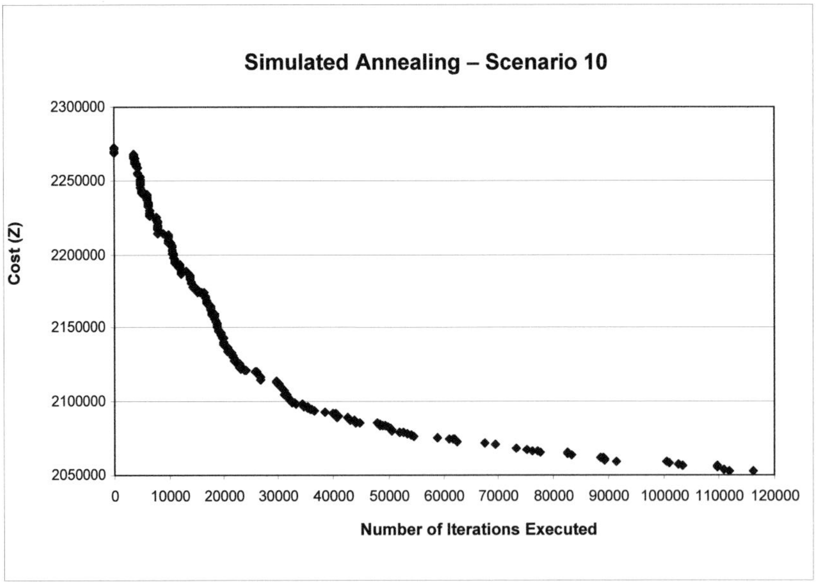

Figure 6. Best solution for scenarios 10

It should be noted that the initial solution reflects the heuristics used by the planner in a real environment. Additionally, in relation to the lower bound value, there is a clear convergence between the value of the solution and the limit. In this case, the better solution lies just 5.48% from the value obtained for the lower bound and may be considered a satisfactory result.

8. Conclusion

Considering the type of industry studied, the results show the clear importance of a structured approach to improving the efficiency in problems of lot sizing and scheduling operations. In the real case study, the developed tool allowed for an improvement of approximately 10%.

It is possible, based on the developed work, that when this is extrapolated to other industrial environments, it may also bring significant benefits.

The generalisations made for the initial problem allow an innovative formulation compared with the known approaches to problems of lot sizing and scheduling of uniform parallel machines. Specifically, the following generalisations were useful: i) consider different setup times for the same sequence of tasks depending on the lot size and the task position in the queue of the machine; ii) the machines, in addition to being grouped together according to their processing speed, should be further sub grouped according to technical features, including the machine width.

From a more conceptual perspective, the work presented also provides significant advantages in terms of tactical planning. The study identified a planning philosophy in the textile industry based on their Material Requirement Planning (MRP) that generates production needs from the backlog.

It is expected, although difficult to quantify, that if the material requirements are generated from the nearly optimal sequence obtained with the developed program, additional gains beyond the direct gains demonstrated in the work will exist in the upstream production sections because it is a system-dependent demand.

Author details

F. Charrua Santos
Electromechanical Department of University of Beira Interior, Calçada Fonte do Lameiro, Covilhã, Portugal

Francisco Brojo
Department of Aeronautical Sciences of University of Beira Interior, Calçada Fonte do Lameiro, Covilhã, Portugal

Pedro M. Vilarinho
Departament of Ecomics, Management and industrial Engeneeringl University of Aveiro, Campo Universitário de Santiago, Aveiro, Portugal

9. References

Allahverdi, Gupta and Aldowaisan (1999), "A review of scheduling research involving setup considerations", International. Journal. Management Science, Vol. 27, pp 219-239

Allahverdi, Ng, Cheng, Mikhail and Kovalyov (2008), A survey of scheduling problems with setup times or costs, European Journal of Operational Research, 187 pp. 985–1032.

Anagnostopoulos and Rabadi (2002), A simulated annealing algorithm for the unrelated parallel machine scheduling problem, Robotics, Automation and Control and Manufacturing: Trends, Principles and Allpications, pp15-120.

Anghinolfi D and M. Paolucci (2007), Parallel machine total tardiness scheduling with a new hybrid metaheuristic approach, Computers and Operations Research, 34 (11), pp. 3471-3490.

Baker, K. R. (1974), Introduction to Sequencing and Scheduling New York: Wiley.

Behnamian, J; Zandieh, M; Ghomi, SMTF 2010 A multi-phase covering Pareto-optimal front method to multi-objective parallel machine scheduling International Journal Of Production Research 48 pp: 4949-4976

Blazewicz; Drozdowski; Formanowicz; Kubiak and Schmidt (2000), Scheduling preemptable tasks on parallel processors with limited availability, Parallel Computing, 26, pp. 1195-1211.

Chaudhry, IA; Drake, PR 2009Minimizing total tardiness for the machine scheduling and worker assignment problems in identical parallel machines using genetic algorithms International Journal Of Advanced Manufacturing Technology pp 581-594

Chen, CL; Chen, CL 2009 Hybrid metaheuristics for unrelated parallel machine scheduling with sequence-dependent setup times International Journal Of Advanced Manufacturing Technology 43 pp161-169

Chen, J.F. (2006), Minimization of maximum tardiness on unrelated parallel machines with process restrictions and setups, The International Journal of Advanced Manufacturing Technology, 29, pp. 557-563.

Chen, JF 2009 Scheduling on unrelated parallel machines with sequence- and machine-dependent setup times and due-date constraints.International Journal Of Advanced Manufacturing 44 Issue: 11-12 pp1204-1212

Chuang, MC; Liao, CJ; Chao, CW 2010Parallel machine scheduling with preference of machines International Journal Of Production Research 48 pp 4139-4152

Curry and Peters (2005), Rescheduling parallel machines with stepwise increasing tardiness and machine assignment stability objectives, International Journal of Production Research, 43, pp. 3231-3246.

Edis, EB; Ozkarahan, I 2011 A combined integer constraint programming approach to a resource-constrained parallel machine scheduling problem with machine eligibility restrictions Engineering Optimization 43 pp: 135-157

Eom, D.H., H.J. Shin, I.H. Kwun, J.K Shim, and S.S. Kim (2002), Scheduling jobs on parallel machines with sequence-dependent family set-up times, International Journal of Advanced Manufacturing Technology, 19, pp. 926–932.

Eren, T; Guner, E 2009A bicriteria parallel machine scheduling with a learning effect International Journal Of Advanced Manufacturing Technology 40 pp 1202-1205

Fanjul-Peyro, L; Ruiz, R 2010Iterated greedy local search methods for unrelated parallel machine scheduling European Journal Of Operational Research 207 pp: 55-69

França, Gendreau, Laport, and Muller (1996), A tabu search heuristic for the multiprocessor scheduling problem with sequence dependent setup times, International Journal of Production Economics, 43, pp. 79-89.

Gacias, B; Artigues, C Lopez, P 2010Parallel machine scheduling with precedence constraints and setup times Computers & Operations Research 37pp: 2141-2151

Gupta, Hariri and Potts (1997), Scheduling a two-stage hybrid flow shop with parallel machines at the first stage, Annals of Operations Research, 69, pp. 171 – 191.

Haouari, Gharbi and Jemmali (2006), Tight bounds for the identical parallel machine scheduling problem, International Transactions in Operational Research, 13, pp. 529-548.

Hsu, C.J.; Kuo, WH; Yang, D.L., 2011. Unrelated parallel machine scheduling with past-sequence-dependent setup time and learning effects. Applied Mathematical Modelling 35, 1492-1496

Huang, SM; Cai, LN; Zhang, XY 2010 Parallel dedicated machine scheduling problem with sequence-dependent setups and a single server Computers & Industrial Engineering 58 pp165-174

Huo, Y; Leung, JYT 2010 Parallel machine scheduling with nested processing set restrictions European Journal Of Operational Research 204 pp229-236

Jaehwan and Marc, Scheduling Parallel Machines for the Customer Order Problem, Journal of Scheduling, Vol. 8, 2005, pp. 49-74

Jiang, Yiwei; Tan, Zhiyi; He and Yong (2005), Preemptive Machine Covering on Parallel Machines, Journal of Combinatorial Optimization, 10, pp. 345-363.

Jouglet, A; Savourey, D 2011 Dominance rules for the parallel machine total weighted tardiness scheduling problem with release dates Computers & Operations Research 38 pp1259-1266

Kang, YH; Shin, HJ 2010 An adaptive scheduling algorithm for a parallel machine problem with rework processes International Journal of Production Research 48 pp95-115

Kim, Na, Jang, and Chen (2006). Simulated annealing and genetic algorithm for unrelated parallel machine scheduling considering set-up times. International Journal of Computer Applications in Technology 26, 28-36.

Kim, S.I., H.S. Choi and D.H. Lee, (2006) Scheduling algorithms for parallel machines with sequence-dependent set-up and distinct ready times: minimizing total tardiness, Proceedings of the Institution of Mechanical Engineers Part B-Journal of Engineering Manufacture, 221, pp. 1087-1096

Lee, Wu and Chen (2006), A simulated annealing approach to makespan minimization on identical parallel machines, International Journal of Advanced Manufacturing Technology, 31, pp. 328-334.

Liao and Lin, (2003), Makespan minimization for two uniform parallel machines, International Journal of Production Economics, 84, pp. 205-213.

Lin, L; Lin, YX; Zhou, XW, 2010parallel Machine Scheduling With A Simultaneity Constraint And Unit-Length Jobs To Minimize The Makespan Asia-Pacific Journal Of Operational Research 27 pp: 669-676

Lin, YK; Pfund, ME; Fowler, JW 2011 Heuristics for minimizing regular performance measures in unrelated parallel machine scheduling problems Computers & Operations Research 38 pp901-916

Logendran, R., B. McDonell and B Smucker (2007), Scheduling unrelated parallel machines with sequence-dependent setups, Computers & Operations Research, 34, pp. 3420-3438.

Kirkpatrick, S., Gelatt, C. D. and Vecchi, M. P., 1983, Optimization by simulated annealing. Science, 220, 671± 680.

Madhavan, K., (1993). Weighted earliness-tardiness minimization for jobs with sequence dependent set-up times using simulated annealing, Master's thesis, Department of Industrial and Manufacturing Engineering, The Pennsylvania State University, USA

Mellouli, R; Sadfi, C; Chu, CB, et al. 2009Identical parallel-machine scheduling under availability constraints to minimize the sum of completion times European Journal Of Operational Research 197 pp 1150-1165

Metropolis, N., Rosenbluth, A., Rosenbluth, M., Teller, A. and Teller, E., 1953, Equation of state calculations by fast computing machines. Journal of Chemical Physics, 21, 1087± 1092

Meyr (2002), Simultaneous lotsizing and scheduling on parallel machines, European Journal of Operational Research, 139, pp. 277–292.

Mokotoff (2001), Parallel Machine scheduling problems: A survey, Ásia-Pacific Journal of Operational Research, 18, pp. 193-242.

Monma and Potts (1989), On the complexity of scheduling with batch setup times, Operations Research, 37, pp. 798-804

Moradi, E; Zandieh, M 2010 Minimizing the makespan and the system unavailability in parallel machine scheduling problem: a similarity-based genetic algorithm International Journal Of Advanced Manufacturing Technology 51 pp 829-840

Nessah, Yalaoui and Chu (2006), A branch and bound algorithm to minimize total weighted completion time on identical parallel machines with job release dates, Proceedings of International Conference on Service Systems and Service Management, 2, pp. 1192-1198.

Oh and Kim (2008), A branch and bound algorithm for an identical parallel machine scheduling problem with a job splitting property, Computers & Operations Research, 35, pp. 863-875.

Okolowski, D; Gawiejnowicz, S 2010Exact and heuristic algorithms for parallel-machine scheduling with DeJong's learning effect Computers & Industrial Engineering 59 pp:272-279

Picard, and Queyranne (1978), The time dependent traveling salesman problem and its application to the tardiness problem in one-machine scheduling, Operations Research, 26, pp. 86-110.

Potts and Kovalyov (2000), Scheduling with batching: A review, European Journal of Operational Research, 120, pp. 228-249.Drexl and Kimms (1997), Lot sizing and scheduling - Survey and extensions Invited Review, European Journal of Operational Research, 99, pp. 221-235.

Potts and Wassenhove (1992), Integrating scheduling with batching and lot-sizing: a review of algorithms and complexity, Journal of the Operational Research Society, 43, pp. 395-406.

Rabadi, Moraga and Al-Salem (2006), Heuristics for the Unrelated Parallel Machine Scheduling Problem with Setup Times, Journal of Intelligent Manufacturing, 17, pp. 85-97.

Radhakrishnan and Ventura (2000), Simulated annealing for parallel machine scheduling with earliness tardiness penalties and sequence-dependent set-up times, International Journal of Production Research, 38, pp. 2233- 2252.

Raja, K., V. Selladurai, R. Saravanan and C. Arumugam, (2008), Earliness-tardiness scheduling on uniform parallel machines using simulated annealing and fuzzy logic, Proceeding of the Institution of Mechanical Engineers part B-Journal of Engineering Manufacture, 222, pp. 333-346

Rocha P. L., M. G. Ravetti and G. R. Mateus (2008), Exact algorithms for a scheduling problem with unrelated parallel machines and sequence and machine-dependent setup times, Computers and Industrial Engineering, 35, pp. 1250-1264

Shakhlevich and Strusevich (2005), Pre-Emptive Scheduling Problems with Controllable Processing Times, Journal of Scheduling, 8, pp. 233-253.

Shie-Gheun, Pyung, Jae and Woon (2004), Scheduling parallel batch processing machines with arbitrary job sizes and incompatible job families, International Journal of production Research, 42, pp. 4091-4107.

Shim and Kim (2007), Minimizing total tardiness in an unrelated parallel-machine scheduling problem, Journal of the Operational Research Society, 58, pp. 346-354

Shin and Leon (2004), Scheduling with product family set-up times: an application in TFT LCD manufacturing, International Journal of Production Research, 42, pp. 4235-4248.

Sivrikaya-Serifoglu, F. and G. Ulusoy, "Parallel machine scheduling with earliness and tardiness penalties", Computers and Operations Research, Vol. 26,1999, pp. 773–787.

Su, LH 2009Scheduling on identical parallel machines to minimize total completion time with deadline and machine eligibility constraints International Journal of Advanced Manufacturing Technology, 40 pp 572-581

Sumichrast and Baker (1987), Scheduling parallel processors: an integer linear programming based heuristic for minimizing setup time, International Journal of Production Research, 25, pp. 761- 771.

Tang and Luo (2006), A new ILS algorithm for parallel machine scheduling problems, Journal of Intelligent Manufacturing, 17, pp. 609-619.

Unlu, Y; Mason, SJ 2010Evaluation of mixed integer programming formulations for non-preemptive parallel machine scheduling problems Computers & Industrial Engineering 58 pp785-800

Wang, XL; Cheng, TCE 2009Heuristics for parallel-machine scheduling with job class setups and delivery to multiple customers International Journal Of Production Economics 119 pp199-206

Wen-Chiunge, Chin-Chia and Chen (2006), A simulated annealing approach to makespan minimization on identical parallel machines, The International Journal of Advanced Manufacturing Technology, 31, pp. 328-334

Chapter 9

Use of Simulated Annealing Algorithms for Optimizing Selection Schemes in Farm Animal Populations

Edo D'Agaro

Additional information is available at the end of the chapter

http://dx.doi.org/10.5772/50665

1. Introduction

The design of optimal mating schemes is a mean to improve farm animal performances. During the last decades, breeding strategies and techniques addressing both genetic improvement and inbreeding control have been well documented and applied in several countries [1]. The detrimental effects of inbreeding have been reported in farm animals and, in the recent years, many selection and mating strategies were proposed to restrict inbreeding in selection programmes [2]. Recent advances in animal breeding theory have clearly shown the importance of mating design optimization by means of new analytical models as the optimum contribution selection method [3] and simulated annealing algorithms [4]. Stochastic simulation programs are generally used to create farm animal populations under artificial selection, and, by this way, genetic and inbreeding effects are easily modelled and studied for several generations. In this study, a stochastic simulation (Monte Carlo method) was used to evaluate and optimize different mating schemes of farm animals under a restricted inbreeding rate.

2. Breeding strategies in farm animal populations

2.1. Selection based on phenotype

The selection of individuals based on the phenotype has been a breeding strategy widely practiced over time as it allowed to obtain significant benefits in the economic sector. Most of these selection schemes have been applied by isolating a specific phenotype, ignoring the genetic structure of selected traits [5]. However, in practice, selection programs based only on phenotypes have shown effects both positive and negative. In fact, although some significant improvements were observed in animal production and reproduction performances, at the same time, several undesirable characteristics were also selected.

2.2. Marker assisted selection

In recent years, the importance of molecular genetics to understand the genetic nature of quantitative characters has worldwide been recognized, identifying specific regions of genes or chromosomes that affect production and reproduction traits [5]. For example, there are some traits, such as the resistance to a specific pathogen, that can be studied only by a selection method based on the genotype or studying the correlation between the phenotype and genotype. Different types of molecular markers can be used to identify specific gene variants and, a marker assisted selection scheme or MAS (Marker Assisted Selection) implemented in a population. The MAS is a direct selection technique which is based on the association between a trait and several molecular markers. This technique allows to select at a very early stage of development, since it is not necessary to wait for the phenotypic expression of the trait. To date, for many species, a very large number of molecular markers or sequences of DNA are available. An important goal, realized for many species of commercial importance and under way for others, is to set up a map of the genome, identifying several molecular markers and then use them in association mapping, linkage analysis or QTL (Quantitative Trait Loci) studies. The technique of QTLs mapping, based on quantitative genetic laws and molecular methods, allows to associate, for a quantitative trait one or more genetic markers. In this way, for example, it is possible to find markers associated with resistance to a certain pathology or high growth rates. However, in order to obtain a successful QTL analysis, it is necessary to use a large number of polymorphic molecular markers linked to measurable and heritable traits. So, the ideal situation is to perform the analysis in a population showing a high degree of polymorphism and high variability in the genes that control the expressed phenotypes. All individuals of the segregating population are identified for both the molecular markers and quantitative traits.

2.3. Genomic maps

Genomic maps are used in order to get more information concerning the genome of individuals of a given species, describing the order in which genetic loci or markers are displaced and the distance between them on each chromosome. There are two ways to map the genome, using physical or genetic (linkage) methods. Maps are a useful tool for the isolation and cloning of genes of interest.

2.4. Physical maps

A physical map is set up to show the position of specific genes. A physical map consists on a set of markers or physically identifiable regions of DNA and is constructed without using the recombination analysis between genes. The main role is to measure the order and distance between two markers. Physical maps can have different resolutions. For example, the location of a marker on a specific chromosome is given by the hybridization technique of somatic cells. By this way, it is possible to produce a chromosome map in which each chromosome is characterized by a particular banding, observable after staining under a microscope. Another type of physical map has a medium-high resolution, allowing to make eukaryotic metaphase fluorescent chromosomes or specific DNA sequences, and DNA specific fluorochrome-labeled (FISH, fluorescence in situ hybridization). The third type of physical maps has a high resolution of thousands of STS (Sequence-Tagged Site), that defines unique portions of the genome. The STS are short segments of DNA, long approximately 60-1000 bp, which represent points of reference in the genome.

2.5. Maps of association or linkage

Association maps maps show the distances between various genes, their position and other features. Distances between genes are determined by the frequency by which two markers, located on the same chromosome, are inherited together. Alleles which are very closed, they have a higher probability of being transmitted together than those found on distant loci. A unit of genetic map or cM (centimorgan) represents the distance between two genes (1% of recombination). A linkage map, then, defines the distance between markers and their positions on the genome, determining the frequency by which two markers are associated. Maps that use genes as markers show generally a low density and therefore are not always informative. In the construction of a linkage map, DNA sequences should be preferably used.

2.6. Genomic selection

More recently, the availability of high-throughput sequencing techniques has allowed to discover in several livestock species, thousands of Single Nucleotide Polymorphisms (SNPs) spread across the whole genome. Currently, beadchips for genotyping bovines at more than 750,000 marker loci are commercially available. Such a map density is enough to find Linkage Disequilibrium (LD) between markers and QTLs and, by this way looking for associations between traits and markers without specific knowledge of the population structure. These new techniques give rise to new perspectives for the genetic evaluation of farm animals with a so called genome-wide approach. On one hand, this new advance allows to explore the genome looking for QTLs and associations between SNPs and phenotypes. On the other hand, it allows to use directly the marker information to estimate the genomic breeding value (GEBV). In the former case, we talk about the genome-wide association (GWA) studies, while in the latter, the term genomic selection (GS) is generally adopted. Briefly, the GS rely on the segmentation of the genome using a dense marker map in several thousands of bits, each contributing to the explanation of part of the genetic variance of a quantitative trait. The effect of each segment is estimated in a reference population (animals with known phenotypes and genotypes). SNPs effects are then used to predict the breeding values of another set of genotyped animals (prediction population) without phenotypes. Meuwissen et al. in 2001 [6] proposed to use dense marker information to predict the breeding values of animals. Afterwards, several of models and approaches – mainly on simulated dataset – have been proposed to solve the main statistic issue of practical implementation of GS: the great asymmetry of data matrix i.e., the number of effects (single markers or haplotypes) which is much greater than the number of phenotypic records available. In brief, potential advantages of using high density markers in genomic evaluation are the following: i) each QTL is expected to be in LD with at least one marker; ii) all the genetic variance is taken into account in the estimation of breeding values; iii) the animals can be genotyped early in life, and this may guarantee a reduction of generation interval; iv) a better estimation of mendelian sampling (deviation of the individual from the average family effect) term may give rise to a lower inbreeding rate. Genotypes of a particular marker provide a direct information on variability at the locus and frequently at the closely linked loci. When the marker map is not very dense, we may get a biased measure of the variation of the non-genotyped part of the genome. There is an active development of new molecular genetic technology allowing for high-throughput genotyping. Dense marker maps cover the entire genome giving a detailed picture of the genetic variability. This technology has also facilitated the detection of important regions or loci with adaptive effects [7]. Loci could be studied further over breeds and individuals using a technique called re-sequencing [8]. As the technology in molecular genetics advances, it is very likely that sequencing of the whole genome of individuals will soon replace the marker typing. This would result in increments in the accuracy of the estimation of genomic variation and, correspondingly, in the power of strategies devoted to the management of the genetic diversity (and also in selection efficiency). Over generations, alleles at different loci are recombined. If population size has stayed large over a long period, there has been time to produce recombinations even over a very narrow genome area. On the other hand, in a very small population, variants tend to be transmitted over longer genome stretches. Such blocks would therefore indicate a small population size (bottleneck) in the recent history of the population [7]. Furthermore, considering different populations within species, allele frequency differences are used to quantify relationships (through the calculation of different genetic distances) among all groups [9-10].

3. Calculation of inbreeding and additive relationship coefficients in farm animals and small populations

3.1. Calculation of the inbreeding coefficient

Alleles at one locus can be classified into two categories: alleles identical in structure (IS) and identical by descent (ID). The inbreeding coefficient (F) of an animal is defined as the probability that both alleles at a locus are identical by descent (copies of the same allele present in a common ancestor) [11]. The presence of a common ancestor is a key element. Figure 1 shows a diagram of half sibs mating.

F_u coefficient can be computed as the probability F of the animal U to get two copies of an allele from a common ancestor. In the example, animals W and V can have each four different genotypes and inherit the same allele (A_3 or A_4) from the common grandfather (Z) but not from the grandmothers (X and Y). Animal U receives one allele from each parent and so we can get 16 different genotypes (A_1A_3, A_1A_4, A_1A_5, A_1A_6, A_2A_3, A_2A_4, A_2A_5, A_2A_6, A_3A_3, A_3A_4, A_3A_5, A_3A_6, A_3A_4, A_4A_4, A_4A_5, A_4A_6). In the example of half sibs, the condition in which both alleles at a locus are identical by descent is satisfied only for the genotypes A_3A_3 and A_4A_4. The probability to obtain these genotypes is equal to two out of 16 possible combinations (2/16 or 1/8).

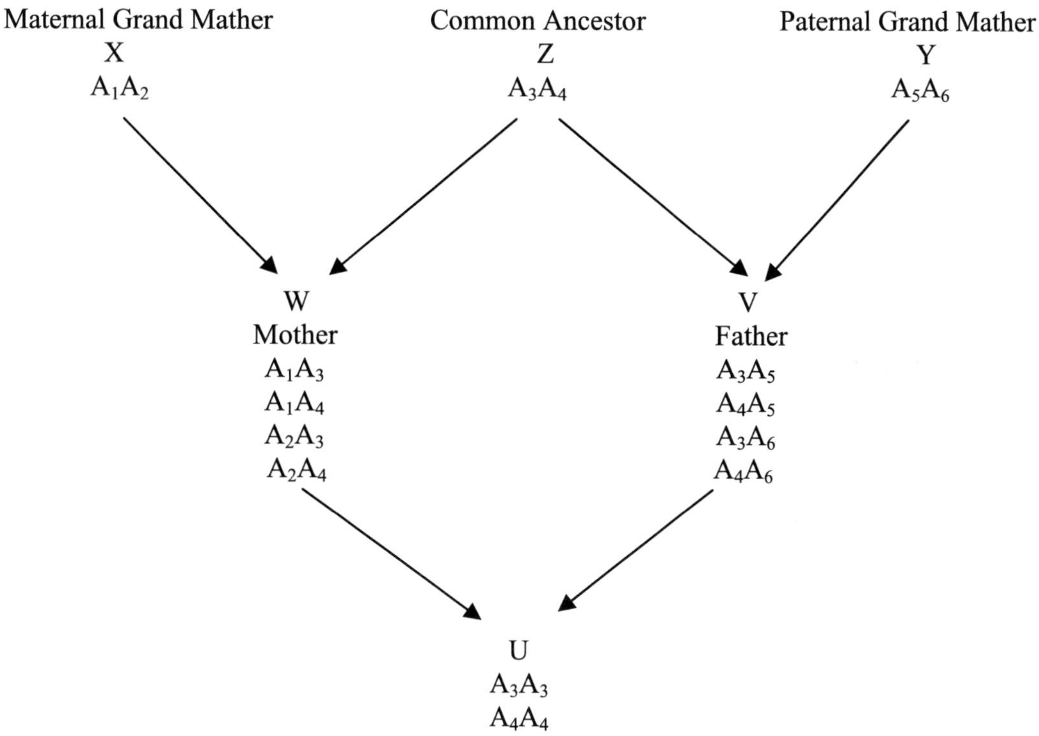

Figure 1. Example of half sibs mating

F coefficient can be computed as follows:

1. for one offspring, the probability to inherit one allele from his father is equal to 1/2;
2. the result of gamete segregation is independent to other segregations that occur at the same or in previous generations. Fu coefficients can be computed as: for allele A_3, the probability that the animal U inherits the allele A_3 from Z and W is equal to: 1/2 x 1/2 = 1/4. Similarly, the probability that animal U inherits the same allele via the Z * V * U path is equal to 1/4. The probability that U inherits the A_3 allele from both parents is equal to the product of the two probabilities:

$$P(A_3A_3) = (1/2 \times 1/2) \times (1/2 \times 1/2) = (1/2)^4 = 1/16$$

and the probability that U inherits the allele A_4 is equal to:

$$P(A_4A_4) = (1/2 \text{ x}1/2)\ (1/2 \text{ x } 1/2) = (1/2)^4 = 1/16$$

As the two genotypes A_4A_4 and A_3A_3 are mutually exclusive:

$$P\ (U = ID) = (1/2)^4 + (1/2)^4 = (1/2)^3 = 1/8$$

Fu coefficient is equal to 1/8. Note that this result is equal to 1/2 powered 3.

In general, the coefficient of inbreeding of an animal U is computed using the following formula:

$$F_U = \sum (1 + F_Z) \left(\frac{1}{2}\right)^n \tag{1}$$

where:
n = number of individuals in the path connecting U and Z. Where Z is descendent of U.
1/2: probability that one allele is transmitted to the next generation
F_Z: inbreeding coefficient of Z (common ancestor).

Values of F range between 0 and 1. In the reference population (assuming no homozygous animals), F coefficient is equal to 0.
In the example, there are three ancestors (W, Z and V) which are considered in the transmission of alleles but X and Y are ignored since they don't influence the inbreeding coefficient. If the common ancestor Z is inbred, the Fz is calculated and multiplied by 1/8. So, the inbreeding coefficient is calculated as:

$$\begin{aligned} Fu &= 1/8 + 1/8Fz \\ &= (1 + Fz)\ 1/8 \\ &= (1 + Fz)\ (1/2)^n \end{aligned}$$

If the inbreeding coefficient of the common ancestor is not specified, it is assumed that is 0.

Example 3.1 – One common ancestor (one path)

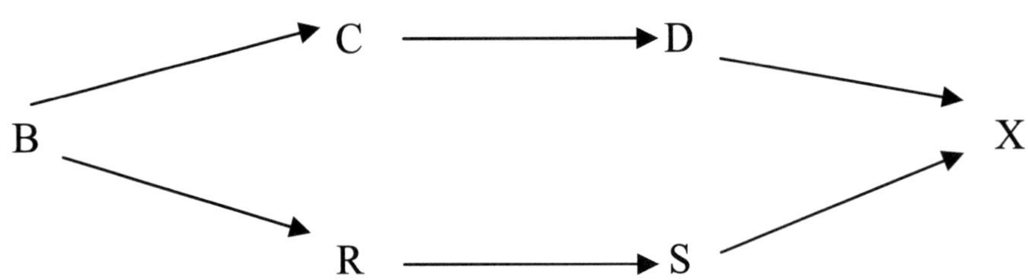

The common ancestor is B: SR<u>B</u>CD; n = 5

$$Fx = (1/2)^5 = 1/32 = 0{,}03125$$

Example 3.2 - One common ancestor (two paths)

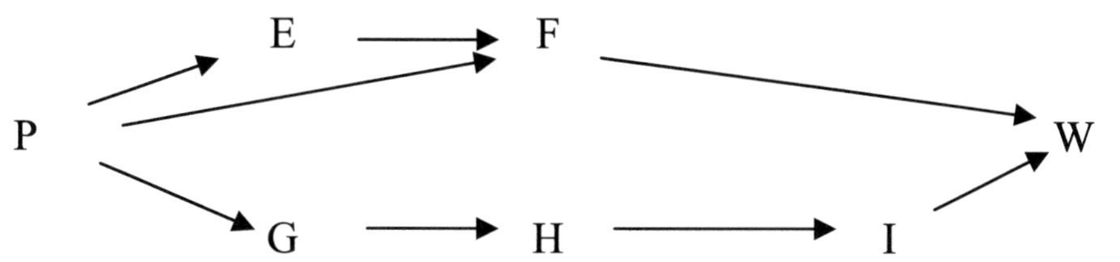

The two paths are:

IHG$\underline{P}$EF	$(1+ F_P)(1/2)^6$
IHG$\underline{P}$F	$(1+ F_P)(1/2)^5$

If P is not inbred:

$$P_W = (1/2)^6 + (1/2)^5=3/64=0,0469$$

Example 3.3 – Two common ancestors (one animal is inbred):

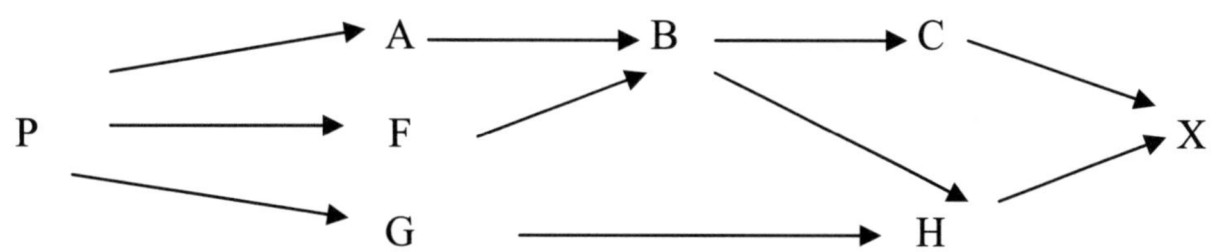

P and B are the common ancestors of C. H, B are inbred but they don't contribute to the inbreeding coefficient of X. The inbreeding coefficient of B is equal to:

$$F_B = (1/2)^3 = 0.125$$

There are three paths:

H$\underline{B}$C	$(1+F_B)(1/2)^3$
HG$\underline{P}$ABC	$(1 + F_P) (1/2)^6$
HG$\underline{P}$FBC	$(1 +F_P) (1/2)^6$

$$F_X = (1 +1/8)(1/2)^3 + (1/2)^6 + (1/2)^6 = 0.172$$

All paths can be easily identified using the following rules:

- one animal appears only once in a path;
- the path has a direct trend;
- all individuals, with the exception of the common ancestors, are ignored in the calculation of the inbreeding coefficient.

Inbreeding occurs in the progeny of related parents increasing the degree of genetic homozygosity, at the expense of heterozygous genes. The increase of inbreeding rate in the population induces two genetic events:

a. a progressive fixation of alleles;
b. a gradual reduction of dominance effects;
c. an increment of inbreeding depression effects due to the higher frequency of recessive genes.

Inbreeding coefficients refer also to the inbreeding level averaged across all individuals living in a population.

3.2. Effective population size

In small populations, the effective number of reproducing animals or effective population size (Ne) determines the expected increment of inbreeding per generation (rate of inbreeding) [11]:

$$\Delta F = \frac{1}{2N_e} \tag{2}$$

Note that equation 2 is appropriate only if the population is in Hardy-Weimberg equilibrium (panmictic population).

3.3. Unequal numbers of males and females

Sometimes, in small populations, the number of males (Mm) and females (Nf) is not 1:1. In this case, sexes are not contributing equally and Ne is calculated as:

$$\frac{1}{N_e} = \frac{1}{2}\left(\frac{1}{2Nm} + \frac{1}{2Nf}\right) = \frac{1}{4Nm} + \frac{1}{4Nf} \tag{3}$$

The equation 3 can be also written as:

$$N_e = \frac{4N_m N_f}{N_m + N_f}$$

The effective population size is primarily determined by the less numerous sex. The increase of F in one generation is computed as:

$$\Delta F = \frac{1}{2N_e} = \frac{1}{8N_m} + \frac{1}{8N_f} \tag{4}$$

Example 3.3.1: 2 males and 50 females

$$1/Ne = 1/8 + 1/200 = 0.13$$

$$Ne = 1/0.13 = 7.7$$

$$\Delta F = 1/2(7.7) = 0.0650$$

In terms of increment of inbreeding per generation, this population of 52 individuals is equivalent to a population of 8 animals: 4 males and 4 females.

Example 3.3.2: the number of males is 2 and the number of females is assumed to be infinite

$$1/Ne = 1/4Nm= 1/4(2) = 1/8$$

$$\Delta F = 1/16 = 0.0625$$

The result obtained in the example 3.3.2 is similar to the value calculated in the previous example.

3.4. Non-random distribution of the family size

The family size is the number of offspring of each family that become parents in the next generation. Under an ideal situation, the size of the population remains constant in successive generations and each of the parents has to be replaced by another animal. In this case, the average number of offspring per parent is equal to 1 with an average size of the family of 2. Ne is function of the variance of the family size:

$$N_e = \frac{4N}{2 + V_k} \tag{5}$$

where:
N = total number of animals in the population
V_K: variance of the family size
Note that if $V_K = 2$ than Ne = N

Example 3.4: Vk = 6 for both sexes. Population size: 25 males and 25 females

$$Ne = \frac{4(50)}{2+6} = \frac{200}{8} = 25$$

In terms of inbreeding rate, this population is equivalent to a population made of 12 males and 12 females.

If each male mate with more than one female, then the number of offspring and the variance of family size will be different within sexes. In this case, equation 5 becomes:

$$N_e = \frac{8N}{4 + V_{km} + V_{kf}}$$

V_{km} and V_{kf} are variances of family size for males and females.

3.5. Variable number of breeding animals over generations

If the number of parents is not constant over generations, the effective population size can be calculated by the harmonic mean as follows [11]:

$$\frac{1}{N_e} = \frac{1}{t}\left(\frac{1}{N_1} + \frac{1}{N_2} + \ldots\ldots + \frac{1}{N_t}\right) \quad (6)$$

where:

t = number of generation.

N_1 = number of reproducing animals at the first generation

Example 3.5: Numbers of parents over four generations: 10, 10,50 and 10 animals

$$\frac{1}{Ne} = \frac{1}{4}\left(\frac{1}{10} + \frac{1}{10} + \frac{1}{50} + \frac{1}{10}\right)$$

$$1/4(0.32) = 0.08$$

$$\overline{N_e} \cong 13$$

After four generations, the expected inbreeding coefficient will be the same as for a population of 13 animals in each generation. Note that, the increase in the number of breeding animals up to 50 in the third generation will not modify the value of inbreeding rate.

3.6. The kinship coefficient and the additive relationship

The kinship coefficient f_{IJ} between two individuals (A and C) is measured by the probability of taking a given allele at a locus of an animal that is identical by descent to another allele on the same locus in a second animal [12]:

Father A $b_1\ b_2$ → C ← Mother B $b_3\ b_4$

C

$b_1\ b_3$ (1/4)

$b_1 b_4$ (1/4)

$b_2 b_3$ (1/4)

$b_2 b_4$ (1/4)

The probability that two alleles taken at random from A and C are identical is equal to:

Possible combinations	Probability (P)	
b_1b_1	1/2 x 1/4	⇒ 1/8
b_1b_2	1/2 x 1/4	
b_1b_3	1/2 x 1/4	
b_1b_4	1/2 x 1/4	⇒ 1/4
b_2b_1	1/2 x 1/4	
b_2b_2	1/2 x 1/4	⇒ 1/8
b_2b_3	1/2 x 1/4	
b_2b_4	1/2 x 1/4	

The probability to get two identical alleles (b_1b_1 or b_2b_2) from A and C (father and son) is equal to ¼ (P = 1/8 + 1/8=1/4). Because each locus contains two alleles, the process must be repeated two times. The coefficient of kinship or additive relationship is defined as:

$$2f_{AC} = a_{AC} = 2 \times 1/4 = 1/2 \tag{7}$$

where:
f_{AC} = kinship coefficient
a_{AC} = additive relationship

The coefficient of inbreeding of an animal C is equal to the coefficient of kinship of his parents (A and B) [11]:

$$F_C = f_{AB} \tag{8}$$

The inbreeding coefficient is equal to half of the additive relationship coefficient of his parents:

$$a_{AB} = 2F_C \qquad F_C = \tfrac{1}{2} a_{AB} \tag{9}$$

3.7. Calculation of the additive relationship and relationship coefficients using the tabular method

The simplest method to determine the additive relationship coefficient between individuals and inbreeding coefficient is the tabular method [13]. This method is called tabular method because, the result takes the form of a table. The tabular method allows to construct a matrix relationship in the following way:

1. the number of columns is equal to the number of animals in the population. In the following population, there are 6 individuals and 6 columns. Animals are sorted by birth date starting from the left;
2. parents are indicated above each individual (-: missing record);
3. the value of 1 (on the diagonal) indicates the relationship of each individual with himself ($a_{xx} = 1$);

4. the calculation of the additive relationship starts on the first animal on the first row (A) and continues with the other individuals in the same line;
5. the additive relationship of each animal is computed as 1/2 of the sum of the additive relationship coefficients of his/her parents at the left on the same row;
6. the inbreeding coefficient is calculated adding to 1 values on the diagonal, 1/2 of the additive relationship of the animal's parents (e.g. see in table 1 the DD cell: 1+1/8).

	- -	A -	A -	B C	D -	D C
	A	B	C	D	X	Y
A	1	1/2	1/2	1/2	1/4	1/2
B	1/2	1	1/4	5/8	5/16	7/8
C	1/2	¼	1	5/8	5/16	13/16
D	1/2	5/8	5/8	1+1/8	1/2	13/16
X	1/4	5/16	5/16	1/2	1	13/32
Y	1/2	7/8	13/16	13/16	13/32	1

Table 1. Example of calculation of the additive relationship and inbreeding coefficient using the tabular method

Parameters obtained from the pedigree analysis provide a useful information for predicting the genetic consequences of a given management scheme or for designing future resources of a conservation programme, where biodiversity has to be maintained. Use of molecular information (combined with pedigree data or alone) may be the most useful for dealing with adaptive variation and to unveil the old history of populations (i.e. before pedigree recording started) [7].

4. Restricted inbreeding strategies in farm animals

There are a number of approaches described in the literature to assess the acceptable rate of inbreeding or conversely the minimum effective population size to maintain a relatively 'safe' population. Regarding the short-term prevention of inbreeding depression problems, there is a consensus among animal breeding researchers that ΔF of 0.5 to 1% is the acceptable rate. Therefore, an effective size of 50-100 could be sufficient to keep a population in a healthy state. Meuwissen and Woolliams in 1994 [14] also considered balancing the depression due to inbreeding, which decreases fitness, against the genetic variation available for natural selection, which improves fitness. Depending on the fitness parameters assumed, the critical effective size varied between 50-100 individuals. When taking into account other criteria (i.e. long-term potential to evolve and accumulation of mutations), figures should be higher, with the value depending on the assumptions about the mutational model (i.e. the mutational rate and the mean effect of spontaneous mutations). Some organisations (e.g. FAO [15]) often use the effective population size to define the level of endangerment. Breeding programmes for mainstream breeds are focused on achieving significant gains in the trait of interest but the programmes should also deal with the problems associated with the loss of diversity. One way to cope with the situation is an efficient monitoring process to detect undesirable changes in fitness traits that are sensitive to inbreeding depression. However, a more reasonable strategy is to incorporate restrictions on the level of expected kinship (or inbreeding) in the animals with the objective of maximising their gain. The maintenance of variation is related to the effective population size or rate of inbreeding. From the definition of Ne itself and the factors maximising Ne (or minimising the genetic drift), some basic recommendations can be extracted. First, we should obviously keep the highest possible number of parents and try to have the same number of sires and dams. Then, we should try to equalise the number of offspring (contributions) to be obtained from every potential parent. The idea behind this is to give the same opportunity to every parent of effectively transmitting their alleles. And finally, we should prolong the generation interval as genetic drift occurs always when parents' alleles are sampled in creating offspring. Note that, this last recommendation and that of using many parents decreases the annual rate of response in a selection scheme. In practice, in many livestock species it is impossible to reach the 1:1 sex ratio. To cope with this situation some hierarchical (several dams mated to each sire) and regular systems have been developed [16,17]. The idea is always to equalise the contributions from each individual to the next generation. Basically, these strategies consist of a more or less optimised form of within-family selection. Hierarchical methods have the advantage of being simple and easy to implement for non specialised personnel and of providing predictions on the evolution of inbreeding over the years. The disadvantage of them is that they are very sensitive to deviations from the assumed conditions (i.e. related founders, mating failures, number of females not being an exact multiple of the number of males, fluctuating population size) as shown by Fernandez et al. [18] and, therefore, they are not applicable in most real situations. When no pedigree is available there are two options. To begin with, we could use molecular information to complete or replace the genealogical information. In its simpler form, it is very common to carry out a paternity analysis, useful for determining the probabilities for the sire candidates (and sometimes also for the dams) in free range animals, and consequently filling the gaps in the pedigree. In more complex situations, we could determine the general relationships in a group of animals through a set of available kinship estimators [19,20] or a IBD (Identical By Descent) matrix is constructed. Fernàndez et al. (18) studied the accuracy of molecular kinship in maintaining the genetic diversity in a conservation programme when replacing or complementing the genealogy with molecular genetic information. The study relied on the use of microsatellites and conclusions should be re-evaluated in the context of dense SNP maps. The genomic information could also be utilised for comparing the genetic value of individual animals for quantitative traits. The pedigree-based relationships can be augmented or even replaced by marker-based information. This is probably easier to envisage by considering a new genomic selection method [6], where the genetic value of an animal is determined by summing the effects of tens or hundreds of thousands markers over the whole genome. Marker effects are estimated from a sufficiently large reference population. Management of variation is very important in genomic selection because, as a very efficient method, it is expected to lead to a long-term depletion of variation with a higher risk compared than conventional methods [21].

5. Use of the optimum contribution selection (OCS)

The kinship between individuals is directly related to the genetic diversity of the population (measured as the expected heterozygosity) and also related to the expected inbreeding in the next generation. The kinship between individuals also reflects the proportion of common genes and, thus, the redundancy of the alleles in the individuals. From this, it follows that a good methodology should consist of finding the combination of contributions from available parents to minimise the expected average kinship in the next generation. This is achieved by applying the OCS [22]. Long-term selection schemes also benefit from it by restricting the average kinship to a desired level in the in the objective function (with a negative sign), directed at maximising the gain [3]. There are interesting similarities behind the two terms. In finding the best candidates for selection, the comparison of genetic values also use the information from relatives. The well-known additive relationship matrix, used in such an evaluation using the BLUP methodology, equals twice the kinship matrix. In conclusion, with OCS one can either minimise the rate of inbreeding (ΔF), or constrain it into a predefined value and maximise genetic gain simultaneously. Recently, software has been developed for choosing the sires and dams and allocating the contributions for them both in conservation and selection programmes. GENCONT [23] is able to perform OCS selection for a given rate of inbreeding. EVA [24] produces a similar kind of outcome but puts cost weights against the kinship instead of restricting the rate of inbreeding. Once the parents and the optimal number of offspring from each of them have been decided, we should determine the mating scheme. It should be noted that the optimisation of contributions is the main task in the management, leaving little margin for any improvement in the mating design. With a one generation horizon, the genetic level and average kinship do not depend on the way the parents were mated. Inbreeding is greatly influenced, because the inbreeding of the descendants is, by definition, the kinship between the mating pairs. If we are worried about inbreeding, it is sensible to implement strategies that prevent matings between close relatives [22]. In a general non-regular population, this methodology is called the minimum kinship mating and consists of finding the combinations of couples that yield the minimum average kinship between each pair of individuals to be mated. As pointed out by some authors [25], the prevention of mating between relatives is not the best method in the long term but the method they proposed implies a large increase in inbreeding in the short term, which would not be acceptable in most conservation programmes. Other strategies like compensatory mating [8] have been proposed. This methodology works by mating the most related females with the least related males, and vice versa, trying to balance the genetic contributions from under- and over-represented lineages. However, performances are not really very different from that of the minimum kinship mating, so the former may be recommended. Henryon et al. [26] proposed to reduce the covariance between ancestral contributions (MCAC mating), showing that lower levels of inbreeding can be reached when performing truncation selection. When physiologically feasible, some authors [27] have proved that performing a factorial mating design (i.e. mating each parent to several mates) would reduce the levels of inbreeding achieved due to the reduced correlation between the contributions of mates.

Moreover, factorial mating increases the flexibility in breeding schemes for achieving the optimum genetic contributions. Sometimes, for practical reasons (e.g. a female is not able to mate with more than one male), and results from the OCS methodology cannot be fitted into a realistic mating design. In that situation, we would like to determine, at the same time, not only how many offspring an animal should have, but also with which animal it should be mated. The simultaneous optimisation of selection and mating is called 'mate selection' and, instead of deciding just on the number of offspring to be had from each candidate, it also looks into the number of offspring produced from every possible couple. It is easy to include some restrictions on the number of matings per particular animal or the maximum number of full-sibs to generate among the progeny.

6. A stochastic simulation program (Matlab), based on a simulated annealing algorithm, for optimizing farm animal breeding schemes under restricted inbreeding

In species with large families, the management of the pedigree to minimize ΔF can be combined with appropriate selection techniques within families. The high reproductive potential, in some commercial species (pig, chicken, fish), allows high genetic gains by applying high selection intensities. This means that, a very small number of individuals are used to generate successive generations and hence the rate of inbreeding can be high [27]. The detrimental effects of inbreeding are well documented in several commercial species. In recent years, many selection and mating strategies have been proposed to restrict inbreeding in selection programmes [27]. In this study, a stochastic simulation model was used to simulate and optimize mating schemes of farm animals using different genetic parameters and under restricted inbreeding. The structure of the simulated breeding scheme was that of a closed nucleus. An animal population under artificial selection was modelled by stochastic (Monte Carlo) simulation using the Matlab software. Selection was applied for a single trait measured on both sexes and based on estimated breeding values (EBVs) using the ASREML2 statistical package. Generations were discrete (equal number of sires and dams were selected at each generation). The trait under selection was assumed to be determined by an infinite number of unlinked additive loci, each with an infinitesimal effect. The trait was considered to be standardized, so the initial phenotypic variance is unity. Phenotypes of unrelated base population animals (generation 0) were generated as the sum of a normally distributed environmental and genetic effects. Phenotypic values of the offspring born every generation were generated as:

$$P_i = \mu + \left(\sigma_A RND(0,1)_S + \sigma_A RND(0,1)_D\right)/2 + \sigma_E RND(0,1) + \\ + \left(0.5\left(1-\left(F_s+F_d\right)/2\right)\right)^{1/2} \sigma_A * RND(0,1) \quad (10)$$

where:

$\sigma_A = \sigma_{A(o)}/(1+kh^2)$

$k=(0.5)(k_m+k_f)$
$k_y= i_y(i_y-x_y)$ y=male or female
i= selection intensity
RND = random number

Phenotypic values (Pi) were calculated as $Pi = \mu + \sigma_{Gi} + \sigma_{Ei}$ where σ_{Gi} is the genetic effect and σ_{Ei} is the environmental effect, which were sampled from N (0,1) making the base phenotypic variation (σ^2_P) equal to 1.0. The base generation additive genetic variance, σ^2_A was 0.1, or 0.25, or 0.50 corresponding to a heritability, h^2, of 0.1 or 0.30 or 0.50, respectively. Later generations were obtained by simulating progeny genotypes from $\sigma_{Gi} = 0.5\ \sigma_{Gs} + 0.5\ \sigma_{Gd} + mi$, where s and d denote sire and dam of progeny i, respectively, and mi = mendelian sampling component, which was sampled from $(0.5(1-(F_s+F_d)/2\ ))^{1/2}\ \sigma_A$ *RND(0,1), where $(F_s+F_d)/2$ is the average of the inbreeding coefficients of the sire and the dam. ΔF was restricted to 0.010 per generation, which is an indication of the maximum acceptable rate of inbreeding. Selection was directional upwards and by truncation. Total number of offspring born per generation and numbers of selected males and females were constant (10 or 20 offspring per mating) over generations and varied according to the mating schemes. The simulated breeding schemes are described in Table 2.

Number of selection candidates per generation	90 or 180 or 360 or 720
Number of generations	10
Number of replicated simulations	100
Mating schemes	factorial 3 x 3; 6 x 6; nested 6 (males) x 18 (females)
Heritability coefficient	0.1 or 0.3 or 0.5

Table 2. Parameters of the closed nucleus scheme

The OCS and simulated annealing was used to select animals. The OCS theory maximises the genetic gain while constraining the rate of inbreeding or the relationships among selection candidates. These methods choose the selected parents and assign genetic contributions to the next generation for each selected candidate. This method maximises the genetic level of the next generation of animals:

$$G_{t+1} = c_t'\ EBV_{t\,1} \tag{11}$$

c_t is a vector of genetic contributions of selected candidates to the generation t_{+1};

EBVt is a vector of best linear unbiased prediction (BLUP) estimates of candidates in generation t.

The objective function, $c_t'EBV_t$, is maximized for ct under two restrictions: the first is on the rate of inbreeding and the second is on the contribution per sex. The desired rate of inbreeding, ΔF is obtained by constraining the average kinship of the selection candidates to:

$$C_{t+1} = 1 - (1 - \Delta F)t \tag{12}$$

The actual contributions of the individuals are then obtained in such a way that they fulfil the constraint:

$$C_{t+1} \geq ct' A t c t/2 \quad (13)$$

where:
A_t is a ($n \times n$) relationship matrix among the selection candidates. Note that the level of the constraint C_{t+1}, can be calculated for every generation before the breeding scheme commences.

The contribution of each sex is constrained to ½ :

$$Q' c t = 1/2 \quad (14)$$

where:
Q is a ($n \times 2$) incidence matrix of the sex of the selection candidates (the first column yields ones for males and zeros for females, and the second column yields ones for females and zeros for males). The contributions of the male and those of the female candidates will sum to ½.

In order to obtain the optimal ct that maximize G_{t+1}, Lagrangian multipliers were used. An additional restriction was to select only one full sibs per family.

Using the lagrangian method for restricted optimization, the optimum solution is obtained as follows:

$$c = A^{-1} \ (EBV - Q\lambda) / 2\lambda_0) \quad (15)$$

where:
λ and λ_0) are lagrangian multipliers.

The minimum kinship mating (reduce the average relationship of sires and dams and therefore also the inbreeding of their progeny is minimized) is obtained by applying the simulated annealing algorithm according to Press et al.[4]. The output from the selection method is a vector with genetic contributions for each selection candidate, *ct*. The ultimate goal, in this mating tool, is to reduce the average inbreeding coefficient in the following generation. Input parameters included all possible relationships between pairs of selected dams and selected sires. The scheme with the lowest average inbreeding coefficient in the next generation is considered as the optimal one. The essential steps of the simulated annealing algorithm can be summarized as follows: 1) sires and dams are mated at random according to their frequencies in vector **c**, and than the resulting average inbreeding coefficient is stored as reference value *00*; 2) change of mating partners and comparison of the new resulting average inbreeding value *01* with *00*; 3) if the value *01 is* < *00* than *it* is replaced with *01* and so for all possible matings. By using simulated annealing, inbreeding is avoided as much as possible. The rate of inbreeding (ΔF) and the genetic gain (increase in animal performance through a genetic programme, ΔG) for the three mating designs are reported in Table 3.

	Number of offspring per family					
	10			20		
Heritability	0.1	0.3	0.5	0.1	0.3	0.5
	Full factorial (3males x 3 females)					
ΔF	3.96	5.00	4.55	4.50	4.70	4.65
ΔG	0.29	0.60	0.75	0.37	0.60	0.97
	Full factorial (6 males x 6 females)					
ΔF	2.25	2.20	2.10	2.05	1.80	1.85
ΔG	0.38	0.64	0.76	0.47	0.69	0.94
	Nested (6 males x 18 females)					
ΔF	2.60	2.25	2.36	2.06	2.90	2.05
ΔG	0.43	0.72	0.86	0.57	0.66	0.89

Table 3. Rate of inbreeding (ΔF)(x100) and genetic gain (ΔG))(σ_P) for different mating schemes and genetic parameters.

The full factorial design gives the best results in terms of ΔF and ΔG (1.85 and 0.94 or 1.80 and 0.69) using a higher number of sires and dams (6 x 6), family size per mating ,a family size per mating of 720 offspring and heritability coefficients of 0.3 or 0.5. According to Sorensen *et al.* [2], the superiority of the factorial mating compared to hierarchical scheme can be explained in terms of the different genetic structure of populations obtained showing, in the factorial design, small full-sibs families, more paternal half-sibs and a group of maternal halfsibs. At a lower heritability (0.1) the nested design become competitive with the full factorial mating (6 x 6). This selection approach have been already evaluated in practice for several domestic species such as dairy cattle [7], the Hanoverian horses for show jumpers [19], fish and pigs.

7. Conclusions

Additive relationships among individuals are generally used for weighting records of relatives in the genetic evaluation of farm animals and to calculate inbreeding coefficients. The tabular method, used for computing the additive relationships and inbreeding coefficients of farm animals is the most efficient and widely used method. According to the present simulation study, the best mating scheme of farm animals under a restricted inbreeding rate is a full factorial mating (6 males and 6 females) with full-sibs families of 20 animals. Furthermore, the present work has clearly shown that, the most suitable approach for long-term selection activities under inbreeding restrictions, is to use together the optimum genetic contribution and simulated annealing methods.

Author details

Edo D'Agaro
Faculty of Veterinary Medicine, University of Udine, Udine, Italy

8. References

[1] Woolliams J.A. Genetic contributions and inbreeding. In: Oldenbroek, K. (ed.) Utilisation and conservation of farm animal genetic resources. Wageningen: Academic Publishers; 2007 p 147-165.

[2] Sorensen A.C., Berg P. and Woolliams J.A. The advantage of factorial mating under selection is uncovered by deterministically predicted rates of inbreeding. Genetics Selection Evolution 2005;37 57-81.

[3] Meuwissen, T.H.E.,. Maximizing the response of selection with a predefined rate of inbreeding. Journal of Animal Science 1997;75 934-940.

[4] Press W.H., Flannery B.P., Teukolsky S.A., Vetterling W.T. Minimization or maximization of function. In: Numerical recipes. The act of scientific computing. Cambridge: University of Cambridge Press;1989 pp. 274-334.

[5] Dekkers J.C.M., Hospital F. The use of molecular genetics in the improvement of agricultural populations. Nature Rev. Genet. 2002;3 22-32.

[6] Meuwissen T.H.E., Hayes B.J., Goddard, M.E.. Prediction of total genetic value using genome-wide dense marker maps. Genetics 2001; 157 1819-1829.

[7] Toro M.A., Màki-Tanila, A. Genomics reveals domestication history and facilitates breed development. In: Oldenbroek, K. (ed.) Utilisation and conservation of farm animal genetic resources. Wageningen: Wageningen Academic Publishers; 2007 p 75-102.

[8] Sellner E.M., Kim J.W., McClure M.C., Taylor K.H., Schnabel R.D., Taylor J.F. Board-invited review: Applications of genomic information in livestock. Journal of Animal Science 2007; 85 3148-3158.

[9] Caballero A., Toro, M.A. Characterization and conservation of genetic diversity in subdivided populations. Phil. Trans. R. Soc. B 2005;360 1367-1378. Caballero, A., Santiago, E. and Toro, M.A., 1996. System of mating to reduce inbreeding in selected populations Animal Science 2005;62 431-442.

[10] Toro M.A., Fernàndez J., Caballero A. Molecular characterization of breeds and its use in conservation. Livestock Science 2009; 120 174-195.

[11] Falconer DS Introduction to quantitative genetics. New York: Longman Inc.; 1981

[12] Malecòt G. Les mathèmatiques de l'ereditè. Paris: Masson and Cie ; 1948

[13] Van Vleck L.D., Pollack E.J., Oltenacu E.A.B. Genetics for the animal sciences. New York: Freeman and Company; 1987

[14] Meuwissen T.H.E., Woolliams, J.A.. Effective sizes of livestock populations to prevent a decline in fitness. Theoretical and Applied Genetics 1994; 89 1019-1026.

[15] FAO Secondary Guidelines on the Management of Small Populations at Risk. UNEP, Rome;1998 p 201-204.

[16] Wang J. More efficient breeding systems for controlling inbreeding and effective size in animal populations. Heredity 1997; 79 591-599.

[17] Sànchez-Rodrfguez L., Bijma P., Woolliams J.A. Reducing inbreeding rates by managing genetic contributions across generations. Genetics 2003; 164 1589-1595.

[18] Fernàndez J., Toro M.A., Caballero, A.. Fixed contributions designs versus minimization of global coancestry to contrai inbreeding in small populations. Genetics 2003;165 885-894.

[19] Fernàndez J. and Toro M.A.. A new method to estimate relatedness from molecular markers. Molecular Ecology 2006; 15 1657-1667.

[20] Oliehoek P.A., Windig J.J., Van Arendonk J.A.,Bijma, P., Estimating relatedness between individuals in generai populations with a focus on their use in conservation programs. Genetics 2006;173 483-96.

[21] Hayes, B.J., Bowman, P.J., Chamberlain, A.J. and Goddard, M.E. Invited review: Genomic selection in dairy cattle: progress and challenges. Journal of Dairy Science 2009;92 433-43.

[22] Meuwissen T.H.E. Operation of conservation schemes. In: Oldenbroek, K (ed.) Utilisation and conservation of farm animal genetic resources. Wageningen: Wageningen Academic Publishers; 2007 p 167-193.

[23] Meuwissen T.H.E.. GENCONT: An operational tool for controlling inbreeding in selection and conservation schemes. Proceedings of 7th World Congress on Genetics Applied to Livestock Production, Montpellier,19-23 August 2002; 33 769-770.

[24] Berg P., Nielsen I., and Sorensen M.K. Computing realized and predicting optimal genetic contributions by ÈVA. Proceedings of 8th World Congress on Genetics Applied to Livestock Production, Belo Horizonte; 2006 p 120-121.

[25] Crow J.F. and Kimura, M., 1972 An introduction to population genetics theory. Harper and Row; New York. Daetwyler, H.D., Villanueva, B., Btjma, P., Woolliams, J.A.. Inbreeding in Genome-Wide Selection. Journal of Animal Breeding and Genetics 2007; 124 369-376.

[26] Henryon M., Sorensen A.C. and Berg P. Mating animals by minimising the covariance between ancestral contributions generates more genetic gain without increasing rate of inbreeding in breeding schemes with optìmum-contribution selection. Animal Production 2009;3 1339-1346.

[27] Sonesson A.K., Gjerde B., Meuwissen T.H.E. Truncation selection of BLUP-EBV and phenotypic values in fish breeding schemes. Aquaculture 2005; 243:61-68.

Section 3

Hybrid SA Applications

Chapter 10

Simulated Annealing Evolution

Sergio Ledesma, Jose Ruiz and Guadalupe Garcia

Additional information is available at the end of the chapter

http://dx.doi.org/10.5772/50176

1. Introduction

Artificial intelligence (AI) is a branch of computer science that seeks to create intelligence. While humans have been using computers to simplify several tasks, AI provides new options to use computers. For instance, voice recognition software uses AI to transform the sounds to the equivalent text words. There are several techniques that AI includes. An artificial neural network (ANN) is one of these techniques.

Humans use their intelligence to solve complex problems and perform daily tasks. Human intelligence is provided by the brain. Small processing units called neurons are the main components of the human brain. ANNs try to imitate partially some of the human brain behavior. Thus, artificial neurons are designed to mimic the activities of biological neurons.

Humans learn by experience: they are exposed to events that encourage their brains to acquire knowledge. Similarly, ANNs extract information from a data set; this set is typically called the training set and is organized in the same way that schools design their courses' content. ANNs provide an excellent way to understand better biological neurons. In practice, some problems may be described by a data set. For instance, an ANN is typically trained using a data set. For some problems, building a data set may be very difficult or sometimes impossible as the data set has to capture all possible cases of the experiment.

Simulated annealing (SA) is a method that can be used to solve an ample set of optimization problems. SA is a very robust technique as it is not deceived with local minima. Additionally, a mathematical model is not required to apply SA to solve most optimization problems.

This chapter explores the use of SA to train an ANN without the requirement of a data set. The chapter ends with a computer simulation where an ANN is used to drive a car. Figure 1 shows the system architecture. SA is used to provide a new set of weights to the ANN. The ANN controls the acceleration and rotation speed of the car. The car provides feedback by sending vision information to the ANN. The distance traveled along the road from the *Start* is used by the method of SA. At the beginning of the simulation the ANN does not know how to drive the car. As the experiment continues, SA is used to train the ANN. Each time the

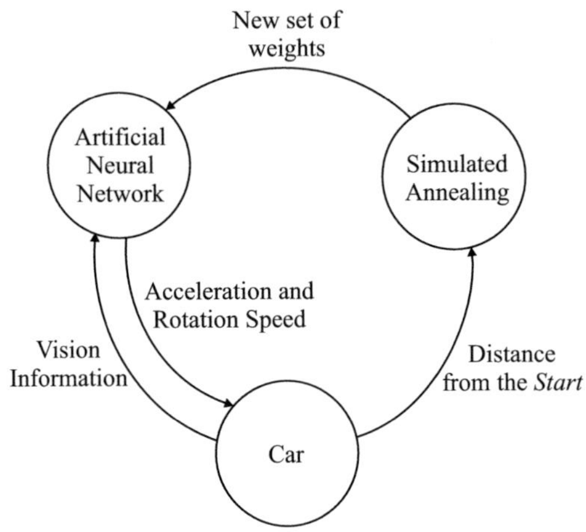

Figure 1. System architecture.

temperature decreases, the ANN improves its driving skills. By the end of the experiment, when the temperature has reached its final value, the ANN and the car have evolved to the point that they can easily navigate a maze.

2. Artificial neural networks

An ANN is a computational method inspired in biological processes to solve problems that are very difficult for computers or humans. One of the key features of ANNs is that they can adapt to a broad range of situations. They are typically used where a mathematical equation or model is missing, see [4]. The purpose of an ANN is to extract, map, classify or identify some sort of information that is allegedly hidden in the input, [13].

2.1. Neuron

The human brain is composed of processing units called neurons. Each neuron is connected to other neurons to form a neural network. Similarly, the basic components of an ANN are the neurons. Neurons are arranged in layers inside the ANN. Each layer has a fixed number of neurons, see [5]. For instance, the ANN shown in the Figure 2 has three layers: the input layer, the hidden layer, and the output layer.

An ANN accepts any kind of input that can be expressed as a set of numeric values; typical inputs may include: an image, a sound, a temperature value, etc. The output of an ANN is always dependent upon its input. That is, a specific input will produce a specific output. When a set of numeric values is applied to the input of an ANN, the information flows from one neuron to another until the output layer generates a set of values.

2.2. Activation function

The internal structure of an artificial neuron is shown in Figure 3(a). The output value z is given by:

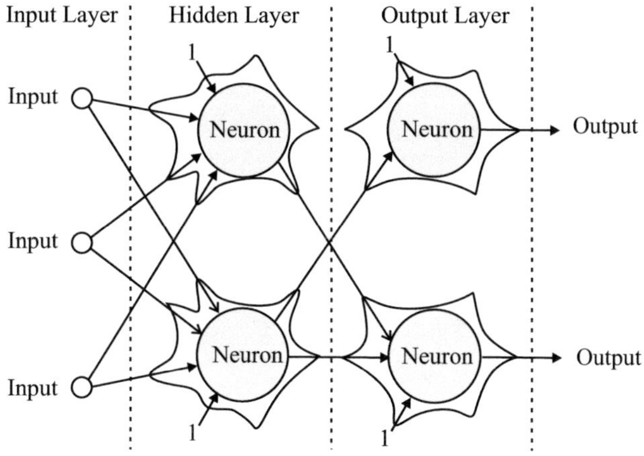

Figure 2. An artificial neural network.

$$z = f(y) = f(w_b + \sum_i x_i w_i), \tag{1}$$

where each w_i is called weight. A fixed input, known as *Bias*, is applied to the neuron, its value is always 1 and w_b is the respective weight for this input. The neuron includes also an activation function denoted by $f(y)$ in Figure 3(a). Without the *Bias*, the output of the network would be $f(0)$ when all inputs are zero. One common activation function used in multilayer ANNs is:

$$f(y) = logsig(y) = \frac{1}{1+e^{-y}}. \tag{2}$$

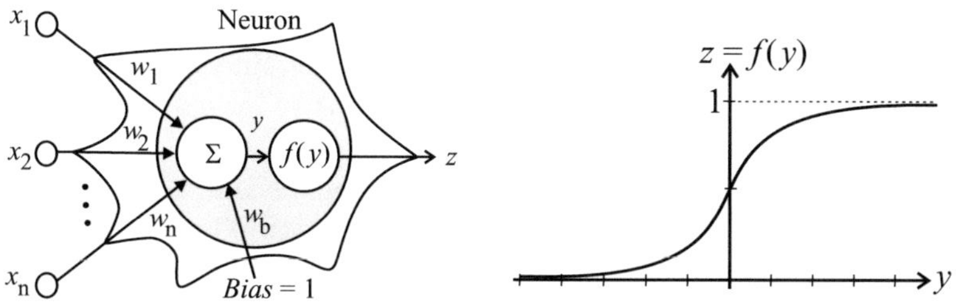

(a) Neuron structure (b) Activation function

Figure 3. Artificial neuron.

The activation function of Equation 2 is plotted in Figure 3(b). The activation function (in this figure) is real, continuous, limited and has a positive derivative.

A neuron can be active or inactive, when the neuron is active its output is 1, when it is inactive its output value is 0. Some input values activate some neurons, while other values may activate other neurons. For instance, in Figure 4, the sound of the word *Yes* will activate the first output neuron, while the sound of the word *No* will activate the second neuron at the output of the network. The structure of the ANN shown in this figure is very simple with only two neurons.

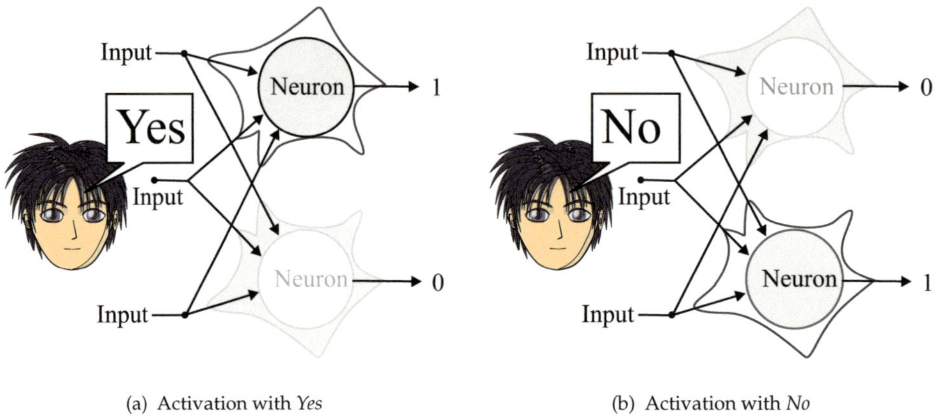

Figure 4. ANN activation.

3. Learning

Before an ANN can be used for any practical purpose, it must be trained. An ANN learns during its training. For the duration of the learning process the ANN weights are recurrently adjusted.

3.1. Structured learning

In some instances, ANNs may learn from a data set. This set is typically known as the training set, and it is used on a new ANN (as its name indicates) for training. The training set has two parts: the input and the target. The input contains the set of inputs that must be applied to the network. The target includes the set of desired values at the output of the ANN. In other words, each sample (case) in the training set completely specifies all inputs, as well as the outputs that are desired when those inputs are presented, see [7]. During the training, each case in the training set is presented to the network, and the output of the network is compared with the desired output. After all cases in the training set have been processed, an epoch or iteration has completed by updating the weights of the network. There are several methods for updating the weights at each epoch or iteration. All these methods update the weights in such a way that the error (measured between the actual output of the network and the desired output) is reduced at each epoch, see [7].

Some training methods are based on the gradient of the error (they are called gradient based methods). These methods quickly converge to the closest local minima. The most typical gradient based methods to train an ANN are: the variable metric method, the conjugate gradient method, and method of Levenberg-Marquardt. To make the training of an ANN robust, it is always recommended to combine gradient based methods with other optimization methods such as SA.

When an ANN is trained using a data set, typically the set includes many training cases, and the training is done at once. This kind of training is called structured learning because the knowledge is organized in the data set for the network to learn. One of the main disadvantages of structured learning is that the training set has to be prepared to describe the problem at hand. Another disadvantage of structure learning is that if more cases are added to the training set, the ANN has again to be trained starting from scratch. As ANN training is time consuming, structured learning may be inadequate for problems that require continuous adaptation.

3.2. Continuous learning

In continuous learning, an ANN does not require a data set for training, the ANN learns by experience and is able to adapt progressively by incorporating knowledge gradually. Because some problems cannot be appropriately described by a data set, and because training using a data set can be time consuming, continuous learning is important for real-time computing (RTC) where there is a "real-time constraint".

3.3. Validation

After the ANN training has been completed, the network performance has to be validated. When using ANNs, the validation process is extremely important, as a matter of fact, the validation is as important as the training, see [7]. The purpose of the validation is to predict how well the ANN will behave in other conditions in the future. The validation process may be performed using a data set called the validation set. The validation set is similar to the training set but not equal. Under normal circumstances (when the ANN is properly used), the error obtained during training and during validation should be similar.

4. Simulated annealing

SA is an optimization method that can be used to solve a broad range of problems, [11]. SA is recommended for complex optimization problems. The algorithm begins at a specific temperature; as time passes the temperature gradually decreases following a cooling schedule as shown in Figure 5. The solution is typically described by a set of variables, but it can be described by other means. Once the algorithm has started, the solution approaches progressively the global minimum that presumably exists in a complex error surface, see [16] and [15]. Because of its great robustness, SA has been used in many fields including the training of ANNs with structured learning, [9].

One of the key features of SA is that it always provides a solution, even though the solution may not be optimal. For some optimization problems that cannot be easily modeled, SA may provide a practical option to solve them.

5. Simulated annealing evolution

Simulated annealing evolution includes the use of: ANNs, continuous learning and SA. In simulated annealing evolution, an ANN does not require a training set; instead the ANN

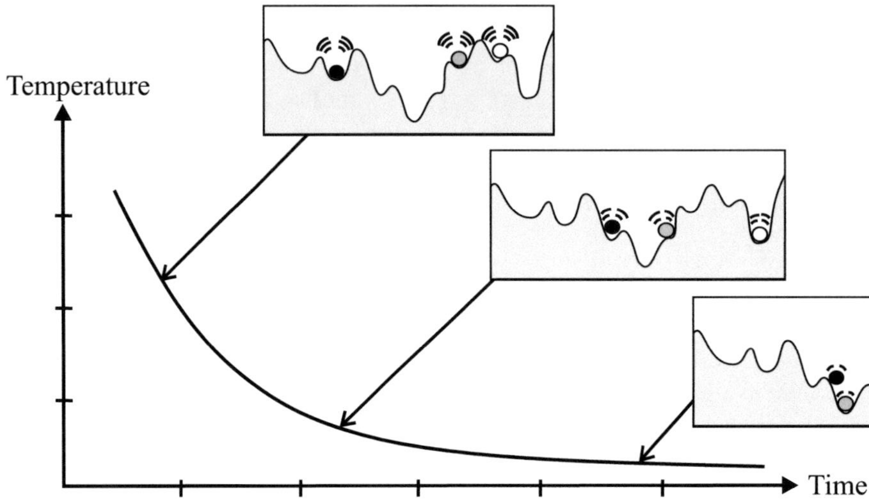

Figure 5. Cooling: three ANNs learning.

gradually learns new skills or improves existing ones by experience. Figure 5 shows how SA evolution works. In this figure, a typical cooling schedule used in SA is displayed. Suppose that there is a 2D landscape with valleys and hills as shown in this figure. Suppose also that it is desired to find the deepest valley on this landscape. Each of the balls, in this figure, represents an ANN. At the beginning of the simulation, the high initial temperature produces a state of high energy; the balls shake powerfully and are able to traverse easily through the high hills of the 2D terrain. In other words, each ANN is exploring, that is, the ANN is in the initial step of learning. As the temperature decreases, the energy of the balls decreases, and the movement of the balls is more restricted than at high temperatures, see [1]. Thus, as the temperature diminishes, the ANN has less freedom to explore new information as the network has to integrate new knowledge with the previous one. By the end of the cooling schedule, it is desired that one of the balls reached the deepest valley in the landscape, in other words, that one ANN learned a set of skills.

At each temperature, an ANN (in the set) has the chance to use its knowledge to perform a specific task. As the temperature decreases, each ANN has the chance to improve its skills. If the ANNs are required to incorporate new skills, temperature cycling can be used, see [6]. Specifically, an ANN may learn by a combination of SA and some sort of hands-on experience. Thus, simulated annealing evolution is the training of an ANN using SA without a training set.

Each ANN may be represented by a set of coefficients or weights. For illustrative purposes only, suppose that an ANN may be described by a set of two weights w_{11} and w_{12}. In Figure 6, there are three individuals, each is represented by a small circle with its two weights: w_{11} and w_{12}. At every temperature, each ANN is able to explore and use its abilities. At high temperatures, each ANN has limited skills and most of the ANNs will perform poorly at the required tasks. The gray big shadow circle in the figure indicates how the ANNs are considering different set of values w_{11} and w_{12} for testing and for learning. As temperature decreases, the ANNs get closer to each other, illustrating the fact that most of them have learned a similar set of skills. By the time the temperature has reached its final value, the skills of each ANN will be helpful to the degree that the network is able to perform the task at hand. At this moment, the simulation may end or the temperature may increase, if there are new skills that need to be incorporated.

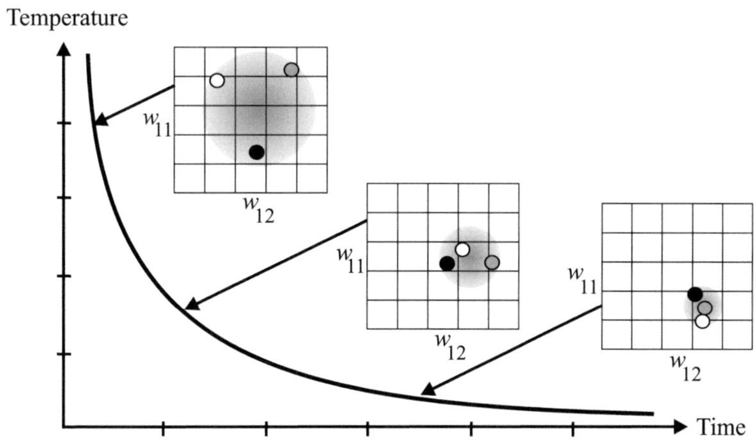

Figure 6. Each ANN tries to learn the same set of skills.

6. Problem description

To illustrate how to use simulated annealing evolution, this section presents a simple learning problem. The problem consists of using an ANN to drive a car in a simple road. Clearly, the problem has two objects: the road and the car. The road includes a *Start* point and a *Finish* point as shown in Figure 7. The road includes several straight lines and turning points. The car is initially placed at the *Start*. The driving is performed by an ANN that was integrated with the car. Specifically, the ANN manipulates indirectly several parameters in the car such as position, speed and direction. The purpose of the simulation is to train the ANN without a training set. At the beginning of the simulation, the ANN does not know how to drive; as the evolution continues the ANN learns and improves its skills being its goal to drive the car quickly from the *Start* to the *Finish* without hitting the bounds of the road (that is, the car must always stay inside the road). To solve this problem simulated annealing evolution was used to train the ANN.

The simulation was performed using object oriented programming (OOP); the respective UML diagrams for the simulation are shown in Figures 8 and 9. The two basis classes are shown in the diagram of Figure 8. This diagram includes two classes: Point and Bounds. The Point class in the diagram represents a point in a 2-Dimensional space, the diagram indicates that this structure includes only two floating point variables: x and y. The Bounds class, in the same UML diagram, is used to describe the bounds of an object when the object is at different positions and rotations angles. The main purpose of the Bounds class is to detect collisions when one object moves around other objects.

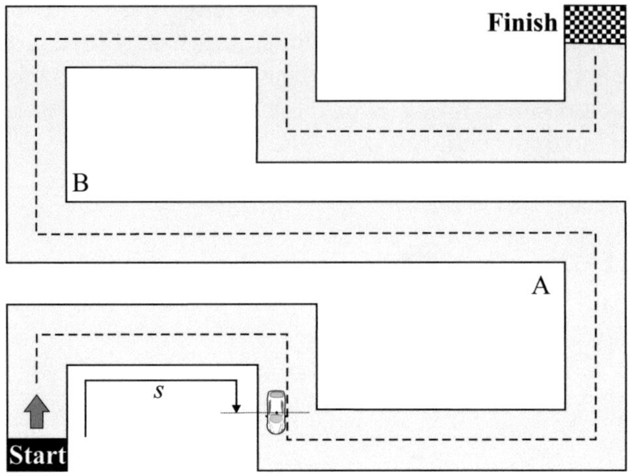

Figure 7. The car and the road.

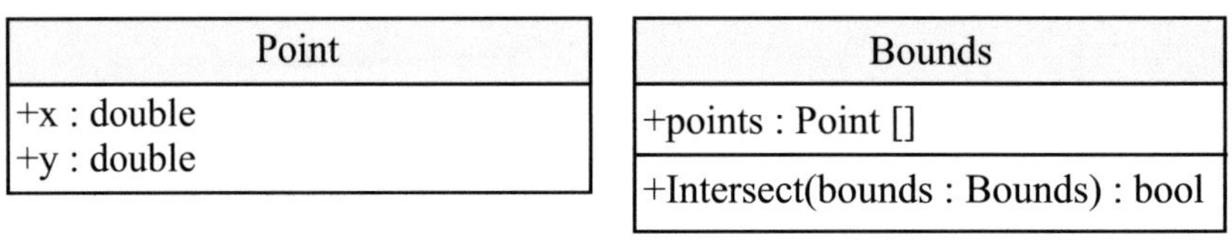

Figure 8. UML diagram showing the basic classes.

6.1. The object class

Figure 9 shows the respective UML diagram for the Object class. This class represents a static object in a 2-Dimensional space. The class name is displayed in italics indicating that this class is abstract. As it can be seen the *Render()* method is displayed in italics, and hence, it is a virtual method and must be implemented by the non-abstract derived classes. If the experiment includes some sort of visualization, the *Render()* method may be used to perform custom drawing. There are many computer technologies that can be used to perform drawing, some of them are: Microsoft DirectX, Microsoft GDI, Microsoft GDI+, Java Swing and OpenGL. From the UML diagram of Figure 9, it can be seen that each object has a position, a rotation angle (theta) and a set of bounds. The method **IsCollision()** takes another object of type Object and returns true when the bounds of one object collide with the bounds of another object. This method was implemented using a simple version of the algorithm presented by [14].

6.2. The mobile class

This class represents a moving object in a 2-Dimensional space. Each *Mobile* object has a speed, an acceleration and a rotation speed as shown in the UML diagram of Figure 9. This class is derived directly from the *Object* class, and therefore, is also abstract as the *Render()* method is not implemented. The **UpdateBounds()** method for this class takes the number of seconds at which the object bounds will be computed. This method is extremely useful when moving an object around other objects, for instance, if the bounds of one object intersect with

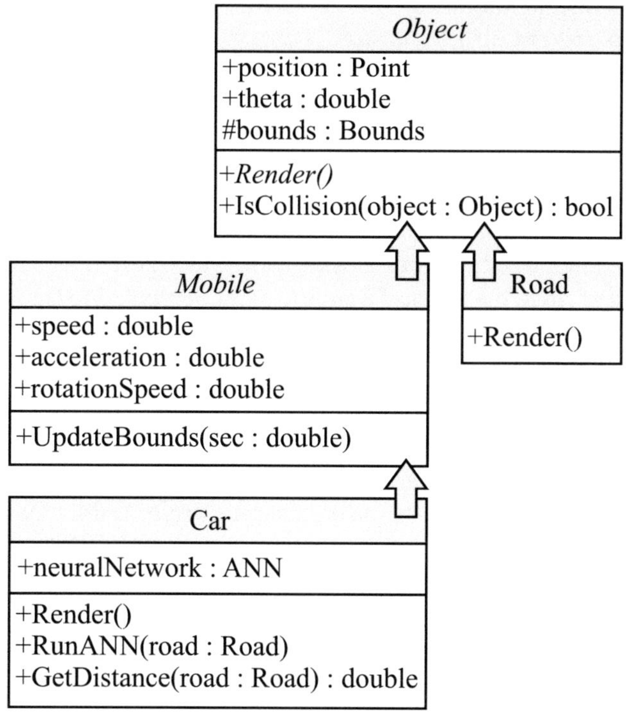

Figure 9. UML diagram showing the Car and the Road classes.

the bounds of another object, the object cannot move, this is implemented internally in the method **IsCollision()**. To update the bounds of a Mobile object, the speed of the object may be computed using Equation 3,

$$v_2 = v_1 + at, \tag{3}$$

where v_1 is the initial velocity of the object, v_2 is its final velocity, a is the acceleration, and t stands for the time for which the object moved. Similarly, the position of the object can be updated using another Kinematic equation. To compute the new position of the object Equation 4 can be used,

$$d = v_1 t + \frac{1}{2}at^2, \tag{4}$$

where the symbol d stands for the displacement of the object. In most cases, the object moves along a line described by its rotation angle and its position. Thus d has to be accordingly projected in the coordinate system as shown by Equations 5 and 6,

$$x_2 = x_1 + v_1 t \cos\theta + \frac{1}{2}at^2 \cos\theta, \tag{5}$$

$$y_2 = y_1 + v_1 t \sin\theta + \frac{1}{2}at^2 \sin\theta. \tag{6}$$

In some cases, the object may be rotating at a constant speed and, hence, the rotation angle has also to be updated at each period of time.

During the experiments, the methods **UpdateBounds()** and **IsCollision()** of the *Mobile* class are used together to prevent object collision. First, the simulator calls the **UpdateBounds()** method to compute the bounds of the object at some specific time, and then will call the **IsCollision()** method to check for potential collisions with other objects.

6.3. The road

The simulation experiments were performed using only two classes: the Road class and the Car class. The UML diagram of the Road class is shown in Figure 9. The Road class is derived directly from the *Object* class; the method **Render()** is implemented to draw the road displayed in Figure 7 using Microsoft GDI and Scalable Vector Graphics (SVG). When the car leaves the *Start*, the ANN has to make its first right turn at 90 $^\circ$, as the car is just accelerating this turn is easy. The following turn is also to the right at 90 $^\circ$ and the ANN should not have any trouble making this turn. Then, if the ANN wants to drive the car to point **A**, it has to make two turns to the left.

The straight segment from **A** to **B** should be easy to drive; unfortunately, the ANN may continually accelerate, and reach point **B** at a very high speed. The turn at point **B** is the most difficult of all the turns, because the car has to make a right turn at 90 $^\circ$ at high speed. Because the simulation is over when the car hits the bounds of the road, as soon as the ANN can see the turn of point **B**, it has to start reducing the speed of the car. Once the ANN has managed to drive to point **B**, reaching the *Finish* should not be difficult.

6.4. The car

The car used for the simulation is shown in Figure 10. The car has a position represented by x and y in Figure 10; the car rotation is represented by θ. The speed and acceleration vectors are represented by v and a respectively. The arrow next to the rotation speed in Figure 10 indicates that the car is capable of turning. The car has several variables to store its state (position, theta, speed, acceleration, rotationSpeed and neuralNetwork) as shown in the UML diagram of Figure 9. The Car class derives directly from the *Mobile* class and implements the method **Render()** to draw the car of Figure 10 using Microsoft GDI and SVG. When $v = 0$ and $a = 0$ the values of x and y do not change. When $v \neq 0$ and $a = 0$, the values of x and y will change while v remains constant. When $a \neq 0$, the values of x, y and v will change. The method **GetDistance()** computes the distance that the car has traveled along the road from the *Start*, this distance is represented by s in Figure 7.

Figure 11 illustrates how the car is able to receive information about its surroundings. The car had seven vision points illustrated by the arrows in the figure. To prevent the ANN from driving backwards, no vision lines were placed in the back of the car. Each value of d_1, d_2, ... d_7, represents the distance from the center of the car to the closest object in the direction of the vision line. These values were computed using the bounds of the road. To create a more interesting environment for the ANN, the values of d_1, d_2, ... d_7 were computed at low resolution and the car could not see objects located away from it.

In real life, a car driver is not able to modify directly the position or velocity of the car, the driver only controls the acceleration and the turning speed. As mention before, each car in the

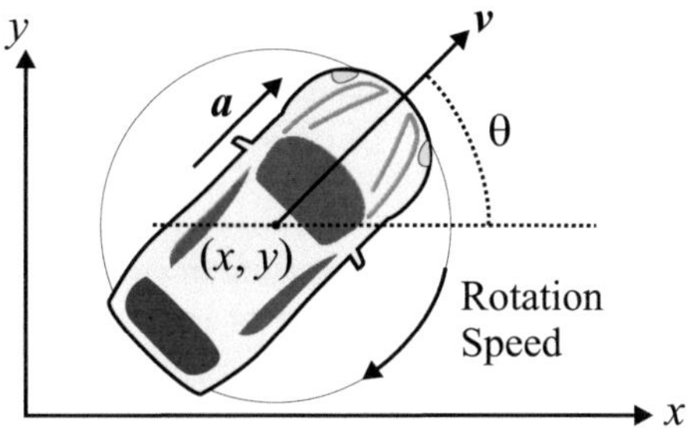

Figure 10. The car.

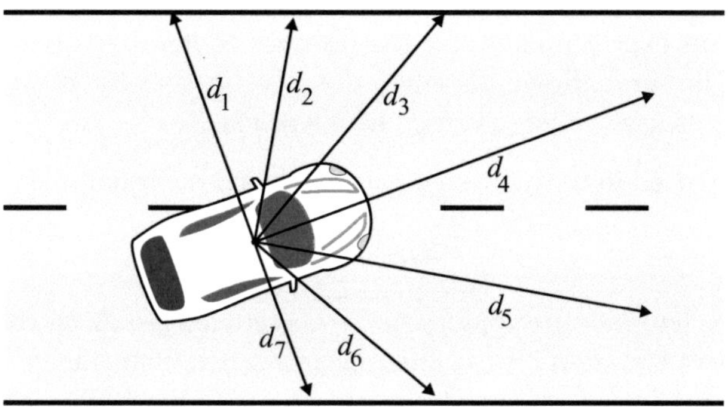

Figure 11. The car vision lines.

simulation has an ANN to do the driving. At each period of time, the ANN receives the vision information from the surroundings and computes the acceleration and the rotation speed of the car. Figure 12 shows the ANN of the car, the ANN has 8 inputs and two outputs. As it can be seen from this figure, the speed of the car is also applied to the input of the network; this is very important because the ANN needs to react differently depending on the current speed of the car. For instance, if the car moves a high speeds and faces a turn, it needs to appropriately reduce its speed before turning. As a matter of fact, the ANN needs to be ready for an unexpected turn and may regulate the speed of the car constantly.

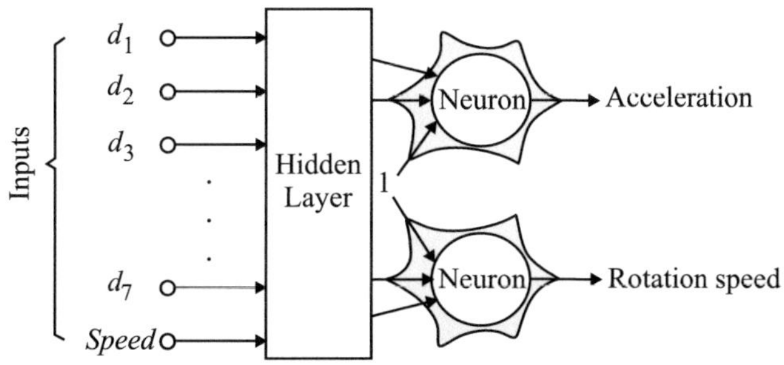

Figure 12. Artificial Neural Network for driving the car.

7. Experimental results

This section explains how SA was used to train the ANN. The implementation of SA was divided in three steps: initialization, perturbation and error computation.

The ANN training process using SA is illustrated in Figure 13. The simulation starts by randomly setting the network weights using a uniform probability distribution $U[-30, 30]$. In the second block, a copy of the weights is stored in a work variable. In the third block, the temperature is set to the initial temperature. For the simulation experiments, the initial temperature was set to 30. In the fourth block (iteration = 1), the optimization algorithm begins the first iteration. Then, the work variable (a set of weights) is perturbed. After the perturbation is completed, the ANN weights are set to these new weights. At this moment, the ANN is allowed to drive the car and the error is computed as shown in the figure. The temperature decreases exponentially and the number of iterations is updated as shown in the flow diagram. The simulation ends when the error reaches the desired goal or when the temperature reaches its final value (a value of 0.1 was used).

The cooling schedule used in the simulations is described by Equation 7,

$$T_{j+1} = cT_j, \tag{7}$$

where T_{j+1} is the next temperature value, T_j is the current temperature, and $0 < c < 1$. Clearly, the cooling schedule of Equation 7 is exponential and slower than a logarithmic one, therefore Simulated Quenching (SQ) is being used for the training of the ANN, see [3].

Observe, that each time the ANN weights are perturbed, the ANN is allowed to drive the car. Then, the error is computed and the oracle makes a decision about whether the new set of weights is accepted or rejected using the probability of acceptance of Equation 8, see [5] and [11]. Some implementations of SA accept a new solution only if the new solution is better than the old one, i.e. it accepts the solution only when the Error decreases; see [7] and [10]. The probability of acceptance is defined as

$$h(\Delta E) \approx exp(-\Delta E/T) \tag{8}$$

where

$$\Delta E = Error_{new\ solution} - Error_{currentsolution}.$$

A uniform probability distribution was used to generate states for subsequent consideration. At high temperatures, the algorithm may frequently accept an ANN (a set of weights) even if the ANN does not drive better than the previous one. During this phase, the algorithm explores in a very wide range looking for a global optimal ANN, and it is not concerned too much with the quality of the ANN. As the temperature decreases, the algorithm is more selective, and it accepts a new ANN only if its error is less than or very similar to the previous solution error following the decision made by the oracle.

7.1. SA initialization

Because of the properties of the activation function of Equation 2, the output of an ANN is limited. As mentioned before, an ANN is trained by adjusting the weights that connect the

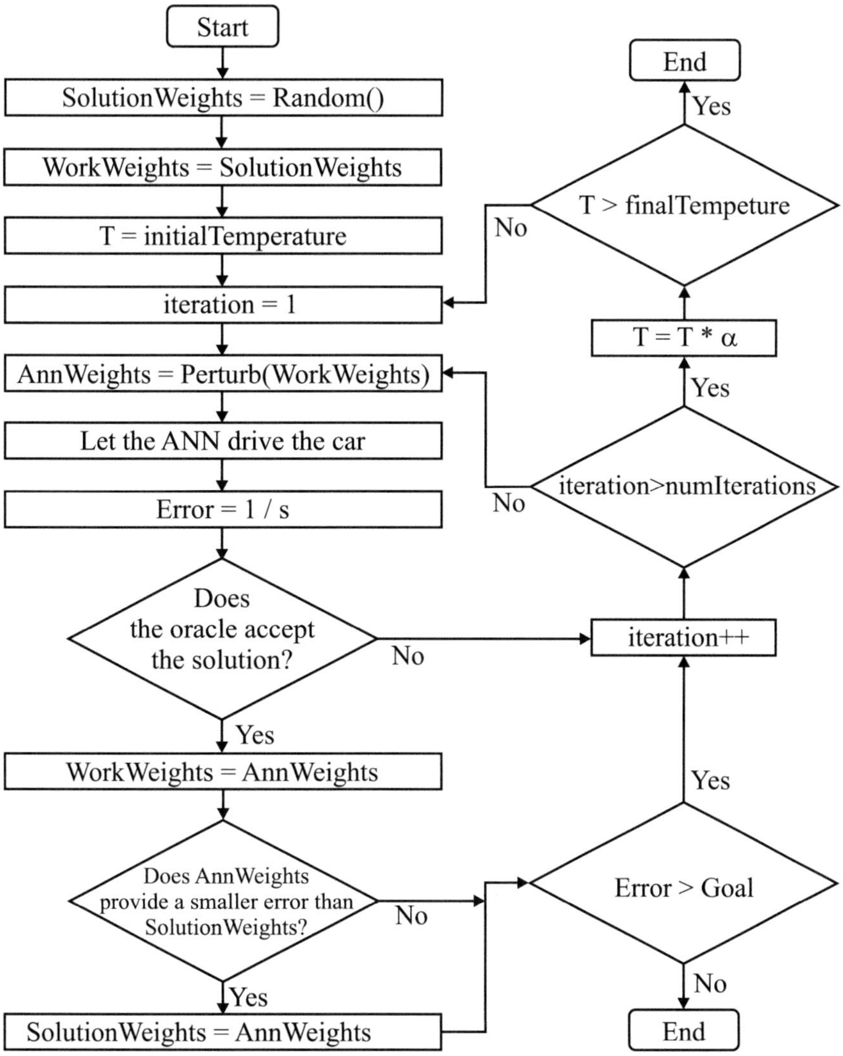

Figure 13. Flow diagram for simulated annealing evolution.

neurons. The training of an ANN can be simplified, if the input applied to the ANN is limited. Specifically, if the input values are limited from −1 to 1, then the ANN weights are limited to approximately from −30 to 30, [8] and [12]. To simplify the simulation, the input values of the ANN were scaled from −1 to 1. Therefore, the SA initialization consisted in simply assigning a random value from −30 to 30 using a uniform probability distribution to each of the ANN weights as shown in the C++ code shown in Figure 14. Observe that the random number generator uses the (*ISO/IEC TR 19769*) C++ Library Extensions **TR1**: *default_random_engine* and *uniform_real*. In this case, the ANN has two sets of weights: the hidden weights and the output weights. Each set of weights was stored in a matrix using the **vector** template from the Standard Template Library (STL); each matrix was built using a vector of vectors.

```
void Solution::SAInitialize()
{
   std::tr1::default_random_engine randGen;
   std::tr1::uniform_real<double> dist(-30.0, 30.0);
   //________________________ Hidden weights
   for(int i=0; i<hidRowCount; i++)
      for(int j=0; j<hidColCount; j++)
         hidWeight[i][j] = dist(randGen);
   //________________________ Output weights
   for(int i=0; i<outRowCount; i++)
      for(int j=0; j<outColCount; j++)
         outWeight[i][j] = dist(randGen);
}
```

Figure 14. Implementation of SA initialization using the C++ language.

7.2. SA perturbation

The code of Figure 15 shows the implementation of the SA perturbation using the C++ language. First, each ANN weight was perturbed by adding a random value from $-T$ to T using a uniform probability distribution (*tr1::uniform_real*), where T is the current temperature. Second, if the perturbed weight was outside the valid range from −30 to 30, the value was clipped to ensure that the weight remained inside the valid range.

7.3. SA error computation

In order to measure the driving performance of the ANN, an error function E was defined as shown in Equation 9,

$$E = \frac{1}{s}, \tag{9}$$

where the variable s represents the distance along the road measured from the *Start* to the current position of the car as shown in Figure 7. As it can be seen the value of the error decreases as the car drives along the road. The smallest error is accomplished when the car reaches the *Finish*.

The code of Figure 16 illustrates the implementation of the error function. The function starts by setting the ANN weights. The variable **deltaTimeSec** is used to refresh the simulation, a value of 16.7 milliseconds was used; it provides a refreshing frequency of 60 Hz (so that the simulation could be rendered on a computer display at 60 frames per second). Next, the function begins a **while** block, at each iteration the car bounds are updated and the simulation checks for a collision between the car and the road. If there is a collision the simulation stops and the error is computed. If there are not collisions, the ANN computes vision information and updates the acceleration and rotation speed of the car.

7.4. Results

Several experimental simulations were performed using different configurations to analyze the behavior of the ANN and the car.

```
void Solution::SimAnnealPerturb(Ann& preAnn, double temperature)
{
    std::tr1::default_random_engine randGen;
    std::tr1::uniform_real<double> dist(-temperature, temperature);
    //________________________ Hidden weights
    for(int i=0; i<hidRowCount; i++)
    {
        for(int j=0; j<hidColCount; j++)
        {
            hidWeight[i][j]= preAnn.hidWeight[i][j]+dist(randGen);
            if (hidWeight[i][j]>30.0) hidWeight[i][j]=30.0;
            if (hidWeight[i][j]<-30.0) hidWeight[i][j]=-30.0;
        }
    }
    //________________________ Out weights
    for(int i=0; i<outRowCount; i++)
    {
        for(int j=0; j<outColCount; j++)
        {
            outWeight[i][j]= preAnn.outWeight[i][j]+dist(randGen);
            if (outWeight[i][j]>30.0) outWeight[i][j]=30.0;
            if (outWeight[i][j]<-30.0) outWeight[i][j]=-30.0;
        }
    }
}
```

Figure 15. Implementation of SA perturbation using the C++ language.

```
double Solution::SimAnnealGetError()
{
    car.neuralNetwork.SetWeights(hidWeight, outWeight);
    const float deltaTimeSec = 0.0167f;
    float duration = 0.0f;

    while (duration<=300.0f) // 5 minutes max to travel the road
    {
        car.UpdateBounds(deltaTimeSec);
        if (car.IsCollision(road) == true) break;
        car.RunANN(road);
        duration+=deltaTimeSec;
    }
    return 1.0/car.GetDistance();
}
```

Figure 16. Implementation of SA error function the C++ language.

In the first simulation, the speed of the car was not applied at the input of the ANN, in all cases, the ANN was not able to turn at point **B**. At some unexpected point, the ANN was able to see the approaching turn of point **B** and did not have enough time to reduce the speed of the car.

In the second simulation, the number of neurons in the hidden layer was varied from 0 to 5. When the number of neurons in the hidden layer was zero, the ANN was able to drive the car to the **Finish** in 90% of the cases. When the number of neurons in the hidden layer was increased to one, the car was always able to get to the **Finish**. It was also observed that the ANN was driving faster when using more neurons in the hidden layer, thus, the car was getting to its destination quicker. When the number of neurons in the hidden layer was increased to 5, there were not any noticeable changes in the performance of the car than when the ANN had 4 neurons in this layer.

The third experiment consisted in varying the number of vision lines described in Figure 11. The number of vision lines was varied from 3 to 7. When using 3 vision lines, the ANN was able to reach 50% of the times to point **A**, 10% of the cases to point **B** and it was never able to get to the **Finish**. When the number of vision lines was set to 4, the ANN was able to drive the car to the **Finish** in 50% of the cases. When the number of vision lines was set to 5, 6 or 7, the ANN was always able to drive the car to the **Finish**. However, the ANN was always driving faster when using more vision lines.

The SA parameters were set to be compatible with the ANN weights. The initial temperature was 30, the final temperature was 0.1. Some experiments were performed by using lower final temperatures, but there were not any noticeable changes in the performance of the ANN. The number of temperatures was set to 10 using 20 iterations per temperature. Some tests were performed using more numbers of iterations, but there were not improvements. All the simulations were run using an exponential cooling schedule.

To validate the training of the ANN, another road similar to the shown in Figure 7 was built. In all cases, the ANN behaved similar in both roads: the road for training and the road for validation.

8. Conclusion

An ANN is a method inspired in biological processes. An ANN can learn from a training set. In this case, the problem has to be described by a training set. Unfortunately, some problems cannot be easily described by a data set. This chapter proposes the use of SA to train an ANN without a training set. We call this method simulated annealing evolution because, the ANN learns by experience during the simulation.

Simulated annealing evolution can be used to train an ANN in an ample set of cases. Because human beings learn by experience, simulated annealing evolution is similar to human learning.

An optimization problem was designed to illustrate how to use SA to train an ANN. The problem included a car and a road. An ANN was used to drive the car in a simple road. The road had several straight segments and turning points. The objective of the ANN was to drive the car from the **Start** to the **Finish** of the road. At the beginning of the simulation, the car was placed at the **Start** and the ANN weights were set to random values. Obviously, the ANN could not drive too far the car without hitting the bounds of the road, and stopping the simulation. By the time the temperature reached its final value, the ANN was able to drive successfully to the **Finish** of the road as it will be briefly described.

During the simulation, the car had a set of vision lines to compute the distance to the closest objects. The distance from each vision line (measured from the car to the closest object) was applied to the input of an ANN. It was noticed that the ANN performed much better when the speed of the car was also applied to the input of the ANN.

The number of neurons in the hidden layer of the ANN was varied during the simulations. It was observed that when the number of neurons in the hidden layer was increased, the ANN was able to reach quicker the **Finish**. It was observed also that when using 5 or more neurons in the hidden layer, the performance of the ANN did not improve. It was noticed, however, that when using zero neurons in the hidden layer, the ANN could not always drive the car to the **Finish**.

The car vision consisted in a set of lines. Experimental simulations were performed varying the number of vision lines form 3 to 7. The experimental results indicated that when 3 vision lines are used, the ANN does not have enough information and cannot drive successfully to the **Finish**. It was observed also that when the number of vision lines was increased, the driving of the ANN was smoother. Finally, it was noticed that when the number of vision lines is increased to 8 or more, the ANN did not improve its performance (meaning that there were not observable changes in its driving).

Author details

Ledesma Sergio, Jose Ruiz and Guadalupe Garcia
University of Guanajuato, Department of Computer Engineering, Salamanca, Mexico

9. References

[1] Bandyopadhyay S., Saha S., Maulik U. & Deb K. A Simulated Annealing-Based Multi-objective Optimization Algorithm: AMOSA, IEEE Transactions on Evolutionary Computation 2008;12(3) 269-283.

[2] Buckland M. AI Techniques for Game Programming, Ohio USA: Premier Press; 2002.

[3] Ingber L. Simulated annealing: Practice versus theory, Mathematical and Computer Modelling 1993;18(11) 29-57.

[4] Jones M. T. AI Application Programming, Massachusetts: Charles River Media; 2005.

[5] Jones M. T. Artificial Intelligence: A Systems Approach, Massachusetts: Infinity Science Press LLC; 2008.

[6] Ledesma S., Torres M., Hernandez D., Avina G. & Garcia G. Temperature Cycling on Simulated Annealing for Neural Network Learning, In: A. Gelbukh and A.F. Kuri Morales (Eds.): MICAI 2007, LNAI 4827: proceedings of the Mexican International Conference on Artificial Intelligence MICAI 2007, 4-10 November 2007, Aguascalientes, Mexico, Springer-Verlag Berlin Heidelberg 2007.

[7] Masters T. Practical Neural Network Recipes in C++, San Diego, USA: Academic Press Inc.; 1993.
[8] Masters T. Signal and Image Processing with Neural Networks, New York, USA: John Wiley & Sons Inc.; 1994.
[9] Masters T. Advanced Algorithms for Neural Networks, New York, USA: John Wiley & Sons Inc.; 1995.
[10] Metropolis N., Rosenbluth M. N., Rosenbluth A. H. & Teller E. Equation of State Calculations by Fast Computing Machines. Journal of Chemical Physics 1953;21(6) 1087-1092.
[11] Press W. H., Teukolsky S. A., Vetterling W. T., & Flannery B. P. Numerical Recipes in C++: The Art of Scientific Computing (Third Edition), Cambridge, New York, Melbourne, Madrid, Cape Town, Singapore and Sao Paulo: Cambridge University Press; 2007.
[12] Reed R. D. and Marks R. J. Neural Smithing, Cambridge, Massachusetts, USA: The MIT Press; 1999.
[13] Russel S. J. and Norvig P. Artificial Intelligence: A Modern Approach (3rd edition), Upper Saddle River, NJ USA: Prentice Hall; 2009.
[14] Shamos M. & Hoey D. Geometric intersection problems, 17th Annual Symposium on Foundations of Computer Science, October 1976, IEEE, Houston, Texas, USA; 1976.
[15] Smith K. I., Everson R. M., Fieldsend J. E., Murphy C., & Misra R. Dominance-Based Multi-objective Simulated Annealing, IEEE Transactions on Evolutionary Computation 2008;12(3) 323-342.
[16] Stefankovic D., Vempala S. & Vigoda E. Adaptive simulated annealing: A near-optimal connection between sampling and counting, Journal of the ACM 2009;56(3).

Chapter 11

Design of Analog Integrated Circuits Using Simulated Annealing/Quenching with Crossovers and Particle Swarm Optimization

Tiago Oliveira Weber and Wilhelmus A. M. Van Noije

Additional information is available at the end of the chapter

http://dx.doi.org/10.5772/50384

1. Introduction

Electronics have received great advance in manufacturing technology and in design automation in the last decades. Systems-on-chip (SoC) integrating complex mixed-signal devices with multi-million transistor circuits have been developed. This fast increase in complexity has been reflected on the need for more engineers and tools to increase productivity.

In the digital domain, the tools are comprehended inside a well-defined design flow that allows the designer to segment the problem in various abstraction levels and then solve it in a straightforward approach. There are industry standards for digital tools in this domain and new tools can be proposed to solve specific problems and be allocated properly in the flow. In analog domain, despite of the efforts in the development of design tools by the scientific community, there is still no well-defined flow due to the great complexity and the interrelation between the several phases of design. Therefore, there is a gap in the automation level of flows between digital and analog domain. A greater automation of the analog flow would allow smaller time-to-market for analog design and also for mixed-signal designs, as analog sections are usually the bottleneck in time required to design.

Analog circuits have an important role in most of modern ICs. They are used in the interface between real world signals (which are analog) and the digital world, covering applications from analog-to-digital and digital-to-analog converters to filters and radio frequency circuits. Therefore, productivity in the analog domain reflects on the IC industry as a whole.

The difficulties of automating analog designs result from the lack of clear separation of stages in the analog flow. Altought it is possible to describe the flow using some abstract steps, in practice the flow is strongly connected through feedback and low-level design tasks need to be considered even when the design is at a high-level stage. Therefore, a complete analog design is usually not a straightforward process, requiring comprehension of the whole process in order to perform a single task.

Other issue that makes the automation in the analog domain more complex than in the digital is the great number of degrees of freedom in the design variables, as well as the lack of clear interrelation among them. Also, analog variables are within an infinite range of values and the measurements performed need to take into account second and third-order effects. This means that the trade-offs that exist are not always evident and therefore very experienced designers are needed to have the correct insights during the sizing of analog blocks.

Furthermore, once the design is accomplished for a selected technology, new effort and time is required to design it for a new technology.

As changing technology usually means that the project has to start from scratch, an expert analog designer (or a team of designers) is required once again as a change of technology usually results in redesign from scratch. Despite all difficulties, analog design is typically accomplished by engineers with no more than a mathematical software, an electrical simulator, intuition and energy to perform several trials and errors, as no hand calculation delivers precise results in one attempt.

The use of CAD tools is a way to aid the designer to bridge the productivity gap between the analog and digital domains. Synthesis is optimization procedure which does not rely on an initial good solution. It has the objective of achieving the specifications without the requirement of an initial solution close to the global minimum provided by the designer.

Initial studies in circuit level synthesis were done in the 1980's and were mostly knowledge-based approaches [5], [10]. These approaches tried to capture the design knowledge required to perform some circuit implementation and implement it in algorithms. Optimization-based approaches started to be made in the late 1980's and are still an active area of research [6], [29], [27]. These approaches provide tools with greater flexibility and smaller overhead in comparison with knowledge-based approaches.

Optimization-based approaches are further classified by the way they evaluate a new solution. Simulation-based evaluation rather than equation-based evaluation was progressively being adopted in order to reduce the preparatory effort and to provide higher accuracy. This type of evaluation is benefited from the advances in computer power which allow faster simulations and therefore can return precise synthesis results within a tolerable amount of time.

This chapter will introduce the basics involved in using Simulated Annealing/Simulated Quenching as the optimization core to synthesize analog integrated circuits, as well as the use of this algorithm together with population-based algorithms in order to improve the effectiveness of the results. The use of multi-objective information to combine Simulated Quenching with Genetic Algorithms as well as the use of Simulated Quenching followed by Particle Swarm Optimization is performed to show the advantages of these techniques. Population-based algorithms are better in dealing with multi-objective problems as they can use pareto-dominance to compare solutions. Combining SA/SQ with population-based algorithms allow the use of SA/SQ qualities with the possibility of overcoming issues related to the use of aggregate objective functions (used to convert multi-objectives into one) . Multi-objective information can guide the algorithm when it is locked in a local minimum and also used combined with SA to explore the pareto-front. In this work, all synthesis were performed using the software *MATLAB* for processing and *HSPICE* for electrical simulations.

The chapter is divided as follows. In section 2, the definition of the analog design problem is presented. Section 3 presents a brief history of the analog synthesis approaches which used

Simulated Annealing. In section 4, the algorithms for SA/SQ and the combinations with GA and PSO are shown. Section 5 presents synthesis of two amplifiers and of a voltage reference using the developed algorithms. Finally, section 6 presents the conclusions.

2. Analog design overview

Analog design is a complex task which involves knowledge at different levels of abstraction and the use of one or more appropriate tools for each level. At the system level, the designer must select the architecture that best fits the analog system that will be developed. This level describes the lower levels simply as functional blocks, which have not yet direct connection with the electrical devices that will implement them. The circuit level is responsible for converting these functional blocks in electronic devices. Each high level block will have a circuit level. The topology and size of the devices in this level will define the specifications which can be used on the system level. The layout level is responsible for converting the dimension values from the circuit level to the mask designs which will later be used in the fabrication process. Figure 1 illustrates the abstraction levels involved in analog design.

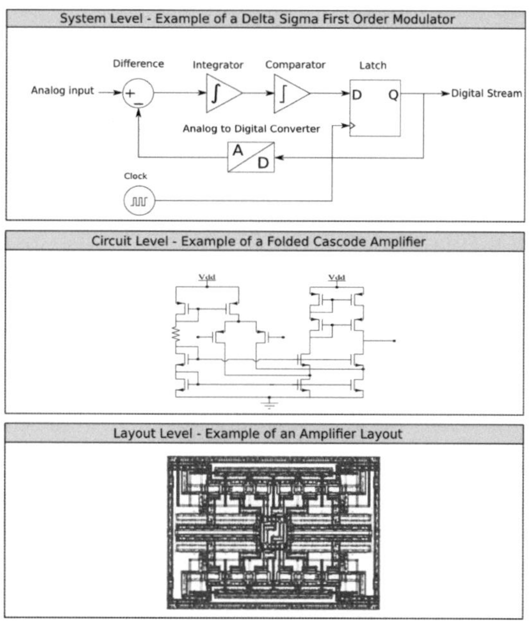

Figure 1. Different Levels (abstractions) of Analog Circuit Design.

The focus of this chapter will be restrained to analog design at circuit level, which is the problem of finding the best device dimensions (transistors, resistors, capacitors,...) and electrical variables (voltages and currents) in order to achieve the requirements for each block. In this perspective, the parameters used in the system level for each block are the specifications in the circuit level. The performance metrics of the circuit must be attended through the manipulation of the dimension variables of the devices.

This is a very challenging problem as it has multi-objectives, has several local minimums and the time required to evaluate each solution is often not negligible if a simulation-based approach is used. The objectives of an analog synthesis and the performance constraints are usually non-linear functions of circuit parameters. In [2] is presented a list of desirable features in an analog synthesis tool. This list includes small preparatory overhead, small synthesis time, starting point independence, accuracy, variation-tolerance, generality and fabrication process independence.

The preparatory overhead is the time spent before the optimization starts in which the designer converts the problem to the format understandable by the optimization tool. In equation-based evaluations, this overhead is usually high (weeks to months) and demands an experienced engineer to be performed properly. In simulation-based evaluations the overhead is reduced as the simulator will be the responsible for extracting the measurements from the proposed solutions. The designer usually only have to provide the circuits in a spice-like format.

Once all setup is ready, synthesis time is much faster in equation-based approaches. However, with the rapid increase in computer power the time required for simulation-based approaches is also being strongly reduced.

Starting point independence refers to being able to achieve good solutions without the requirement of an initial good solution provided by the designer. This is a fundamental requirement for characterizing a procedure as synthesis and justifies the use of global search techniques instead of using less complex local search techniques without capability of avoiding local minimums.

Accuracy is often a problem only on equation-based evaluation techniques, as the design model needs to capture all important effects on the devices and still provides equations manageable by the optimizer. Simulation-based approaches use transistor models like BSIM3 and BSIM4 which offer a good trade-off between accuracy and evaluation time.

Variation-tolerance is the capability to provide solutions that are not only good in the nominal device operation, but also considering the variations that might occur during fabrication process. Generality refers to being able to synthesize a wide range of circuit types. This is also a characteristic at which simulation-based approaches perform better as the only requirement for this approach is the simulator capability of processing that type of circuit.

Fabrication process independence is the possibility of changing the fabrication technology after the problem is formulated. This is a very important requirement as is not unlikely for a design center to change the fabrication process being used. Then, redesign of its intellectual property (IP) blocks become necessary.

As it can be seen by the desired characteristics, a synthesis tool for analog design needs to address most or all of them in order to become useful to the industry. A tool that requires too much time and effort to be properly used or one that does not returns robust solutions will not fit the designers' expectations.

In figure 2 the circuit level analog synthesizer can be seen as a black-box with its inputs and outputs. The designer provides the topology to be synthesized (e.g., folded cascode, miller amplifier,...), the foundry technology, the specifications (e.g., gain, area, power supply,...), the testbenches which will be used to make the measurements of the circuit and finally the synthesis configurations. The synthesizer interacts several times with the electrical simulator in order to evaluate the solutions that are proposed during the optimization procedure. The outputs are a dimensioned and biased circuit, the specifications achieved by this solution and graphics showing an approximation of its pareto front.

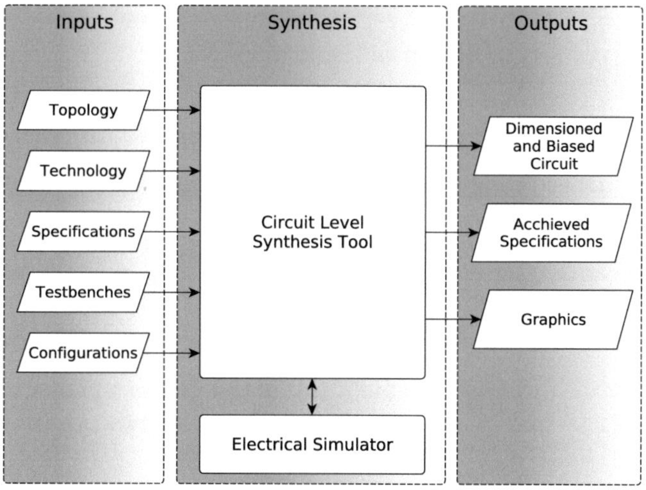

Figure 2. Inputs and Outputs of the Circuit Level Synthesis Tool.

3. Problem definition

Before we deal with the optimization of analog circuit designs, it is important to spend some effort in correctly defining the problem. An oversimplified definition, considering only the most superficial measurements would either take the solver to a non-practical, non-robust solution or lack enough hints during optimization to direct the solution to the designer specifications.

Analog design at circuit level typically have geometric variables and electrical variables. Geometric variables are the dimensions of the devices present inside a given block, such as the width and length of the transistor. They are important to determine the device characteristics such as the value of a resistor, the value of a capacitor, I/V curve of a transistor and others.

The influence of these geometric variables on the circuit measurements is usually non-linear. The effect of the geometric variables in a transistor current, which will affect the measurements, will depend not only in its dimensions but also on the voltages in its terminals. The equations of a MOS transistor can be seen in equation (1) for long-channel approximation (Schichman-Hodges model):

$$I_D = \begin{cases} 0, & \text{if in cut-off region} \\ \mu_n C_{OX}\dfrac{W}{L}[(V_{GS}\text{-}V_{TH})V_{DS}\text{-}\dfrac{V_{DS}^2}{2}], & \text{if in linear region} \\ \dfrac{\mu_n C_{OX} W}{2L}(V_{GS}\text{-}V_{TH})(1+\lambda V_{DS}), & \text{if in saturation region} \end{cases} \quad (1)$$

where I_D is the drain current, W is the width of the transistor channel, L is the length, μ_n is the electron mobility, C_{OX} is the gate oxide capacitance, V_{GS} is the gate to source voltage, V_{TH} is the threshold voltage, V_{DS} is the drain to source voltage and λ is the channel length modulation effect parameter. From these variables, W and L are the ones which the designer can manipulate while C_{OX}, μ_n and V_{TH} are technology dependent, and the voltages are consequence of the whole circuit equations (the designer will have indirect control upon these values by setting the correct currents in all transistors).

Electrical variables used by the synthesizer are voltages or currents that are usually in the interface of the block with its external environment, such as bias currents or voltages.

The choice of the variables is not arbitrary. As they will result in physical devices, there are constraints that need to be addressed in order the design to be feasible. These limitations are called design constraints [22]. They can be of two types: behavior (or functional) and geometric (or side) constraints. Behavior constraints are usually provided by the designer to define limitations that guarantee that the circuit will work properly, such as transistor operation regions or relations between different dimensions across transistors.

The operation region of a transistor is the result of the voltages between its terminals as it can be seen in eq. (2), (3) and (4). To get a transistor in a specific operation region is often a problem constraint as this will affect the operation and robustness of the circuit.

$$\text{cut-off} \quad V_{GS} < V_{TH} \tag{2}$$

$$\text{triode} \quad V_{DS} < V_{GS} - V_{TH} \;\; and \;\; V_{GS} > V_{TH} \tag{3}$$

$$\text{saturation} \quad V_{DS} \geq V_{GS} - V_{TH} \;\; and \;\; V_{GS} > V_{TH} \tag{4}$$

Geometric constraints are usually provided by the foundry which will fabricate the chip and refer to limitations which narrow the possible widths and lengths. The most common geometric constraint in analog design is the minimal channel length of the transistors.

3.1. Multi-objective problem

There are multiple objectives to address in analog synthesis. Each type of analog block has its type of specification and usually a great number of them are competitive. This means that usually improving one characteristic of the circuit decreases one or multiple others, resulting in the so called *design trade-offs*. In [23], an example of the trade-offs involved in the design of analog amplifiers is shown, such as power, gain, output swing, linearity and others.

On the other hand, optimization methods usually make use of only one cost function to measure the quality of a solution. However, on multi-objective problems one cost function for each objective is required. These cost functions can be later integrated into one through an Aggregate Objective Function (AOF) or optimized by a method that can accept multiple cost functions. This type of problem can be described by:

$$Min\ \mathbf{F}(\vec{x}) = \{f_1(\vec{x}), f_2(\vec{x}), ..., f_k(\vec{x})\} \tag{5}$$

Subject to:

$$g_i(\vec{x}) \leq 0 \quad i = 1, 2, ..., m \tag{6}$$

$$h_j(\vec{x}) = 0 \quad j = 1, 2, ..., p \tag{7}$$

where $\vec{x} = [x_1, x_2, ..., x_n]^T$ is a vector containing the decision variables, $f_i : \Re^n \longrightarrow \Re$, $i = 1, ..., k$ are all the objective functions, and $g_i, h_j : \Re^n \longrightarrow \Re$, $i = 1, ..., m$, $j = 1, ..., p$ are the constraint functions of the problem.

As there are many objectives, it is more complicated to compare different solutions than it is in single-objective problems. If one solution is better than the other in one objective but worse

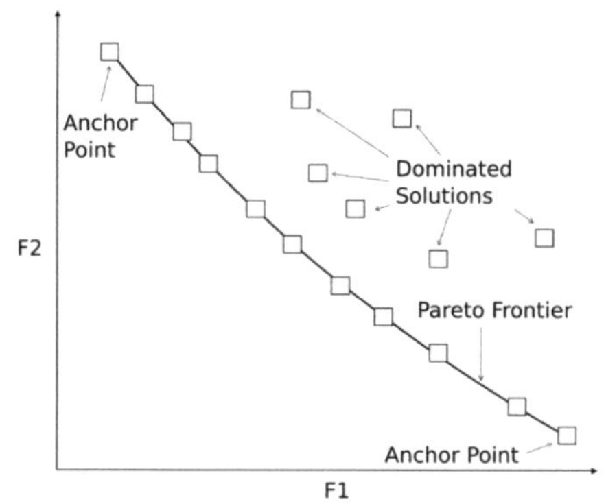

Figure 3. Example of a pareto front for a 2-dimensional design space.

in another, one can't choose which one is the better. However, there are cases in which one solution outperforms another in all specifications, which means one solution dominates the other.

The pareto set is the set of all non-dominated solutions. They represent all the trade-offs involved in the synthesis. Some definitions are used in [24] to formalize the concept of pareto front:

Definition 1: Given two vectors $\vec{x}, \vec{y} \in \Re^k$ we say that $\vec{x} \leq \vec{y}$ if $x_i \leq y_i$ for $i = 1, ..., k$, and that $\vec{x}$ dominates $\vec{y}$ if $\vec{x} \leq \vec{y}$ and $\vec{x} \neq \vec{y}$.

Definition 2: We say that a vector of decision variables $\vec{x} \in \chi \subset \Re^n$ is non-dominated with respect to χ, if there is not another $\vec{x}' \in \chi$ such that $\vec{f}(\vec{x}')$ that dominates $\vec{f}(\vec{x})$.

Definition 3: We say that a vector of decision variables $\vec{x}^* \in \mathcal{F} \subset \Re^n$ ($\mathcal{F}$ is the feasible region) is the Pareto-optimal if it is non-dominated with respect to $\mathcal{F}$.

Definition 4: The Pareto Set P^* is defined by:

$$P^* = \{\vec{x} \in \mathcal{F} | \vec{x} \ is \ Pareto - optimal\}$$

Definition 5: The Pareto Front PF^* is defined by:

$$PF^* = \{\vec{f}(\vec{x}) \in \Re^k | \vec{x} \in P^*\}$$

An illustration of a pareto front for a two-dimensional problem can be seen in figure 3. In this figure, both objectives are of minimization, which means a point close to (0,0) would be preferable if it was possible. The anchor points are the extreme optimal results of the specifications, while the other points in the pareto front are usually trade-offs between several specifications.

3.2. Continuous design space

Simulated Annealing was initially proposed to solve combinatorial problems which use discrete variables. Therefore, the generation of a new solution was accomplished by simply selecting one variable to modify and increasing or decreasing one step in the discrete space.

However, transistor, resistor and capacitor dimensions are usually better treated as continuous variables, as they can assume a large range of values. In problems with continuous variables, the minimum variation ΔX of a variable X becomes subjective and at the same time fundamental to the precision and speed of the annealing process. Choosing a too much small ΔX will result in an excessive number of iterations until the algorithm finds a minimum. On the other hand, a too much great value will decrease the precision in which the design space is explored and therefore it will be difficult to gradually find the best solution.

Another issue is that different types of variables may need different variations. The minimum change in resistor's resistance ΔR which will best satisfy the speed and precision criteria is different than the minimum change in transistor's channel length ΔW. The same type of variable in different types of technologies may also require the same distinction. For example, the ΔL in the channel length of a transistor in $0.18\mu m$ transistor which will be better for finding results is different from the ΔL that will achieve the same results for a $0.5\mu m$ transistor.

Other differentiation that needs to be made is based on the optimization phase. The optimization may be divided in an exploration phase and a refinement phase (which may be mixed). If a variable is far from its optimal final value, the exploration phase is more appropriate to test new values. When the variable has its value near the optimal final value, the refinement phase is better to approach the optimal solution. Modifications magnitude should be greater in the exploration phase than in the refinement phase.

Therefore, it becomes clear the need to change the variable values in an adaptive approach. The approach used for the simulations in this chapter is described in the Simulated Annealing section.

3.3. Generation of cost functions

The construction of a cost function for each specification is very important as it is a way to analyze how close the optimization is from the objective and also to compare which objectives should be prioritized in each stage of the optimization.

When approaching the problem through a single-objective optimizer, the several objectives are aggregated in only one using a weighted sum of each cost function as in eq. (8).

$$C = \sum_{i=1}^{n} c_i \cdot w_i \tag{8}$$

where C is the total cost, i is the objective index, c_i is a single-objective cost and w_i is the weight of the single-objective cost.

The optimizer will sense the variations on the total cost and decide if a solution must be kept or discarded. Therefore, the cost functions that compose this AOF must be designed in a way to have a greater Δc_i for results that are far from the specifications and therefore must be optimized earlier. This greater gradient for measurements distant from the objective (e.g. less than 70% of the specification in an objective maximization) helps the optimizer to develop the circuit as a whole.

Although several types of functions can be used to create the cost function, piecewise linear functions were used in this work in order to precisely control the regions in which a greater $\Delta c_i / \Delta m_i$ should be applied, where m_i is the measurement. The objective of each specification in analog design can be minimization, maximization or approximation:

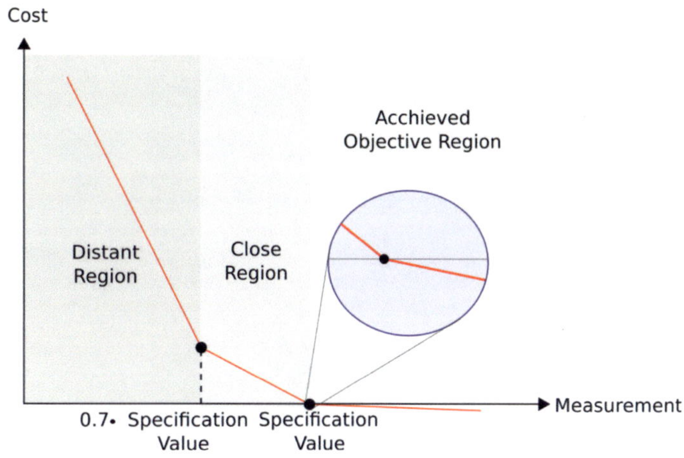

Figure 4. Cost function for a specification with objective maximization.

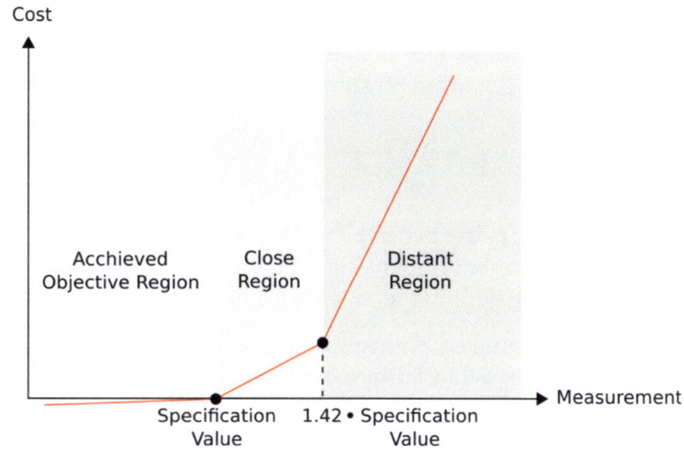

Figure 5. Cost function for a specification with objective minimization.

- **Maximization:** solution measurements must be equal or greater than the specification value. Examples are DC Gain, Cut-off frequency and Slew rate. Figure 4 shows how this work implements this type of function.
- **Minimization:** solution measurements must be equal or less than the specification value. Examples are area, power consumption and noise. Figure 5 shows how this work implements this type of function.
- **Approximation:** solution must approach the specification value. Examples are the central frequency of an voltage controlled oscillator and the voltage output of an voltage reference block. Figure 6 shows how this work implements this type of function.

An indicator containing the number of measures that resulted in failure is used to assign a high cost to these functions in order to provide a gradient even when dealing with non-working designs. Also, the designer can choose to put some or all transistors in a specific operation region. These and other constraints are added to the AOF in order to convert the problem from constrained to unconstrained.

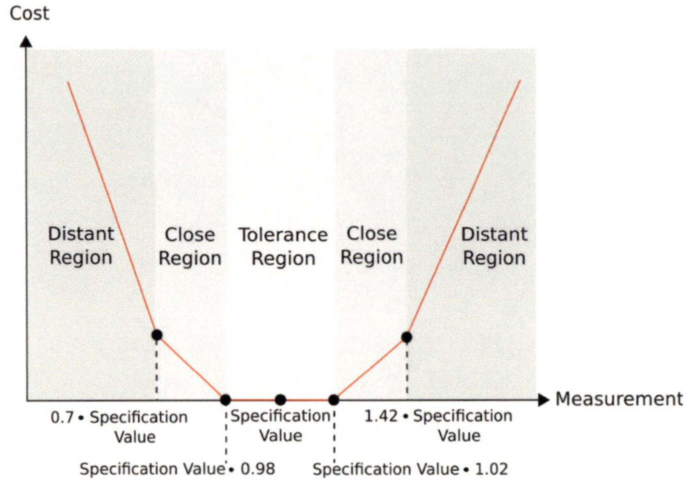

Figure 6. Cost function for a specification with an objective approximation.

Conversion of functional constraints into penalties is performed by adding new terms to the Aggregate Objective Function. Equation (9) shows this conversion:

$$C = \sum_{i=1}^{n} c_i \cdot w_i + \sum_{j=1}^{m} p_j \cdot w_j \tag{9}$$

where j is the penalty index, p_j is the penalty for each of the constraints and w_j is the weight of each penalty and m is the number of penalties. There is the choice of making the penalties more than simple guidelines by defining $w_i >> w_j$ for all objectives and penalties.

Other operation that helps Simulated Annealing to deal with analog design problem is the addition of a boundary check to avoid faulty simulations. The boundary check is performed every-time a new solution is proposed to avoid values out of the geometric constraints. These solutions usually result in simulation errors and take longer than typical ones. As they are irrelevant, the optimizer can skip these values and return to the state before the generation of the solution. The algorithm is kept in loop until a new solution is proposed within the boundaries and at each attempt to generate a wrong solution the sigma of the changed variable is reduced.

4. Simulated annealing approaches in analog design

Simulated Annealing was used in analog design since the early moments of analog synthesis was based on optimization approaches. Still now, it is one of the most popular algorithms to analog design however combined with other techniques.

In [9], a method for optimizing analog circuits using simulated annealing with symbolic simulation of the solutions is proposed. The optimization was accomplished through the use of a program called OPTIMAN and used as input design equations automatically derived using a tool called ISAAC [8]. This tool models the circuit in a set of analytic expressions which can be used by the optimization algorithm to fast evaluate each solution. Later, these tools were integrated into a synthesis environment called AMGIE [6] that covers the complete analog design flow, from specification to layout generation.

Simulated Annealing together with electrical simulations was given in [18]. Non-monotonic and adaptive normalized temperature schedule was used in order to provide fast decrease in temperature followed by reheatings. The temperature schedule could also change as a function of the percentage of accepted movements. This approach allows some level of independence of the optimization regarding the initial and final temperature values. The use of a technique called AWE (Asymptotic Waveform Evaluation) was proposed in the tool ASTRX [19] together with the use of SA. AWE could replace the SPICE simulations as it predicted the small signal behavior of the circuit using a reduced complexity model.

OPTOMEGA [13] performs a nominal optimization in order to search for the global optimal and then performs an optimization aiming robustness regarding process variations. The selection of one of the following optimization algorithms is allowed: Simplex, Simulated Annealing, Enhanced Simulated Annealing and SAPLEX (Simulated Annealing/simPLEX). This method uses Monte Carlo simulations to retrieve a value called Average Quality Index, which is used to evaluate the solutions.

MAELSTROM [15] uses Parallel Recombinative Simulated Annealing, which uses genetic algorithm characteristics with SA. A population of solution is created and at each generation the individuals can have their values changed through recombination with other individuals or through typical SA modification. An approach using SA in combination with evolutionary strategies (ES) was proposed in [1]. The Metropolis criterion from SA is used and the selection mechanism from ES. Recombination and mutation are used to adapt the length of the steps to generate the new solutions. The standard deviation based on which the new solution is made evolutes together with the solution.

In [26], a method to synthesize analog circuits oriented to robustness generating pareto surfaces is proposed. The motivation for generating pareto surfaces was to allow an efficient model of the trade-offs present on the circuit. The model can serve for a high-level synthesis tool as a way to fast evaluate low level blocks. The procedure consisted of a nominal optimization (combining simulated annealing with genetic algorithms) followed by a Monte Carlo approximation. The total time of synthesis considering nominal and process variation optimization takes only 5 to 6 times the nominal synthesis.

Although several other optimization methods can be used for analog problems (some of them can be seen in Table 1), the modern ones such as Simulated Annealing, Genetic Algorithms and Particle Swarm Optimization are specially adequate as they are robust and do not make assumptions about the problem characteristics.

5. SA/SQ algorithm and combination with population-based algorithms

In this section the basic Simulated Annealing/Simulated Quenching algorithm will be described, as well as the required modifications to work with continuous design space. Also, a version of SQ that uses Genetic Algorithm-based concepts and weight adjusting based on multi-objective cost function information. Finally, an approach interfacing SQ with Particle Swarm Optimization is explored.

5.1. SA/SQ algorithm

Simulated Annealing is a local search algorithm proposed in [14]. It has the ability to avoid local minimums by allowing worst solutions to be accepted given a certain probability.

Category	Methods	Published in
Classical Methods	Nonlinear Programming	[16]
	Geometric Programming	[21]
	Sequential Quadratic Programming	[28]
	Linear Programming	[4]
Modern Methods	Simulated Annealing	[14]
	Genetic Algorithms	[11]
	Particle Swarm Optimization	[12]
	Ant Colony Optimization	[3]

Table 1. Some optimization methods and their categories

The name Simulated Annealing is derived from the physical process called annealing in which the algorithm is based. This process is common in metallurgy and consists on heating a crystalline solid to a high temperature and then slowly cool it until its crystal lattice configuration is regular. Due to the high initial temperature, the atom movements follows a random pattern. As the temperature decreases, the movements are slowly reduced to only more regular structures yielding to a surface free of imperfections at the end of the process.

In the algorithm, the initial high temperature allows new generated solutions that are worst than the previous to be accepted. When the temperature decreases, the probability of accepting a worst solution is also decreased until only better solutions are allowed at the end of the process.

The first step of the algorithm is to evaluate the initial solution. Then, a loop as described below is performed.

- **Modification:** a small modification is made on the present solution in order to generate a candidate solution;
- **Evaluation:** the candidate solution is evaluated through the use of a cost function;
- **Decision:** based on the present temperature and on the cost difference between the present solution and the candidate solution, the probability of selecting a worst candidate solution is calculated and a random value is compared with it to make the decision;
- **Control parameter reduction:** the system temperature is decreased in order to reduce the probability of accepting a worst solution.

The stop criteria can be based on a final temperature value, number of iterations, final cost value, among others. There is a great number of variations of the simulated annealing algorithm [20] which aim to accelerate the convergence of the algorithm.

The new solution is selected according to the decision described by equation (10):

$$S_{k+1} = \begin{cases} N, & \text{if} \quad cost(N) < cost(S) \quad \text{or} \quad \tau \leq P(t_k, N, S_k) \\ S_k, & \text{otherwise} \end{cases} \tag{10}$$

where k is the solution index, N is the candidate solution, S_{k+1} is the next solution, τ is an uniform distributed random value within $[0,1]$ and $P(t_k, R, S_k)$ is the probability of accepting a worst solution given a certain temperature t^k and cost difference.

The probability of acceptance of a worst solution, also called acceptance criterion, used in this work is the Metropolis Criterion which can be seen in equation (11):

$$P(t^k, N, S^k) = e^{-\frac{Cost(S^k) - Cost(N)}{t^k}} \tag{11}$$

Theoretical proof of convergence to the global minimum for simulated annealing only exists when using a logarithmic temperature schedule. In order to acchieve solutions faster, new cooling schedules that do not have the proof of convergence derived from the original SA are used in many applications. Despite the lack of proof, these algorithms have the same structure as a simulated annealing algorithm and are usually better for the subset of problems to which they were created. Simulated Quenching is an approach to simulated annealing that allows a quicker cooling schedule and further increases in the temperature. In this work, a Simulated Quenching technique was used.

The control parameter reductions used in this work followed a simple exponential cooling scheme $t_{k+1} = \alpha t_k$, where α is a value close but smaller than one. The only variation used is when the detection of a local minimum occurs, in which the temperature is set as $t_{k+1} = \frac{t_0 + t_k}{2}$ where t_0 is the initial temperature.

5.2. Modification of variables

As mentioned in section 3.2, modification of variables on a continuous design space requires special efforts in relation to classical simulated annealing which deals with discrete variables. In this work we used an auxiliary variable attached to each optimization variable defining the direction and standard deviation of the step using a folded normal distribution. This auxiliary variable defining the step is updated after a new solution is generated based on how well that solution performed based on the previous one. Therefore, an adaptive step for each variable is implemented.

If the generated solution based on the provided step yielded a great improvement of the solution, the standard deviation is kept. Otherwise, if the generated solution resulted on a worse result, the standard deviation of that variable is decreased. At last, if the modification offered only a small benefit, the standard deviation is increased. The idea behind this standard deviation modification is that greater change probability must be given to variables that offer only reduced effect on the circuit when perturbed.

The equation for setting the value for the new variable is shown in equation (12):

$$X_i = X_i \cdot (1 + |f(0,1)| \cdot aux_i) \tag{12}$$

where X_i is the variable being modified, aux_i is the auxiliary variable that combined with X_i determines the standard deviation associated with the variable, $f(0,1)$ is a normal distributed random variable with zero-mean and unitary standard deviation (which will be modified by the auxiliary variable). The standard deviation σ_i will determine the probability of new solutions to be close to the previous one. This standard deviation will be equal to $X_i \cdot aux_i\ X_i$. If the auxiliary variable is smaller than 0.1, all new values within one sigma from the previous value for that variable will be less than 10% different of the previous. Therefore, a shift in the auxiliary variable sign and an increase in its values is performed when it acchieves such a small magnitude. This allows the algorithm to know in which direction a variable must be modified and update that at each iteration in which the variable is modified.

5.3. SA with genetic algorithm

Several algorithms that integrate Simulated Annealing with Genetic Algorithms have been proposed in the optimization field and in the analog synthesis field. These integrations usually apply the acceptance criteria and movements from simulated annealing together with the population-based nature of genetic algorithms. They keep several individuals performing SA and then, at some point, allow them to perform crossover among themselves.

On the proposed algorithm only one individual is used (as in general simulated annealing). The crossovers are made between this individual and the memory-based past solutions. The algorithm uses these past solutions to perform crossover with the present one in order to avoid local minimums when these are detected. This is possible due the existence of multi-objective information despite the fact SA is operating based on an Aggregate Objective Function (AOF). These informations are used to save the solutions that are better than all other in a single objective during the optimization. When using an AOF, the objectives in which the solution is performing worse usually are responsible for the local minimum. When the local minimum is achieved, the crossover together with a weight adjustment helps the algorithm to escape from the minimum and attempt new solutions.

Each of the objectives will have one saved solution associated to it. At each new evaluation, the present solution is compared with the saved ones in order to update the current best in each objective. Figure 7 shows a flow diagram of the algorithm and the most important steps are described above:

- **Saving the Anchor Solutions:** Choosing which solutions to save is an important step since they are going to be the responsible for taking the present solution away from the minimums. Anchor solutions are the ones that are responsible for the extreme points in a Pareto front. They represent the best solution already achieved for each given specification. Selecting directly these anchor points to be kept in the memory is not interesting for the flow of Simulated Annealing. A solution can be extremely good in I_{supply}, for instance, but perform bad in all other specification. Crossover between the present solution and such an anchor solution would result most likely in a non functioning solution. A more interesting approach is to save solutions that are not too far from the overall cost function achieved by the present solution. The saved solution must be better than the present solution in one specific objective but cannot be much worst in the overall AOF. This trade-off can be accomplished through a weighted sum of the cost for the specification and the overall cost.
- **Detecting Local Minimums:** detection of local minimums is based on the number of consecutive failures. If the SA algorithm is not able to accept a solution within a given number of attempts, it is assumed that a local minimum was achieved. As analog synthesis is a continuous problem, sometimes very small improvements are accepted however they

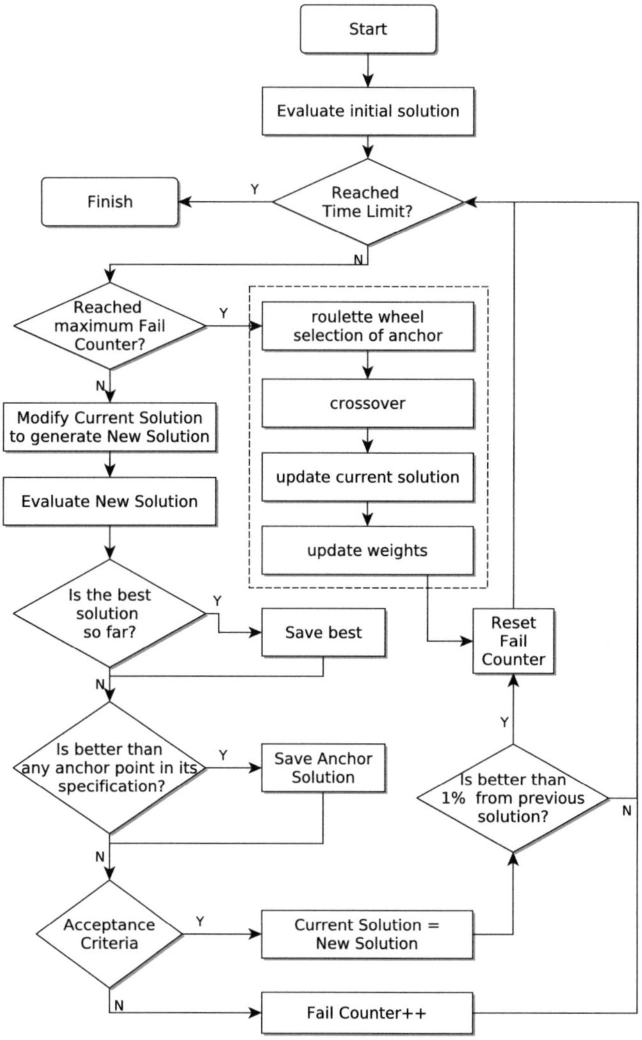

Figure 7. Algorithm flow for the developed SA + GA integration.

do not represent an actual significant improvement in the circuit. Therefore, it is important to reset the failure counter only when an improvement greater than 1% is accomplished.

- **Selection of anchor points to perform Crossover:** the selection of which of the previous solutions will be mixed with the present one is based on how far each of the objectives is from being achieved. The worst present objectives will grant more chance to their respective anchor solutions to be selected to crossover with the present solution. This is accomplished through a roulette selection among the anchor points.
- **Crossover Operator:** the crossovers are the responsible for taking a solution out of a local minimum. This process produces a child from two parents, which in this case are the present solution and the selected anchor solution. It is expected that the crossover between these solutions will create a child that has characteristics of the anchor (which will take the solution away from the local minimum) as well as some characteristics of the present

solution. The crossover operator used in this work is based on conditional probability and can be seen in eq. (13) and eq. (14):

$$P(g_i = present | g_{i-1} = present) = 0.8 \tag{13}$$

$$P(g_i = anchor | g_{i-1} = anchor) = 0.6 \tag{14}$$

where g is the gene, i is the gene index, *present* is a gene from the present solution and *anchor* is a gene from the anchor solution.

- **Weight Adjusting:** change the weights prevents Simulated Annealing from going to the same local minimum after the crossover operator was performed. This is performed by increasing the weight of the selected specification after it was sorted on the roulette wheel. It is important to keep the same scale when comparing solutions collected with different weights for the specifications, therefore the weight modifications are added only as penalties in the AOF, keeping the first terms unaltered. This allows the comparison of only the first preserved terms in order to keep the same criteria.

  Every time a local minimum is found a vector sp receives a new value with specification index sorted on the roulette wheel. Equation (15) shows how the weight adjustment affects the cost after one or more local minimums are found.

$$C_A(x) = C(x) + k \sum_{i=max(1,ms-n)}^{ms} w_{sp_i} f_{sp_i} \tag{15}$$

  where $C_A(x)$ is the total cost, k is a factor used to increase the weight of the adjustments on the total cost, ms is the index of the last value of the sp vector,n is the maximum number of previous local minimums are going to be considered for the adjustment, $C(x)$ is the original total cost, w_{sp_i} is the weight of specification saved on sp_i and f_{sp_i} is the cost value of the specification saved on sp_i.

5.4. SA with particle swarm optimization

Particle Swarm Optimization is a population-based algorithm that simulates the social behavior of birds within a flock. Its usage as an optimization algorithm was introduced by [12]. The individuals of the population are called particles and they interact with each other and the design space in search for the global minimum. The analogy of a flock of birds is the exchange of information between the particles that is converted to factors that influence their flight over the design space.

Each particle update their speed based on the following factors: inertia, experience of the particle and experience of the group. Equation (16) shows how these factors influence the speed.

$$\vec{v}_i(t) = \underbrace{w\vec{v}_i(t-1)}_{\text{inertia factor}} + \underbrace{c_1 r_1(\vec{x}_{lbesti} - \vec{x}_i(t))}_{\text{best local factor}} + \underbrace{c_2 r_2(\vec{x}_{gbesti} - \vec{x}_i(t))}_{\text{best global factor}} \tag{16}$$

where w, c_1 and c_2 are constants that defines the influence of each factor on the final speed, $r1$ and $r2$ are uniform random variables from 0 to 1, $\vec{v}_i$ is the velocity vector, t is the time (or iteration number), $\vec{x}_{lbest}$ is the local best result and $\vec{x}_{gbest}$ is the global or neighborhood best

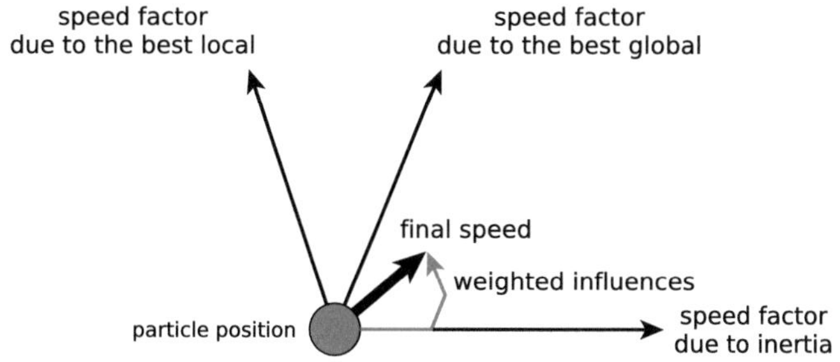

Figure 8. Influence of factors on the particle speed in PSO.

result. An illustration of the influence of the factors in the final speed of the particle can be seen in figure 8.

The connection topology among the particles define how a neighborhood is formed. In the simulations performed in the experiment section, the fully connected neighborhood topology was selected. In this case, all particles are in contact with the whole group so the best neighborhood solution is actually the best global solution.

In [24], a survey of multi-objective particle swarm optimization implementations is made. To treat multi-objective problems, the PSO algorithm must create and update a list of pareto solutions already visited by each particle and by the whole flock. The main goal is to maximize the number of well spread solutions in the list of pareto solutions of the flock in order to have a good approximation of the real pareto front.

For a new solution to enter on the list it must be non-dominated by any other solution already in the list. Also, at each iteration the old solutions on the list have to be checked to see if they were not dominated by new solutions. The selection of the local leader from the list of local paretos and the selection of the global leader from the list of global paretos is based on the density of solutions on the list. The leader is the one that has less close solutions already on the list. This allows the optimization to search for new solutions rather than to have a fast convergence.

Directly using PSO as circuit synthesizer in a problem with many objectives might take a long time and demand a lot of memory to keep the lists updated. Also, the required number of particles necessary to achieve a good result is very high if the particles are randomly initialized. The use of PSO for analog circuit synthesis is shown in [7] in an equation-based evaluation approach for a mono-objective and for a case with two objectives.

The proposed combination of SA and PSO tries to combine the best features of each algorithm. To accomplish that, the search for the solutions is divided in three phases. Figure 9 illustrates how this different phases are connected.

- Phase 1: Simulated annealing for multi-objective optimization using aggregate objective function. As all objectives are combined into one through weighted sum, this is a straightforward phase ideal for simulated annealing. The first objective of this phase is to find a solution that cope with the specifications and weights defined by the designer. Once this solution is found it will set all weights to the same value and start searching for a global minimum. The seed (initial solution) of this phase does not need to be good,

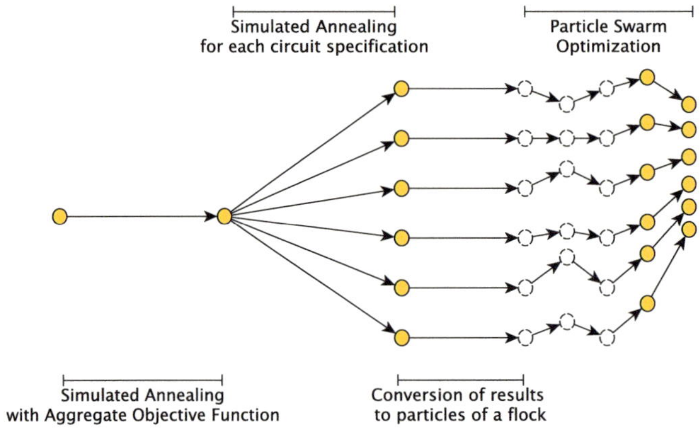

Figure 9. Connection between different optimization phases and methods.

although the better it is the faster this phase will end. On the example illustrated in the results section, the seed was a design with all transistor sizes set to the minimum allowed by the technology.

- Phase 2: Simulated annealing for single-objective optimization for each specification. This phase will take the best achieved circuit from phase 1 and start optimizing it for each of the specifications. All weights of the aggregate objective function are set to zero except the one of the specification being optimized. Actually, to avoid solutions that are out of the minimum specifications, only the section of the piece-wise linear function that represents the achieved specification region is set to zero. The other regions are kept for all specifications. Therefore, if a solution is outside the minimum specifications it is quickly discarded.
- Phase 3: Particle Swarm Optimization for multi-objective optimization. This phase uses the outputs of phase 2 as particles in a PSO. As each particle is already representative of an interesting part of the design space, the number of iterations to achieve an useful representation of the pareto front is reduced. The end flag for this phase is set when a sufficiently number (defined by the designer) of values from the pareto front is on the list of best globals or after a timeout.

After all phases are ended the designer still needs to have a feedback about the pareto front. However, as there might be more than two specifications, the pareto front becomes impossible to be plotted. In this work, one alternative to solve this problem was to allow the designer to select two specifications as fixed axis and plot multiple graphs of the pareto front using the rest of the specifications as the third axis of each graph. Before plotting the pareto front for each graph, a filter procedure is done to select from the n-dimensional pareto solutions which ones continue to be non-dominated if we consider only a the new 3-dimensional objective.

6. Experiments and results

This section will use the presented algorithms to synthesize practical circuits. Two different types of tests were performed. First, a comparison between standard Simulated Annealing and the modified Simulated Annealing allowing crossovers among anchor points and weight

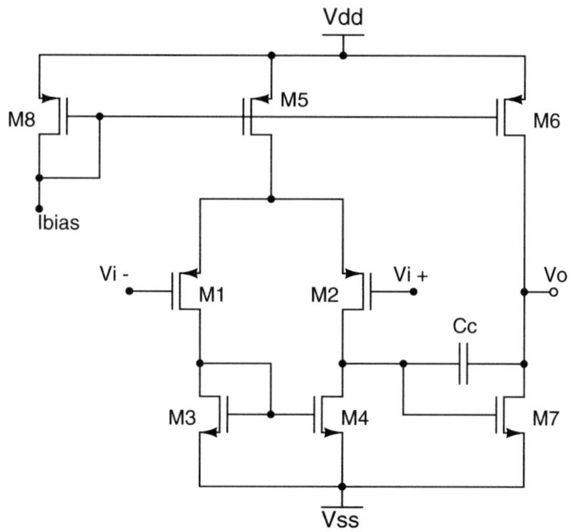

Figure 10. Miller amplifier topology

adjusting will be performed for two topologies. Then, a pareto front exploration using Particle Swarm Optimization together with SA will show the performance of the algorithm to illustrate the trade-offs in circuit synthesis. The computer used for these tests is a Intel(R) Core(TM)2 CPU with a 1.8 GHz frequency and with a RAM memory of 3GB.

6.1. Comparison between SA and modified SA with crossovers

The aim of comparing SA and Modified SA with crossovers and weight adjusting is to analyze the benefit of crossovers among anchor points in the convergence speed to a global optimum. The benchmark for this comparison is the synthesis of two amplifiers within a limited amount of time. Therefore, the optimization algorithms will not be able to continue the synthesis until their own stop criteria are met. The time limit is set to 20 minutes and then the best cost found during optimization is saved.

As the algorithms use random variables to achieve their results, several synthesis procedures are required to state a valid comparison between them. In this work, 10 synthesis for each topology in both algorithms were performed and then statistical metrics were used to compare both techniques. The metrics used are mean, standard deviation, median and also the minimal and maximum cost value for all syntheses.

The topologies used for this test are a miller amplifier (figure 10) and a complementary folded cascode amplifier (figure 11). The miller amplifier uses a voltage supply of +1.65 and -1.65 and a load capacitance of 10 pF while the complementary folded cascode uses a voltage supply of +1.8 and -1.8 and a load capacitance of 20 pF. All the syntheses were performed using the AMS 0.35 μm technology.

The specifications for the miller amplifier as well as the best results of all 10 synthesis found for each of the techniques can be seen in Table 2. Both techniques could achieve most of the minimal specifications within the 20-minute synthesis for at least one synthesis row. However, the statistical measurements taking into account all 10 syntheses in Table 3 show

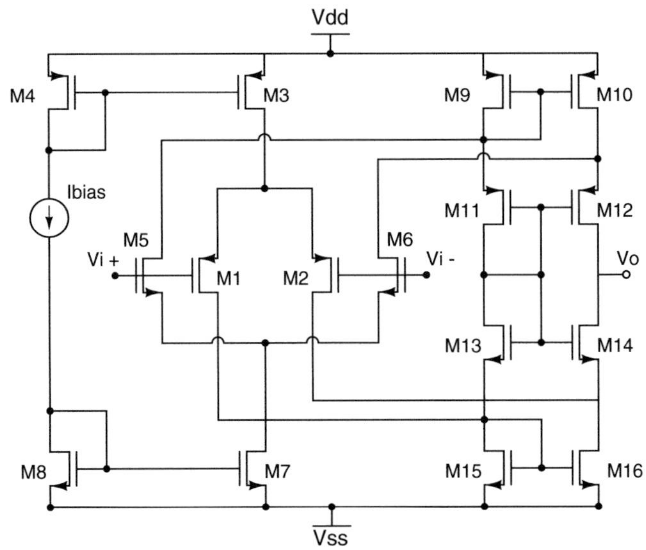

Figure 11. Complementary Folded Cascode amplifier

Measurement	Specification	SA Result	SA with Crossovers Result
UGF	> 15 M [Hz]	15.3 M [Hz]	15 M [Hz]
I_{supply}	< 300 μ[A]	314 μ[A]	328 μ[A]
Phase Margin	> 60 [°]	53 [°]	60 [°]
DC Gain	> 80 dB[V]	82.6 dB[V]	86.0 dB[V]
Slew Rate (pos)	> 20 M [V/s]	20.0 M [V/s]	21.8 M [V/s]
Slew Rate (neg)	< -20 M [V/s]	-26 M [V/s]	-25.5 M [V/s]
ICMR	> 1.5 [V]	2.9 [V]	2.9 [V]
Output Swing	> 1.5 [V]	2.9 [V]	3.0 [V]
CMRR	> 70 dB[V]	64.2 dB[V]	70.7 dB[V]
PSRR	> 70 dB[V]	63.5 db[V]	70.0 dB[V]
Gate Area	< 1000 [μm]^2	307.4 [μm]^2	557.7 [μm]^2

Table 2. Specifications of the Miller amplifier and results of the best 20-minute bounded synthesis. The gray background indicates an achieved specification.

that the modified simulated annealing with crossovers performs closer to the best solutions in mean. The classical SA/SQ algorithm had solutions that were not close to acchieve all specifications and therefore had a higher final cost mean (19.9 for SA/SQ and 1.87 for SA/SQ with crossovers), representing an indicator that the crossovers can effectively improve the results. Also, it can be seen that the maximum cost result (worst result) of the classical SA/SQ is much higher than the proposed algorithm (72 for SA/SQ and 16.5 for SA/SQ with crossovers).

The same analysis was performed for the complementary folded cascode with specifications and results shown in Table 4. Similarly to the miller amplifier case, the best result achieved by each algorithm in the 10 syntheses for the complementary folded cascode are close in cost (-0.01 for SA -0.02 for SA with crossovers). However, the results from the proposed algorithm achieved a mean closer to these best results. The SA/SQ algorithm had several results far from

Measurement	Standard SA	SA with Crossovers
Mean	19.9	1.87
Minimum	0.20	0.0006
Maximum	72.0	16.5
Standard Deviation	29.6	5.2

Table 3. Statistical measurements of cost results from equation (9) for 10 syntheses limited by 20 minutes of a Miller Amplifier in 0.35 μm technology.

Measurement	Specification	SA Result	SA+GA Result
UGF	> 10 M[Hz]	10.3 M[Hz]	10.1 M[Hz]
I_{supply}	< 1 m[A]	0.84 m[A]	0.92 m[A]
Phase Margin	> 90 [°]	87.9 [°]	92.45 [°]
DC Gain	> 50 dB[V]	60.9 dB[V]	56.2 dB[V]
Slew Rate (pos)	> 13 M [V/s]	16.4 M [V/s]	16.6 M [V/s]
Slew Rate (neg)	< -13 M [V/s]	-27.1 M [V/s]	-27.7 M [V/s]
ICMR	> 1.5 [V]	2.5	2.9
Output Swing	> 1.5 [V]	1.5 [V]	1.8 [V]
CMRR	> 80 dB[V]	91.1 dB[V]	84.56 dB[V]
PSRR	> 80 dB[V]	56.5 db[V]	84.8 dB[V]
Gate Area	< 1000 $[\mu m]^2$	210.0 $[\mu m]^2$	527.0$[\mu m]^2$

Table 4. Specifications of the Complementary Folded Cascode amplifier and results of the best 20-minute bounded synthesis.

Measurement	Standard SA	SA with Crossovers
Mean	26.7	3.55
Minimum	-0.01	-0.02
Maximum	93.2	29.7
Standard Deviation	43.2	9.4

Table 5. Statistical measurements of cost result from equation (9) for 10 syntheses limited by 20 minutes of a Complementary Folded Cascode Amplifier in 0.35 μm technology.

achieving the specifications due to local minimuns and the lack of efficient ways of escaping them. This resulted in a difference of the cost mean between the techniques: 26.7 for SA/SQ and 3.55 for SA/SQ with crossovers (results are summarized in Table 5). It is important to perceive that these cost values are in a non-linear scale as the cost functions are piece-wise linear functions.

As this 20-minute synthesis benchmark is limited by time and uses the same computer hardware available for both algorithms, the mean of the results can be used to directly compare the algorithms. In both synthesis cases, the use of crossovers could improve the results effectively. Choosing this algorithm as core optimizer, the next subsection will use it together with Particle Swarm Optimization to explore the Pareto Front.

6.2. Pareto front exploration through modified SA with crossovers + PSO

Using the modified SA as an initial synthesis stage followed by Particle Swarm Optimization (as described in subsection 5.4), this benchmark explores the Pareto Front after trying to achieve the basic specifications. This type of synthesis is useful to highlight the design

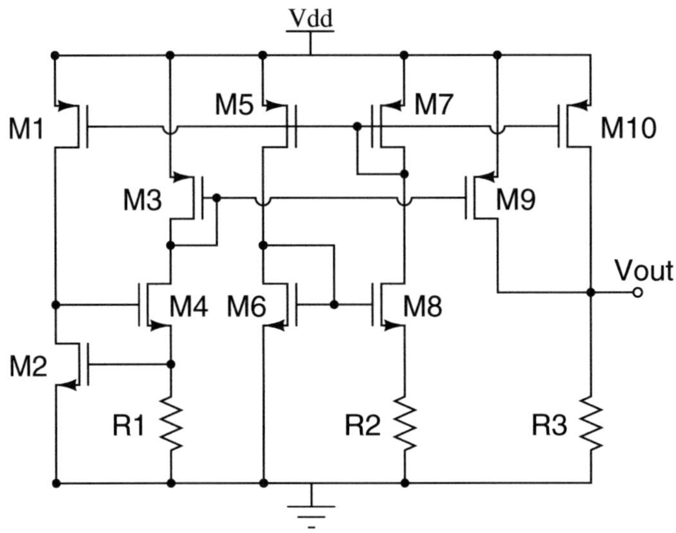

Figure 12. Voltage Reference Topology [17].

trade-offs. In the end of the procedure, graphs showing where the best achieved solutions are in the specification space. The topology used for this synthesis is a voltage reference as described in figure 12 [17] with a 1.25 voltage supply. Resistor values were acchieved using width and lenght dimensions. The used 0.35μm technology has a high resistive module with a 1.2 k$\Omega/\Box$ sheet resistance (R_{sr}). The final resistance is approximatedly $R = R_{sr} \cdot \frac{L}{W}$.

The specifications to be achieved and the results of the global search can be seen in Table 6. The dimensions of the devices achieved after the phase 1 of synthesis was performed can be seen in Table 7. Figure 13 shows the pareto front ploted after the Particle Swarm Optimization and the pareto filters for 3 dimensions were used. These plots were obtained by fixing two axes (Power and Area) and using the remaining specifications (TC, Noise, LR and PSSR) as the third axis of each graph.

The graphs show the relations among these specifications found on the best achieved solutions. It can be seen, for instance, that solutions with greater area and power had smaller LR and output noise than the others. As these graphs show only solutions within

Measurement	Specification	Result
Power [μ W]	< 4	2.95
Voltage at 27°C [mA]	170	170
Temperature Compensation [ppm/°C]	< 10	2.24
Output Noise [μV/Hz]	< 5	1.26
Line Regulation [mV/V]	< 5	4.07
PSSR dB[V]	> 40	42.3
Gate Area [mm]^2	< 0.01	0.0096

Table 6. Specifications of the Voltage Reference and Results after global search.

Device	W [μm]	L [μm]
M_1	16.8	10.7
M_2	392	3.23
M_3	112	3.93
M_4	4.65	4.37
M_5	50	10.7
M_6	9.51	1.49
M_7	50	10.7
M_8	225	1.49
M_9	11.2	3.93
M_{10}	19.3	10.7
$R1$	3.4	625
$R2$	3.76	684
$R3$	1.7	796

Table 7. Dimensions of MOS transistors and resistors of the Voltage Reference Circuit.

the minimum specifications, it can be seen in the first subplot that solutions with less power than 2.85μW usually go out of the desired specifications. However, it can also be seen that if the area is greater than 9400 μm, there are results within specifications but with high TC despite having lower power. It is important to notice that these results do not represent all the

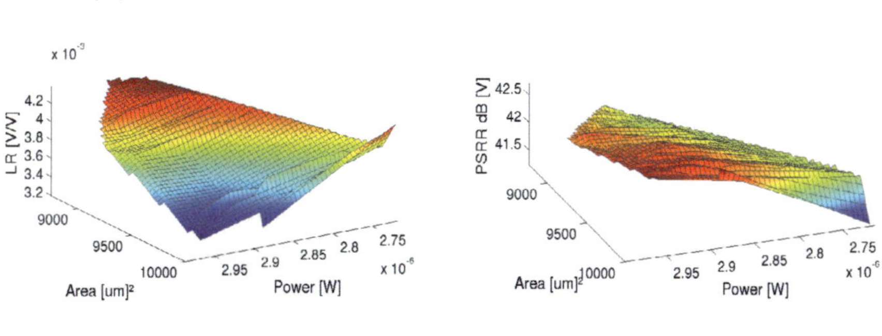

Figure 13. Pareto Front achieved after Simulated Annealing plus Particle Swarm Optimization

pareto solutions as the method is non-exaustive. The objective is to provide sufficient data to alow the user to make design choices.

The time required for phase 1 (global search) was 78 minutes. The time for all optimizations in phase 2 was 15 minutes and the time for the particle swarm optimization was bounded in 1 hour. Results have shown all specifications were met within a small amount of time. Therefore, the use of this technique offers good results for synthesizing and searching the pareto front of analog circuits.

In this experiment, only nominal simulations were performed. The designer can use this technique together with a optimization method taking in consideration mismatch variations [25] to further improve the circuit robustness.

7. Conclusion

This chapter presented the analog design synthesis problem and the use of Simulated Annealing/Quenching combined with other algorithms to solve it. The difficulties in synthesizing analog circuits are result of several factors, including high inter-connectivity among its variables, non-linearity and blurry separation among the different abstraction levels.

A brief history of the analog synthesis approaches which used Simulated Annealing was presented. SA/SQ has been used in analog circuits since the beginning of the optimization approach to analog synthesis, and it is still a popular algorithm as it yields good results and is easy to implement. Modern approaches use SA together with other algorithms in order to improve its results.

A set of experiments using SA/SQ with crossovers and also SA/SQ with Particle Swarm Optimizations were performed. The simulations have shown the combination of SA/SQ with other algorithms can improve the results considerably and also explore the pareto front. All tests were performed using a 0.35 μm technology. The circuits synthesized were a Miller amplifier, a Folded-cascode amplifier and a Voltage Regulator, and for each of them good optimization results were achieved.

Author details

Tiago Oliveira Weber and Wilhelmus A. M. Van Noije
Electronic Systems Engineering Department, Polytechnic School
University of São Paulo, Brazil

8. References

[1] Alpaydin, G., Balkir, S. & Dunkar, G. [2003]. An evolutionary approach to automatic synthesis of high-performance analog integrated circuits, *IEEE Transactions on Evolutionary Computation* 7: 240–252.

[2] Balkir, S., Dundar, G. & Ogrenci, S. [2003]. *Analog VLSI Design Automation*, CRC Press, Abingdon.

[3] Colorni, A., Dorigo, M. & Maniezzo, V. [1991]. Distributed optimization by ant colonies, *Towards a Practice of Autonomous Systems: Proceedings of the First European Conference on Artificial Life*, MIT Press, Cambridge, MA, pp. 134–142.

[4] Dantzig, G. B. [1963]. *Linear programming and extensions*, Princeton University Press, Princeton, N.J.

[5] Degrauwe, M. G. R., Nys, O., Dijkstra, E., Rijmenants, J., Bitz, S., Goffart, B. L. A. G., Vittoz, E. A., Cserveny, S., Meixenberger, C., van der Stappen, G. & Oguey, H. J. [1987]. IDAC: an interactive design tool for analog CMOS circuits, *IEEE Journal of Solid-state Circuits* 22: 1106–1116.

[6] der Plas, G. V., Debyser, G., Leyn, F., Lampaert, K., Vandenbussche, J., Gielen, G., Veselinovic, P. & Leenarts, D. [2001]. AMGIE-A synthesis environment for CMOS analog integrated circuits, *IEEE TCAD, 1037* .

[7] Fakhfakh, M., Cooren, Y., Sallem, A., Loulou, M. & Siarry, P. [2010]. Analog circuit design optimization through the particle swarm optimization technique, *Analog Integrated Cir Process* 63: 71–82.

[8] Gielen, G. G. E., Walscharts, H. C. C. & Sansen, W. M. C. [1989]. ISAAC: a symbolic simulator for analog integrated circuits, *IEEE Journal of Solid-state Circuits* 24: 1587–1597.

[9] Gielen, G. G. E., Walscharts, H. C. C. & Sansen, W. M. C. [1990]. Analog circuit design optimization based on symbolic simulation and simulated annealing, *IEEE Journal of Solid-state Circuits* 25: 707–713.

[10] Harjani, R., Rutenbar, R. A. & Carley, L. R. [1989]. OASYS: a framework for analog circuit synthesis, *IEEE Transactions on Computer-aided Design of Integrated Circuits and Systems* 8: 1247–1266.

[11] Holland, J. H. [1975]. *Adaptation in Natural and Artificial Systems: An Introductory Analysis with Applications to Biology, Control and Artificial Intelligence*, University of Michigan Press.

[12] Kennedy, J. & Eberhart, R. C. [1995]. Particle swarm optimization, *IEEE International Conference on Neural Networks*, IEEE Service Center, pp. 1942–1948.

[13] Keramat, M. & Kielbasa, R. [1998]. OPTOMEGA: an environment for analog circuit optimization, *IEEE ISCAS* .

[14] Kirkpatrick, S., C. D. Gelatt, J. & Vecchi, M. P. [1983]. Optimization by simulated annealing, *Science* 220: 671–680.

[15] Krasnicki, M., Phelps, R., Rutenbar, R. & Carley, L. [1999]. Maelstrom: Efficient simulation-based synthesis for custom analog cells, *Proc. Design Automation Converence* .

[16] Kuhn, H. W. & Tucker, A. W. [1951]. Nonlinear programming, *Proc. Second Berkeley Symposium on Mathematical Statistics and Probability*, University of California Press.

[17] Mateus, J., Roa, E., Hernandez, H. & Noije, W. V. [2008]. A 2.7μa sub 1-v voltage reference, *Proceedings of IEEE Symposium on Integrated Circuits and System Design* pp. 81–84.

[18] Medeiro, F., Fernández, F. V., Domínguez-Castro, R. & Rodríguez-Vázquez, A. [1994]. A statistical optimization-based approach for automated sizing of analog cells, *International Conference on Computer Aided Design*, pp. 594–597.

[19] Ochotta, E., Rubenbar, R. & Carley, L. [1996]. Synthesis of high-performance analog circuits in ASTRX/OBLX, *IEEE TCAD* .

[20] Pereira, A. I. & Fernandes, E. M. [2004]. A study of simulated annealing variants, *XXVIII Congreso de Estadística e Investigación Operativa.*

[21] R. J. Duffin, Bittner, L. [1967]. Geometric programming - theory and application, *ZAMM - Journal of Applied Mathematics and Mechanics / Zeitschrift für Angewandte Mathematik und Mechanik* 47(8): 561–561.

[22] Rao, S. S. [1996]. *Engineering Optimization: Theory and Practice*, New Age Publishers.

[23] Razavi, B. [2001]. *Design of Analog CMOS Integrated Circuits*, 1 edn, McGraw-Hill, Inc., New York, NY, USA.

[24] Reyes-Sierra, M. & Coello, C. A. C. [2006]. Multi-objective particle swarm optimizers: A survey of the state-of-the-art, *International Journal of Computational Intelligence Research* 2: 287–308.

[25] Sáenz Noval, J. J., Roa Fuentes, E. F., Ayala Pabón, A. & Van Noije, W. [2010]. A methodology to improve yield in analog circuits by using geometric programming, *Proceedings of the 23rd symposium on Integrated circuits and system design*, SBCCI '10, ACM, New York, NY, USA, pp. 140–145.

[26] Tiwary, S. K., Tiwary, P. K. & Rutenbar, R. A. [2006]. Generation of yield-aware pareto surfaces for hierarchical circuit design space exploration, *Design Automation Conference*, pp. 31–36.

[27] Weber, T. O. & Noije, W. A. M. V. [2011]. Analog design synthesis method using simulated annealing and particle swarm optimization, *Proceedings of the 24th symposium on Integrated circuits and systems design*, SBCCI '11, ACM, pp. 85–90.

[28] Wilson, R. B. [1963]. *A Simplicial Algorithm for Concave Programming*, PhD thesis, Cambridge.

[29] Yu, G. & Li, P. [2011]. Hierarchical analog/mixed-signal circuit optimization under process variations and tuning, *IEEE Transactions on Computer-Aided Design of Integrated Circuits and Systems* 30: 313.

Chapter 12

Genetic Algorithm and Simulated Annealing: A Combined Intelligent Optimization Method and Its Application to Subsynchronous Damping Control in Electrical Power Transmission Systems

Xiaorong Xie

Additional information is available at the end of the chapter

http://dx.doi.org/10.5772/50371

1. Introduction

Series compensations (SCs) have been widely used in electrical power systems to enhance transmission capability through partial compensation of line reactors. However, they will trigger oscillatory modes inherent in the mass-spring system of turbine-generators, resulting in subsynchronous resonance (SSR), which, if not properly handled, could cause shaft failure. The SSR phenomenon was first discussed in 1937 [1] and until 1971, shaft torsional oscillations were ignored. Two shaft failures at the Mohave power plant in Southern Nevada, USA [2] led to the understanding and development of the theory of interaction between series-capacitor compensated lines and the torsional modes of steam turbine-generators. After the second shaft failure at Mohave, the utility industry devoted considerable effort to the analysis and suppression of SSR. Much has been written on the subject as evidenced by the bibliography [3] and three supplements [4, 5, 6]. Generally, when preliminary analysis shows the system to be at risk of unstable SSR, control measures must be applied [7]. In this chapter, two types of controllers, namely, supplementary excitation damping control (SEDC) and static var compensator based subsynchronous damping control (SVC-SSDC), are used for the mitigation of SSR problem. The emphasis of our work is to achieve an optimal design of the controller parameters with the proposed genetic algorithm and simulated annealing (GA-SA) algorithm.

The SSR issue is generally addressed under three different types [8]: induction generator effect (IGE), torsional interaction (TI), and torque amplification (TA). In all cases, SSR is due to the interaction of a series capacitor compensated system with a turbine-generator.

However, this chapter will focus on the TI type SSR, because it is the main problem we encountered in practice.

TI-SSR is the interplay between the mechanical shaft system (turbine-generator) and the series- capacitor compensated electrical network. Small-signal disturbances in a power system result in simultaneous excitation of all natural modes of the electrical and mechanical systems. The turbine-generator shaft system responds to disturbances with oscillations at its torsional natural frequencies. Therefore, TI-SSR can be viewed as a stability phenomenon of the linearized system model [9-10]. In other words, their stability and design of control parameter can be investigated with the eigenvalues of the small-signal model obtained at a certain working point [10]. In our study, for each of the concerned operation conditions, a detailed linearized system model of the studied system is developed. It can be expressed as

$$\begin{cases} \Delta\dot{\mathbf{X}} = \mathbf{A}\Delta\mathbf{X} + \mathbf{B}\Delta\mathbf{U} \\ \Delta\mathbf{Y} = \mathbf{C}\Delta\mathbf{X} \end{cases} \tag{1}$$

Where $\Delta\mathbf{X}$ is the state vector, $\Delta\mathbf{U}$ is the input vector and $\Delta\mathbf{Y}$ is the output or feedback vector.

In model (1), the following various sub-models are included:

- the lumped mass-spring model of the mechanical shaft system of the turbine-generator set;
- the electromagnetic model of the generator, which is generally represented with the dq0 model with three damper windings [11, 12];
- the excitation system, for instance, the static self-parallel excitation system is modeled with the IEEE ST4B-type automatic voltage regulator (AVR) and PSS2B-type power system stabilizer (PSS);
- the model of the electrical network, incorporating the lines, transformers and the series capacitors.

The controller, either the SEDC or the SVC-SSDC to be discussed, can be also be linearized as

$$\begin{cases} \Delta\dot{\mathbf{Z}} = \mathbf{F}(\boldsymbol{\alpha})\Delta\mathbf{Z} + \mathbf{G}(\boldsymbol{\alpha})\ \Delta\mathbf{Y} \\ \Delta\mathbf{U} = \mathbf{H}(\boldsymbol{\alpha})\Delta\mathbf{Z} \end{cases} \tag{2}$$

where $\Delta\mathbf{Z}$ is the state vector, and $\mathbf{F},\mathbf{G},\mathbf{H}$ are the coefficient matrices with appropriate dimensions and their elements are determined by the control parameter vector $\boldsymbol{\alpha}$, which including the gains and time constants of the controller to be designed.

Therefore the close-loop system model under a certain operation situation can be obtained by combining the open-loop system model (1) and the controller (2), i.e.

$$\Delta\dot{\hat{\mathbf{X}}} = \hat{\mathbf{A}}\Delta\hat{\mathbf{X}} \tag{3}$$

where $\Delta\hat{\mathbf{X}} = \begin{bmatrix} \Delta\mathbf{X} & \Delta\mathbf{Z} \end{bmatrix}^{\mathrm{T}}$ is the extended state vector and $\hat{\mathbf{A}} = \begin{bmatrix} \mathbf{A} & \mathbf{BH} \\ \mathbf{GC} & \mathbf{F} \end{bmatrix}$ is the extended coefficient matrix with appropriate dimensions.

Obviously, the eigenvalues of the matrices $\mathbf{A}$ and $\hat{\mathbf{A}}$ that correspond to the torsional modes represent the stability or damping of these TI modes for the open-loop and closed-loop systems respectively. Hence, by the eigen-analysis method, the relationship between the torsional damping with system variables and control parameters can be traced out under different operating conditions. Generally, for an uncontrolled practical series-compensated power system, several of the eigenvalues related to the TI-SSR problem would be negative or unstable under certain operating conditions, i.e., the SSR problem is a multimodal one. Thus, the key issue is to design an effective control strategy which can meet the following requirements: First, the multimodal SSR must be damped simultaneously and the controller parameters determined in a coordinated way. Second, the controller must be robust enough to stabilize SSR under all possible operating conditions. In addition, the controller should be practical, i.e., easy to implement and operate. Finally, its size and operational loss should be reasonable for reduction of cost, room and maintenance.

In this chapter, the aimed power system and its SSR problem is represented with detailed electromagnetic models and the task of control-design is formulated into the constrained nonlinear optimization problem, which is of the following form:

$$\begin{aligned} &\max f \\ &\text{subject to: } \underline{\boldsymbol{\Gamma}}(\boldsymbol{\alpha}) \le \boldsymbol{\Gamma}(\boldsymbol{\alpha}) \le \overline{\boldsymbol{\Gamma}}(\boldsymbol{\alpha}) \end{aligned} \tag{4}$$

Where f is the fitness function to be optimized, $\boldsymbol{\Gamma}(\boldsymbol{\alpha})$ is the function of the control parameter $\boldsymbol{\alpha}$, which has its lower and upper limits being $\underline{\boldsymbol{\Gamma}}(\boldsymbol{\alpha})$, $\overline{\boldsymbol{\Gamma}}(\boldsymbol{\alpha})$ respectively.

For our concerned multimodal SSR problem, the fitness function is generally defined as [13]

$$\begin{aligned} & f = \sum_{i=1}^{N} w_i \eta_i + w_{N+1} \min\{\eta_1, \eta_2, \eta_3\}, \sum_{i=1}^{N+1} w_i = 1, w_i > 0 \\ & \eta_i = \min_{j=1}^{j=M} \left\{ \sigma_{ij} \right\}, \ \sigma_{ij} = -\mathrm{Re}(\lambda_{ij}) / \left| \mathrm{Im}(\lambda_{ij}) \right|, \\ & \lambda_{ij} = \lambda_i \left\{ \hat{A}_j(\boldsymbol{\alpha}) \right\}; i = 1,2,3; j = 1,\ldots,M \end{aligned} \tag{5}$$

where $\hat{\mathbf{A}}_j$ is the closed-loop coefficient matrix for the j-th operating condition and is jointly determined by the system variables and the control parameters; $\lambda_i\{\cdot\}$ means the calculation of the i-th SSR mode of the included close-loop coefficient matrix; $\mathrm{Re}(\lambda_{ij})$ and $\mathrm{Im}(\lambda_{ij})$ are the real and imaginary parts of the SSR mode; σ_{ij} is the close-loop damping; η_i is the least damping among all operating conditions; w_i are the positive weight coefficients; N is the number of concerned SSR modes; M is the number of small-signal system model with each representing an evaluation condition; the subscripts i, j denote the index of the SSR mode and the operating condition respectively.

The control-design problem can be summarized as follows:

Firstly, for a SSR-threatened system, representative conditions should be selected as the "evaluation conditions", which cover the full range of generator output levels as well as the status of transmission lines. While these evaluation conditions by no means limit the situations under which the power system operates, together they form the edge of practical operating conditions and the most unfavorable situations relevant to the controller to be designed. Therefore, the derived controller can be robust enough to stabilize the system under all possible conditions.

Secondly, for each evaluation condition, a linearized model like (1)-(3) is derived for the nonlinear system as well as the controller via small-signal approximation, and the damping of torsional modes can be calculated by eigen-analysis. Consequently, the parameter sensitivity of the control on the stability of TI-type SSR is quantified.

Thirdly, the control-design task, i.e., simultaneously tuning of the control parameters, was formulated into a standard nonlinear optimization problem as described in (4)-(5), by appropriately determining the fitness function and the constraints on the parameters of the controller. As seen in (5), the ability of the controller to depress SSR is determined by the weighted sum of the damping of concerned torsional modes under all listed conditions. Furthermore, the least damped one is emphasized with additional weight to achieve better damping.

Next, by solving the optimization problem (4)-(5), the control parameters can be tuned simultaneously and an optimal control strategy to mitigate the SSR issue is obtained.

Finally, the designed controller should be verified with simulation study on the nonlinear system model, or even be implemented as hardware equipment and tested in practical applications.

2. The combined intelligent optimization method: Genetic algorithm and simulated annealing (GA-SA)

The control-design problem (4)-(5) is a complex nonlinear optimization problem with many models, each corresponding to an evaluation condition. However, it is difficult for conventional methods due to its nonlinear nature and high dimension of the solution space. Here, we provide the solution by combining genetic algorithm (GA) and simulated annealing (SA), termed GA-SA. As a hybrid and global optimization strategy, GA-SA takes advantage of both GA's parallel-searching capability and SA's probabilistic jumping property [14-17]: The fast and global parallel searching ability of GA is retained, and the diversity is improved by SA state transition. Thus, premature convergence in GA can be avoided. Furthermore, SA renders GA mutation more controllable by its temperature. Thanks to the powerful global searching ability, the criteria for the selection of algorithm parameters are very much relaxed, resulting in improved performance and robust optimization. Since the basic principle of GA-SA has been well documented [14-17], only the particular procedures employed to design the SSR control (i.e., SEDC and SVC-SSDC) are elaborated in the following.

The flow chart of the proposed GA-SA is shown in Fig. 1.

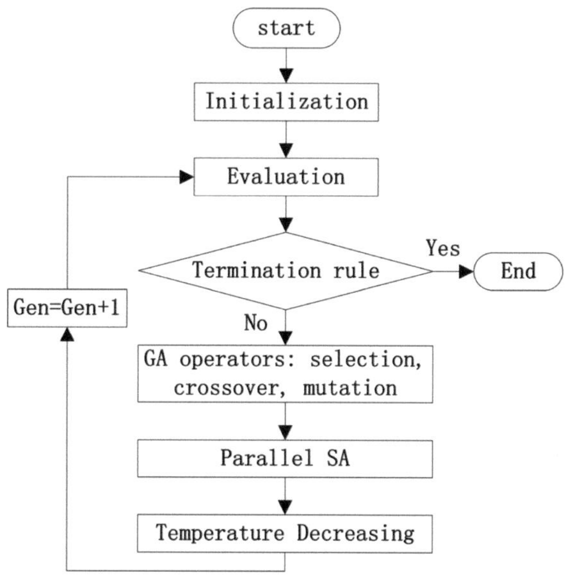

Figure 1. Flow chart of the implemented GA-SA

The GA-SA is executed with the following steps.

Step 1. GA-SA initialization.

i. Define algorithm parameters, including: boundaries of control parameters, GA parameters (population size n, number of generations k_{GA}, number of preferential chromosomes n_b, crossover probability p_c, mutation probability p_m, initial perturbation amplitude ξ_{GA0})and SA parameters (number of state transition k_{SA}, cooling rate r, perturbation amplitude ξ_{SA0}, initial acceptance rate p_r).

ii. Set an initial population by randomly generating n feasible vectors or chromosomes in a reasonable range of the control parameters;

iii. Define the initial temperature as $t_0 = -\Delta_{max} / \ln p_r$, where $\Delta_{max} = |f_{max} - f_{min}|$ and f_{max}, f_{min} are fitness values of the best and the worst chromosomes of the initial population.

Step 2. Evaluate each chromosome by using the fitness function.

Step 3. Apply GA operations to generate a new generation. This is fulfilled with three sub-operators:

i. Selection: Select n_b chromosomes as potential parents from the old population. The selection is simulated by spinning a weighted roulette wheel [18] in such a way that the fittest chromosomes have a higher probability of being selected, so that their genes have a greater chance to be passed on to the next generation.

ii. Crossover: This is carried out between each parent member (p_i, i=1,…,n_b) and the fittest chromosome (p_0) of the current generation. The convex crossover method is adopted to produce the offspring x_{i1}, x_{i2}, where $x_{i1} = \lambda p_i + (1-\lambda)p_0$, $x_{i2} = (1-\lambda)p_i + \lambda p_0$, $0 < \lambda < 1$ [17]. All offspring are added to the original population and then n chromosomes of higher fitness are chosen to form a new generation.

iii. Mutation with one elite reservation strategy: Keep the best chromosome and apply mutation to other chromosomes with a probability of p_m. For each mutation, a new chromosome x_n is generated from an old chromosome x_o by applying a random perturbation, i.e., $x_n = x_o + m\xi_{GA}$, where $m \in (-5,5)$ is a random variable of Cauchy distribution, and ξ_{GA} is the perturbation amplitude.

Step 4. Apply the parallel SA algorithm to enhance the quality of the new population. State transition and acceptance are repeatedly used during the SA.

i. State transition: First, SA selects candidate chromosomes from the GA-generated population according to the roulette strategy [18]. State transition is then performed k_{SA} times for each of these chromosomes. The operation of state transition is identical to that of GA mutation, the only difference being that the former has a much smaller parameter of perturbation amplitude ξ_{SA}.

ii. State acceptance: Let f_i and f_{i+1} be the fitness values of the original state x_i and the state x_{i+1} obtained by state transition, respectively. If $f_{i+1} > f_i$, x_{i+1} is accepted as the starting point for the next state transition; otherwise, x_{i+1} is accepted with probability $p = \exp[(f_i - f_{i+1}) / T_p]$.

Step 5. Cooling: As the optimization proceeds, the temperature T_p is dynamically updated according to the rule $Tp_{k+1} = r\ Tp_k$. Thus, at higher temperatures, SA will accept poorer status with certain probability to avoid plugging into local minimum; while at lower temperatures, it basically turns out to be an optimization function in small random searches.

Step 6. Termination: GA-SA terminates if the maximum number of generations allowed (k_{GA}) is reached or the best result has not seen improvement for a certain number of generations. Otherwise, go back to Step 2.

3. Application of GA-SA for SEDC optimization

3.1. The target system

The Shangdu power plant (SPP), located in the Inner Mongolia Autonomous Region, is about 300 kilometers north of Beijing city. It has four 600 MW steam turbine-generators connected to the North China Power Grid through 500 kV transmission lines, including the 243 km double lines connecting SPP with the Chengde substation and the 130km double lines connecting the Chengde substation with the Jiangjiaying substation. Fig.2 illustrates the one-line diagram of the equivalent transmission system. To improve the transferring capability as well as system stability, fixed series capacitors (FSCs) are applied to the parallel transmission lines between SPP and the Chengde substation with 45% compensation degree. The four turbine-generators are subcritical air-cooled machines with almost the same parameters. Each turbine-generator consists of four rotors, i.e., a high-and-intermediate-pressure (HIP) turbine rotor, two low-pressure (LPA and LPB) turbine rotors, and the generator rotor, thus resulting in three subsynchronous torsional modes. The characteristic frequencies (in Hz) are 15.19~15.33 (mode 1), 26.01~26.12 (mode 2) and 30.25~30.54 (mode 3). To solve the SSR problem, torsional stress relays (TSRs) and supplementary excitation damping controllers (SEDCs) are applied to the SPP generators (as shown in fig. 2).

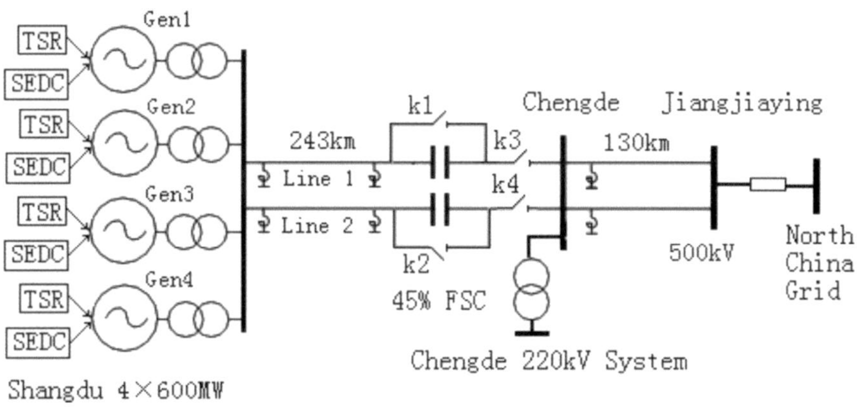

Figure 2. The one-line diagram of the equivalent transmission system

3.2. The SSR problem

A thorough evaluation of the severity of the SSR problem was conducted under all possible system conditions [19, 20], which comprised frequency scanning, eigenvalue analyses and electromagnetic transient (EMT).

With eigenvalue analysis, the modal damping (i.e., the negative real part of the torsional eigenvalue, which is a combined outcome of mechanical and electrical damping) can be obtained with respect to different operating conditions. To illustrate, 24 representative conditions are selected as the "evaluation conditions" (listed in Table 1), which cover the full range of generator output levels as well as the status of the Shangdu-Chengde lines. Table 1 also lists the calculated modal damping. Thus, the severity of SSR is quantified and the most risky situations are identified.

Through the evaluation study, characteristics of the SSR problem are summarized as follows:

i. Mode 1 is well-damped in all operating conditions; however, modes 2 and 3 may be under-damped or even unstable under some conditions. Thus, the SSR problem is a multimodal one.
ii. Mode 2 is the worst damped and tends to be unstable for numbers of operating conditions, especially when only one Shangdu-Chengde line is in service or the generator output is relatively low (corresponding to a lower mechanical damping).

iii. Modal damping is affected by several factors and each torsional mode has a most undesirable (or least damped) condition unique to its own, making it a challenge to design a controller adaptable to all operating conditions.

Oper. Cond. #	Generator output (%)				Status of SPP-Chengde lines		Mode 1 (s^{-1})	Mode 2 (s^{-1})	Mode 3 (s^{-1})
	Gen 1	Gen 2	Gen 3	Gen 4	Line 1	Line 2	No SEDC	No SEDC	No SEDC
1	0	0	0	0	online	offline	0.0312	**-0.5603**	0.0211
2	0	0	0	0	online	online	0.0367	**-0.0429**	**-0.0053**
3	0	0	0	offline	online	offline	0.0338	**-0.1992**	**-0.0764**
4	0	0	0	offline	online	online	0.0380	0.0449	0.0236
5	0	0	offline	offline	online	offline	0.0402	**-0.0235**	**-0.0921**
6	0	0	offline	offline	online	online	0.0442	0.0628	0.0290
7	0	offline	offline	offline	online	offline	0.0506	0.0594	0.0267
8	0	offline	offline	offline	online	online	0.0533	0.0796	0.0325
9	40	40	40	40	online	offline	0.0598	**-0.4887**	0.0511
10	40	40	40	40	online	online	0.0654	0.0383	0.0242
11	40	40	40	offline	online	offline	0.0590	**-0.1882**	0.0229
12	40	40	40	offline	online	online	0.0641	0.0569	0.0460
13	40	40	offline	offline	online	offline	0.0599	**-0.0123**	**-0.0864**
14	40	40	offline	offline	online	online	0.0646	0.0748	0.0567
15	40	offline	offline	offline	online	offline	0.0664	0.0684	0.0529
16	40	offline	offline	offline	online	online	0.0717	0.0916	0.0620
17	100	100	100	100	online	offline	0.0674	**-0.5455**	0.1344
18	100	100	100	100	online	online	0.0716	0.0787	0.1026
19	100	100	100	offline	online	offline	0.0615	**-0.1835**	0.1035
20	100	100	100	offline	online	online	0.0648	0.1015	0.1284
21	100	100	offline	offline	online	offline	0.0555	0.0248	**-0.0192**
22	100	100	offline	offline	online	online	0.0580	0.1226	0.1409
23	100	offline	offline	offline	online	offline	0.0511	0.1163	0.1370
24	100	offline	offline	offline	online	online	0.0540	0.1414	0.1467

Table 1. Torsional damping of the open-loop system under the evaluation conditions

3.3. Supplementary Excitation Damping Controller

SEDC is a real-time control system that works through the excitation system by modulating the field voltage at the torsional frequencies. Fig.3 illustrates the signal relationship of the SEDC, the excitation regulators, the generators and the grid. As a supplementary control loop, SEDC uses the mechanical speed of the HIP turbine (ω_1) as the only feedback signal to generate the subsynchronous control output (u_{SEDC}). u_{SEDC} is then added to the output of AVR (u_C) to form a modulated control signal (u_f), which drives the excitation circuit to yield the field voltage (E_f). There is a time delay (τ) between u_f and E_f, mainly due to signal sampling, data processing and thyristor transport lag. This time delay, generally several to a dozen milliseconds, is important to the practicality of SEDC's design, since it is comparable with the period of torsional modes, which, if not taken into consideration, would deteriorate or even destabilize the SEDC-controlled system.

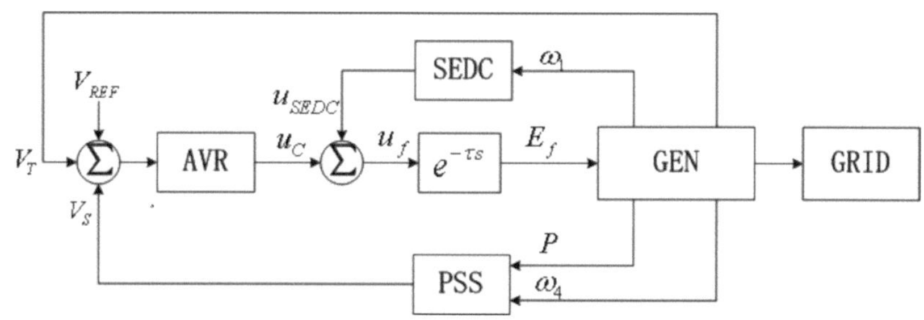

Figure 3. The signal relationship of the SEDC and the excitation regulators
(AVR: automatic voltage regulator, PSS: power system stabilizer)

Fig.4 shows the block diagram of the proposed SEDC. The mechanical speed of the HIP turbine provides the input, which after proper filtering and conditioning becomes the deviation signal $\Delta\omega_1$. It then passes through three separate control paths. Each control path, tuned to a specific mode, comprises a band-pass filter, an amplifier G_k and a unity-gain phase-shifter $(1-T_k s)^2/(1+T_k s)^2$ to generate the control signal for the corresponding mode. The control signals of all torsional modes are summarized, clipped and finally added to the AVR output to modulate the field voltage. Consequently, three subsynchronous components are generated in the excitation current, which in turn produce subsynchronous torque upon the generator shafts. If the gains and phase-shifts are appropriately set, this torque provided by SEDCs will play a role in damping SSR. So proper determination of the gains and time constants, i.e., $G_k, T_k (k=1,2,3)$ in Fig.4, is crucial to mitigate the multimodal SSR.

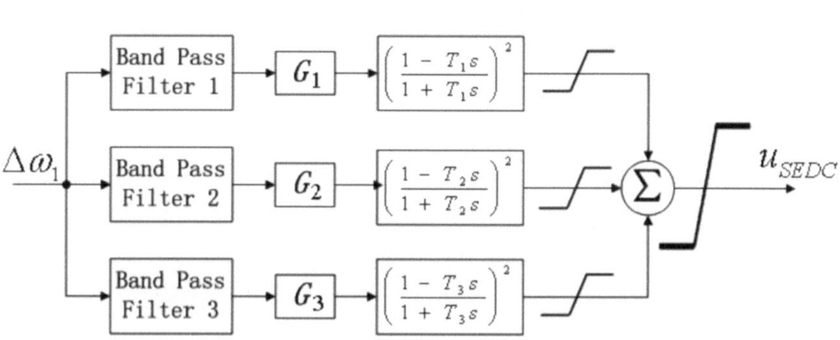

Figure 4. Configuration of SEDC

3.4. System modeling

3.4.1. The linearized open-loop system model

The linearized state equations of the four generators and network around a specific operating point can be expressed by (1). Here the state vector $\Delta\mathbf{X}$, input vector $\Delta\mathbf{U}$ and output vector $\Delta\mathbf{Y}$ are given respectively by $\Delta\mathbf{X}^{\mathrm{T}} = \left[\Delta\boldsymbol{\delta}_{4\times1} \quad \Delta\boldsymbol{\omega}_{4\times1} \quad \Delta\mathbf{i}_{6\times1} \quad \Delta\mathbf{u}_{Cd,q}\right]$, $\Delta\mathbf{U}^{T} = \Delta E_{f}$, $\Delta\mathbf{Y}^{\mathrm{T}} = \Delta\mathbf{X}^{\mathrm{T}}$, where $\boldsymbol{\delta}_{4\times1}$ are the mass angles; $\boldsymbol{\omega}_{4\times1}$ are the mass speeds; $\mathbf{i}_{6\times1}$ are the winding currents; $\mathbf{u}_{Cd,q}$ are the d-axis and q-axis voltages of the series capacitors; E_f is the field excitation voltage and $\mathbf{A},\mathbf{B},\mathbf{C}$ are the coefficient matrices with proper dimensions.

3.4.2. The combined system model

From a combination of the models of the grid, the generators and the excitation control system, the close-loop system model can be derived (shown in Fig.3). For the convenience of further analysis and synthesis, the time delay $e^{-\tau s}$ is replaced with a rational approximation in the frequency domain, i.e. $e^{-\tau s} \cong (1+\tau s/2)^{-2}$. For the target system, the time delay τ is measured to be about 7 milliseconds. Since our SEDC scheme is actually a linear dynamic stabilizer, whose transfer function can be expressed with the general form as in (2). Therefore the close-loop system model (3) can be obtained by combining the open-loop system model and the SEDC controller.

3.5. Optimal design of SEDC based on GA-SA

For the Shangdu system, the evaluation conditions listed in Table 1 are used again for the purpose of parameters-tuning because these selected conditions together not only bound the range of practical operating conditions but also incorporate the most unfavorable situations. An SEDC based on these conditions will be robust enough to stabilize the system under all normal conditions. Then the nonlinear optimization problem can be formulated as in (4)-(5), in which the control parameter set $\boldsymbol{\alpha} = \{G_1, G_2, G_3, T_1, T_2, T_3\}$, N=3, M=24. During the parameter-tuning process, the gains and time constants of SEDC should be restricted within a reasonable range due to the control energy and hardware implementation limits. Thus the constraint can be written as

$$G_{\mathrm{lb,k}} \le G_k \le G_{\mathrm{ub},i}, \quad T_{\mathrm{lb,k}} \le T_k \le T_{\mathrm{ub,k}} \tag{6}$$

where $G_{\mathrm{lb,k}}, G_{\mathrm{ub,k}}, T_{\mathrm{lb,k}}, T_{\mathrm{ub,k}}$ are the lower and upper bounds of the gains and time constants.

Then the GA-SA is adopted to solve the problem. The following algorithm parameters have been chosen after running a number of trials: The weights in the fitness function: $w_1 = w_2 = w_3 = 0.2,\ w_4 = 0.4$; population size = 20; crossover rate = 0.9; mutation rate = 0.1; number of generations = 30; cooling rate=0.95; and initial temperature $= -1 * \Delta f / \log_{10}(0.1)$, where Δf is the maximum margin of the fitness value of the initial population.

In our case, it is discovered that the performance of GA-SA is not very sensitive to these parameters. In fact, if the population size and the number of generations are chosen properly, GA-SA will generally converge to satisfactory results after running 8 to 12 iterations, which lasts only several minutes on a modern computer. The typical convergence characteristic of the GA-SA is displayed in Fig.5. The most time-consuming step during the optimization process is the calculation of the fitness function, which is essentially the close-loop damping of the torsional modes. Therefore, a highly efficient and reliable algorithm should be used to calculate the fitness function. In this aspect, the implemented GA-SA works very well. Table 2 lists the optimized SEDC parameters obtained by the proposed GA-SA.

G_1	T_1	G_2	T_2	G_3	T_3
-378.99	0. 0535	311.45	0.0174	239.92	0.0018

Table 2. The optimized SEDC parameters

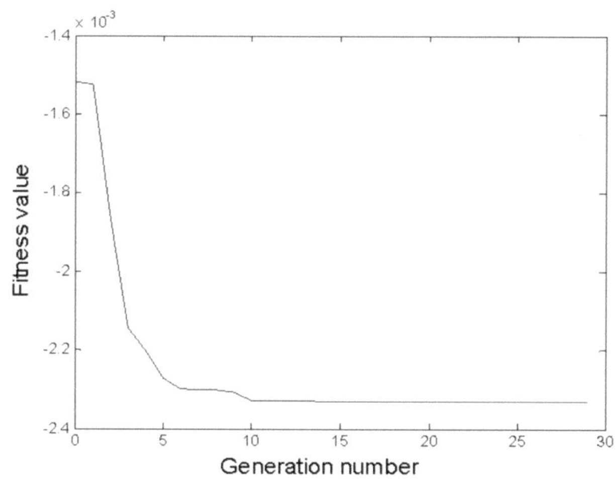

Figure 5. Convergence of the GA-SA algorithm

3.6. Performance verification of the designed SEDC

3.6.1. Model study

To verify the tuned SEDC, both eigenvalue analysis and EMT simulation were fulfilled on the close-loop system.

Eigenvalue analysis was conducted under the evaluation conditions, first in the absence and then in the presence of SEDC. The result is presented in Table 3. It shows that, without SEDC, modes 2 and 3 will incur negative damping, resulting in an unstable system. When SEDC is included into the control loop, the damping of torsional modes under all the specified conditions has been increased considerably and all unstable cases have been well stabilized.

To investigate how well the proposed controller can withstand large disturbances such as short-circuit faults, the nonlinear power system was simulated using EMT software "PSCAD/EMTDC". Step-by-step time-domain electromagnetic simulations were conducted to check the response of the SEDC controlled system during large disturbances.

In both of the two simulation experiments conducted, one Shangdu-Chengde line suffered a serious three-phase short-circuit fault. The faulting line was tripped 70 milliseconds later. In Experiment 1 all generators were 40% loaded while in Experiment 2 all generators were fully loaded. The result is summarized in Fig.6. As is shown in (a) and (c) of Fig. 6, in the absence of SEDC, the diverging delta mechanical speed of the high-pressure turbine indicates that the generators have a growing torsional vibration, which would probably lead to great damage on the shafts. When SEDC is applied in Fig.11 (b) and (d), subsynchronous oscillations in both experiments are successfully damped out.

Besides these two experiments, sufficient simulations have been carried out under other operating conditions and with large but different disturbances. Generally, SEDC is effective in improving the damping of all torsional modes and can satisfactorily mitigate the vibrations caused by SSR in large disturbances.

3.6.2. Practical applications and field tests

After extensive laboratory tests, the proposed SEDC was then applied to the practical SSR problem of the target system. By the end of October 2008, all SEDCs were put into their places. The two FSCs of the Shangdu-Chengde lines were also ready for operation. Then, to further validate the effectiveness of the proposed countermeasure, the project team conducted a series of joint tests on the SEDCs and the series-compensated system. By planned operations of SEDCs, generators, series capacitors and transmission lines, the dynamic characteristics of SSR and the response of the SEDCs were fully investigated. To save space, only some typical test results, to illustrate the function of SEDC in damping TI-SSR, are discussed hereafter.

The initial operating condition of the tested system was as follows: Units 1, 2 and 4 were online and half-loaded; Unit 3 was out of service; both Shangdu-Chengde lines were in operation; FSCs were out of service but ready for operation.

a. Switching on/off FSCs with/without SEDC

First, with all SEDCs in service, the two FSCs were switched on and then they were switched off one by one. During these operations, the torsional oscillation was observed in real time and each operation was initiated only after the dynamics that was triggered by the previous operation completely died away. The next stage involved quitting all SEDCs and repeating the above switch-on/off operations of FSCs. The mode-2 dynamics of this test is depicted in Fig.7. It can be observed that the torsional oscillation of the online machines are convergent under both the SEDC-off and the SEDC-on conditions; while the SEDC can considerably improve the modal damping and make the oscillation converge much more quickly.

Oper. Cond.	Generator output (%)				Status of SPP-Chengde lines		Mode 1 (s^{-1})		Mode 2 (s^{-1})		Mode 3 (s^{-1})	
#	Gen 1	Gen 2	Gen 3	Gen 4	Line 1	Line 2	No SEDC	With SEDC	No SEDC	With SEDC	No SEDC	With SEDC
1	0	0	0	0	online	offline	0.0312	0.0515	**-0.5603**	0.0873	0.0211	0.1078
2	0	0	0	0	online	online	0.0367	0.0571	**-0.0429**	0.1074	**-0.0053**	0.0830
3	0	0	0	offline	online	offline	0.0338	0.0783	**-0.1992**	0.1381	**-0.0764**	0.1286
4	0	0	0	offline	online	online	0.0380	0.0632	0.0449	0.1148	0.0236	0.0864
5	0	0	offline	offline	online	offline	0.0402	0.0827	**-0.0235**	0.1426	**-0.0921**	0.2030
6	0	0	offline	offline	online	online	0.0442	0.0670	0.0628	0.1395	0.0290	0.0690
7	0	offline	offline	offline	online	offline	0.0506	0.0833	0.0594	0.1145	0.0267	0.0782
8	0	offline	offline	offline	online	online	0.0533	0.0694	0.0796	0.1051	0.0325	0.0554
9	40	40	40	40	online	offline	0.0598	0.0897	**-0.4887**	0.1926	0.0511	0.1312
10	40	40	40	40	online	online	0.0654	0.0966	0.0383	0.3859	0.0242	0.1022
11	40	40	40	offline	online	offline	0.0590	0.0910	**-0.1882**	0.4687	0.0229	0.1761
12	40	40	40	offline	online	online	0.0641	0.0973	0.0569	0.3889	0.0460	0.0890
13	40	40	offline	offline	online	offline	0.0599	0.0945	**-0.0123**	0.4788	**-0.0864**	0.2073
14	40	40	offline	offline	online	online	0.0646	0.1002	0.0748	0.3812	0.0567	0.0874
15	40	offline	offline	offline	online	offline	0.0664	0.1039	0.0684	0.4095	0.0529	0.0774
16	40	offline	offline	offline	online	online	0.0717	0.0889	0.0916	0.3418	0.0620	0.0877
17	100	100	100	100	online	offline	0.0674	0.1008	**-0.5455**	0.0958	0.1344	0.2122
18	100	100	100	100	online	online	0.0716	0.1076	0.0787	0.5216	0.1026	0.2274
19	100	100	100	offline	online	offline	0.0615	0.0985	**-0.1835**	0.5564	0.1035	0.2470
20	100	100	100	offline	online	online	0.0648	0.1044	0.1015	0.5413	0.1284	0.2119
21	100	100	offline	offline	online	offline	0.0555	0.0974	0.0248	0.6364	**-0.0192**	0.3600
22	100	100	offline	offline	online	online	0.0580	0.1022	0.1226	0.5553	0.1409	0.2083
23	100	offline	offline	offline	online	offline	0.0511	0.0997	0.1163	0.6081	0.1370	0.2097
24	100	offline	offline	offline	online	online	0.0540	0.1038	0.1414	0.5453	0.1467	0.2076

Table 3. Torsional damping with/without SEDC under the evaluation conditions

b. Tripping and reclosing of one Shangdu-Chengde line

This test was also performed with and without SEDC, respectively. At the initial state, both Shangdu-Chengde lines were in service but only one FSCs was switched on (i.e., k1, k3, k4 were closed and k2 was open, see Fig.2). Breaker k3 was first opened to trip one Shangdu-Chengde line and then reclosed after five seconds. This test was meant to check the system response under the 3-machine and one-line condition. Fig.8 illustrates the dynamics of mode 2 during the operation. It can be seen that during the short period of the 3-machine one-line condition, without SEDC, unit 1 and 2 suffered diverging SSR; while for unit 4 the torsional oscillation converges, but with a very weak damping ratio. The reason is that units 1/2 have a higher mode-2 frequency than units 3/4 (see Table 2) and thus are exposed to a greater SSR risk in this operating condition. If SEDCs were applied, however, the torsional oscillation can be damped out soon and the SSR risk is avoided effectively for all units online.

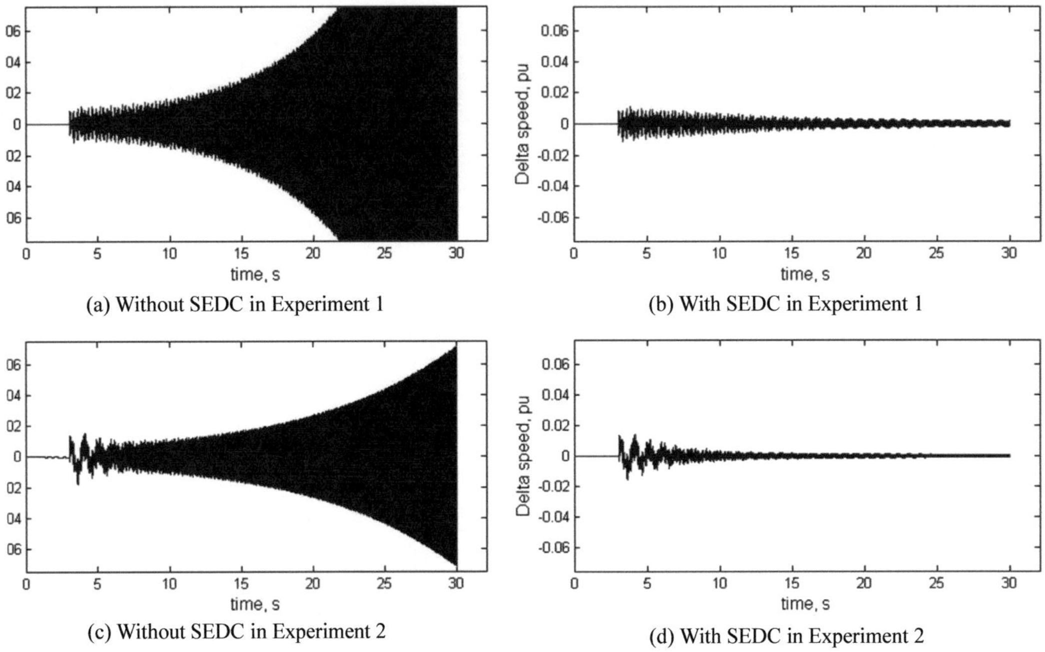

Figure 6. Delta mechanical speed of the high-pressure turbine

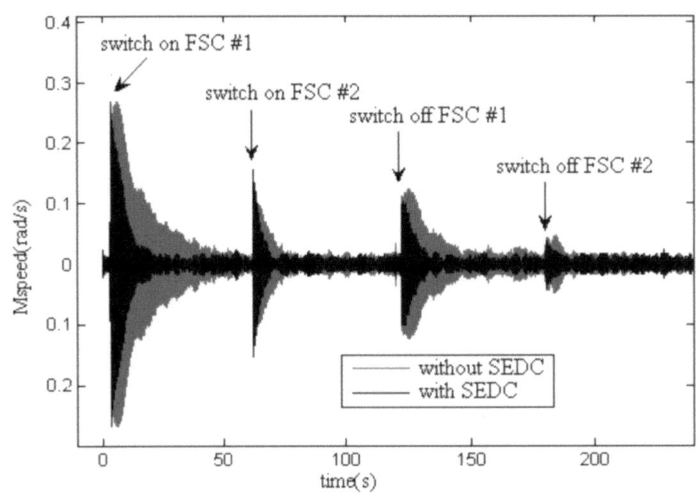

Figure 7. Dynamics of mode 2 during FSCs' switching-on/off operations under the 3-machine condition (unit 4)

From the results of the field tests, it can be concluded that:

i. There is real danger of SSR for the Shangdu series-compensated system, especially for mode 2, which exhibits instability when three or four machines are online while only one Shangdu-Chengde line is in service.

ii. With SEDC applied, the damping of all torsional modes is significantly improved so that the unstable modes become stabilized, which guarantees the safety of the generator shaft and the stability of the system.
iii. The field test includes various operations and covers different working conditions. Especially, the most dangerous conditions previously recognized (i.e., the condition of three/four machine and one Shangdu-Chengdu line) were tested sufficiently. Therefore the effectiveness as well as the robustness of SEDC is very well validated.

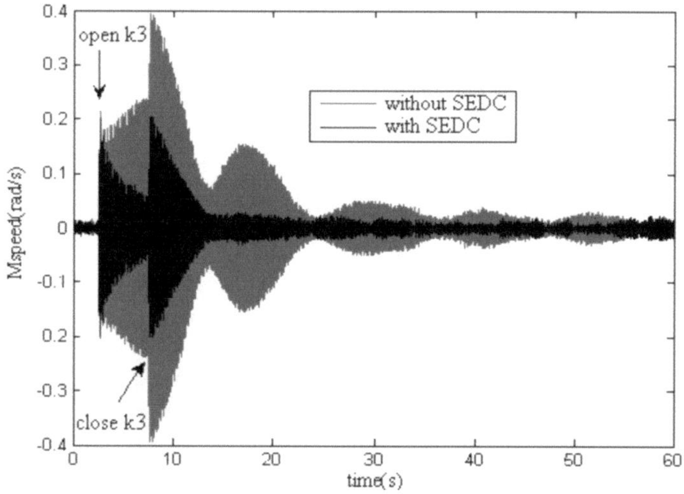

Figure 8. Dynamics of mode 2 during switching operations of Shang-Cheng line #1 under the 3-machine condition (unit 1)

4. Application of GA-SA for SVC-SSDC optimization

4.1. A description of the target system and the multimodal SSR problem

The Jinjie Power Plant is located near Yulin city, Shanxi Province, about 500 kilometers west of Beijing city. As a mine-mouth power plant, it has four 600 MW turbine-generators connected to the North China Power Grid through 500 kV transmissions. Fig. 9 illustrates the one-line diagram of the equivalent system. To improve the transferring capability, 35% SC is applied to the parallel lines between Jinjie Power Plant, Xinzhou substation and Shibei substation. The neighboring Fugu Power Plant, with two 600 MW turbine-generators, is connected to the Xinzhou substation through two uncompensated lines.

Each of Jinjie and Fugu turbine-generators consists of four rotors, i.e., a high-and-intermediate-pressure (HIP) turbine rotor, two low-pressure (LPA/LPB) turbine rotors and the generator rotor, thus resulting in three subsynchronous torsional modes. The modal frequencies were measured via field test and they are about 13.11 Hz (mode 1), 22.77 Hz (mode 2) and 28.22 Hz (mode 3) respectively.

Similarly, a thorough evaluation of the SSR problem was conducted under all possible system conditions for the Jinjie system. As part of the results of eigenvalue analyses, Table 4 gives a list of the real parts, or modal damping, of the three SSR modes under the 4-machine operating conditions (with different load levels and line status, see the "No SVC" column). Thus, the severity of SSR is quantified and the most risky situations are identified.

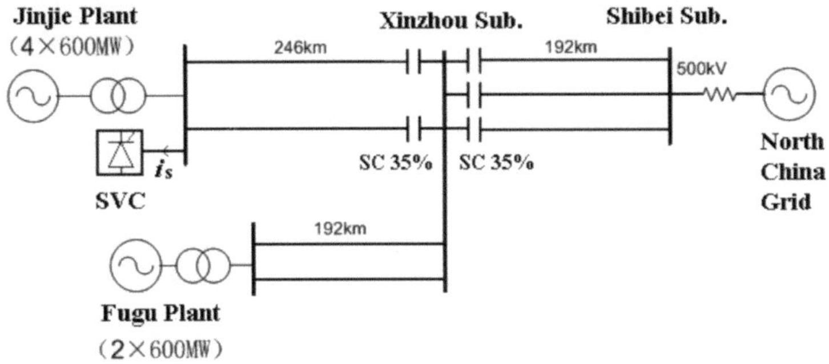

Figure 9. The one-line diagram of the equivalent transmission system

Oper. Con.		Mode 1 (s^{-1})	Mode 2 (s^{-1})	Mode 3 (s^{-1})
Gen. output	lines			
4×0%	2+3	-0.0370	-0.0060	**0.3202**
4×40%	2+3	-0.0378	-0.0262	**0.3099**
4×100%	2+3	-0.0367	-0.0468	**0.3062**
4×0%	2+2	-0.0323	**0.0001**	**0.5708**
4×40%	2+2	-0.0333	-0.0200	**0.5620**
4×100%	2+2	-0.0326	-0.0404	**0.5649**
4×0%	1+3	-0.0156	**0.0223**	**0.3795**
4×40%	1+3	-0.0175	-0.0023	**0.3620**
4×100%	1+3	-0.0182	-0.01760	**0.3480**
4×0%	1+2	-0.0132	**0.0305**	**0.2301**
4×40%	1+2	-0.0152	-0.0105	**0.2108**
4×100%	1+2	-0.0161	-0.0090	**0.1902**

Table 4. Real Parts of SSR modes without and with SVC-SSDC under the 4-machine operating conditions (Note: The "m+n" of the "lines" column means that there are "m" Jinjie-Xinzhou lines and "n" Jinzhou-Shibei lines in service.)

Through the evaluation study, characteristics of the SSR problem are summarized as follows:

i. Mode 1 is well-damped in all operating conditions. Mode 2 is stable in most common operating conditions. However, it becomes weakly-damped or even unstable in some conditions when the generator output is relatively low (corresponding to a lower mechanical damping) or a part of the transmissions are out of service. Mode 3 is the worst damped and tends to be unstable for a number of operating conditions. Thus the SSR problem is a multimodal one.
ii. Modal damping is affected by several factors, including the number of online generators and lines as well as the generators' output. What's more, each torsional

mode has a unique least damped condition, making it a challenge to design a controller adaptable to all operating conditions.

4.2. The SVC-based subsynchronous damping control

Various countermeasures, including the blocking filter, SEDC and SVC were considered to handle the SSR problem of the system. However, SVC was finally chosen as the SSR-depressing device through detailed study and technical versus economic comparison.

4.2.1. *The power-electronic circuit and mathematical model of SVC*

The adopted six-pulse SVC (see Fig. 10) comprises a thyristor controlled reactor (TCR) and a passive LC filter tuned for the 5th harmonics. It is connected to the high-voltage (500kV) side of the power plant through a step-up transformer.

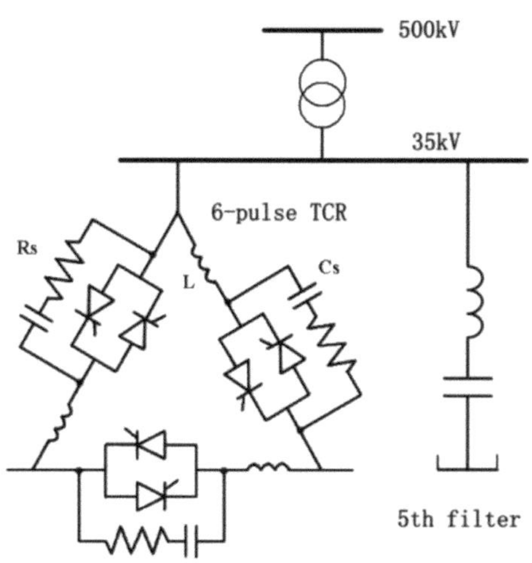

Figure 10. The power-electronic circuit of the six-pulse SVC

Since its MW loss (generally less than 0.6%) is negligible, SVC can be modeled as a controllable inductive admittance [21]. Fig. 11 illustrates the control scheme and mathematical model that describes the dynamics of the fundamental admittance, wherein SVC circuit is represented by combination of a first-order plus dead-time model and a nonlinear function $f(\alpha)$, in series, i.e.

$$f(\alpha)=1/\left[1/(B_{\mathrm{TCR}}+B_{\mathrm{FLT}})-X_{\mathrm{T}}\right],\ B_{\mathrm{TCR}}=\left[2\alpha+\sin(2\alpha)-\pi\right]/(\pi X_{\mathrm{L}}),\ 0\leq\alpha\leq 0.5\pi \tag{7}$$

where α is the firing angle as determined by the reference $B_{1\mathrm{ref}}$ and $f^{-1}(\alpha)$; X_{T} is the transformer reactance, B_{TCR}, B_{FLT} are admittances of the TCR and the filter; X_{L} is the full reactance of the TCR.

With some simplification, the dynamics of fundamental admittance can be expressed concisely by:

$$B_1 = \frac{e^{-sT_d}}{1+sT_s} B_{1ref}, \quad B_1 \in [\frac{-(1-B_{FLT}X_L)}{X_T(1-B_{FLT}X_L)+X_L}, 0] \tag{8}$$

where T_d and T_s represent the dead and delay times of the thyristor circuits. For the six-pulse SVC, generally $T_d = T_0/24$, $T_s = T_0/6$ and T_0=0.02 second.

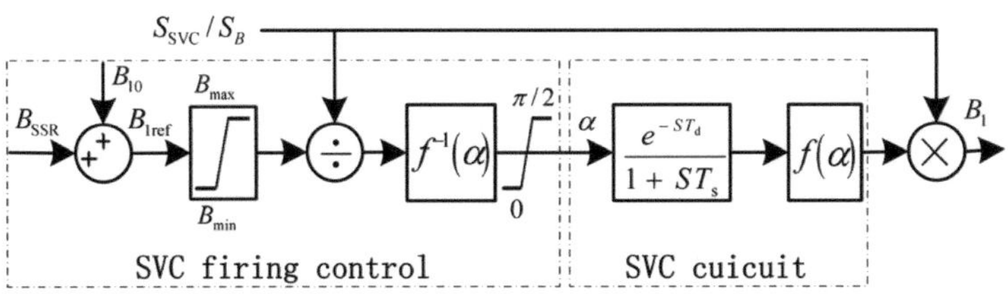

Figure 11. The dynamic model of the SVC

To depress SSR, it is necessary for SVC to generate currents at the complementary subsynchronous frequencies. For this purpose, we propose the idea of modulating the fundamental admittance with subsynchronous frequencies, i.e., to make the reference value of the fundamental admittance (B_{1ref}) vary according to the subsynchronous frequencies, or

$$B_{1ref} = B_{10,ref} + B_{SSR}, \quad B_{SSR} = \sum_{m=1}^{3} B_{1m}(t)\cos(\omega_m t + \varphi_{m,ref}) \tag{9}$$

where $B_{10,ref}$, B_{SSR} denote the DC and subsynchronously modulated components of B_{1ref}; ω_m is the torsional frequency; $\varphi_{m,ref}$ is the initial phase.

Thus, by some derivation, the current of SVC can be derived as

$$i_S = i_1 + \sum_{m=1}^{3} i_{\omega_0-\omega_m} + \sum_{m=1}^{3} i_{\omega_0+\omega_m} + \sum_{k=1}^{\infty}\left(i_{pk\pm1} + i_{(pk\pm1)\omega_0\mp\omega_m}\right) + i_{else} \tag{10}$$

where $i_1 = U_1 B_{10}\cos(\omega_0 t)$ is the fundamental current; $i_{\omega_0\mp\omega_m} = 0.5U_1 B_{1m}(t)\cos\left[(\omega_0 \mp \omega_m)t \mp \varphi_m\right]$ are the sub- and super- synchronous currents; $i_{pk\pm1}$ denotes the characteristic harmonics; $i_{(pk\pm1)\omega_0\mp\omega_m}$ are fractional harmonics caused by the subsynchronous modulating control; i_{else} represents the rest of the negligible components; and U_1 is fundamental voltage.

It is observed that by modulating the fundamental admittance with subsynchronous frequencies, SVC produces sub- and super- synchronous currents with controllable amplitude and phase, which subsequently generate subsynchronous damping torque on generator shaft. This is the basic principle of SSR mitigation using SVC. As shown in (10), this method simultaneously causes fractional currents.

4.2.2. *Configuration of the proposed SVC-SSDC*

The proposed SSDC is a real-time control system that works through the SVC power-electronic circuit to inject sub- and super- synchronous currents into the generator-stators and produces damping torque to ease SSR. Fig. 12 shows the block diagram of the 3-path SVC-SSDC designed for the Jinjie system. The average mechanical speed of HIP turbines of the online machines provides the input, which becomes the standard signal $\Delta\omega_{\mathrm{HIP}}$ after proper conditioning. It is then passed to the low- & high-pass filter. Speed signal filtered next proceeds to three separate control paths. Each, tuned to a specific mode, comprises a modal filter, an amplifier k_i and a unity-gain phase-shifter $(1-sT_k)^2/(1+sT_k)^2$ to generate the control signal for the corresponding mode. The control signals of all torsional modes are summarized, clipped and finally added to form the control signal, or B_{SSR}, which drive the SVC circuit as shown in Fig. 12.

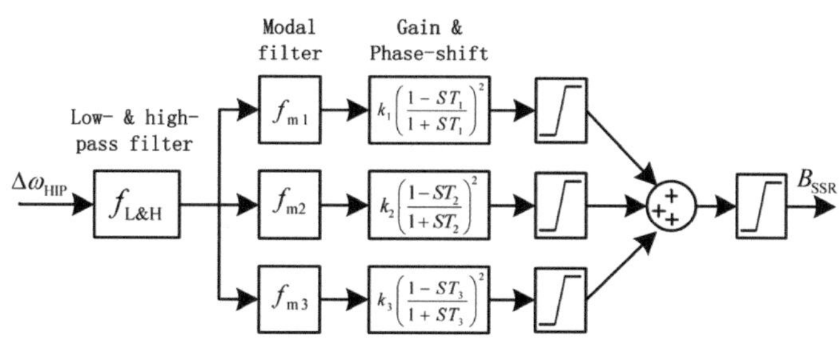

Figure 12. Configuration of SVC-SSDC

The purpose of the low- & high-pass filter, with the transfer functions shown in (11), is to depress the unrelated signals, including the DC, low-frequency (<10Hz) and high-frequency (>40Hz) components, to enable only related subsynchronous signals to pass through.

$$f_{\mathrm{L\&H}}(s)=\frac{1}{1+s/\omega_{\mathrm{L}}+(s/\omega_{\mathrm{L}})^2}\frac{(s/\omega_{\mathrm{H}})^2}{1+s/\omega_{\mathrm{H}}+(s/\omega_{\mathrm{H}})^2},\ \omega_{\mathrm{L}}=2\pi\cdot 40,\ \ \omega_{\mathrm{H}}=2\pi\cdot 10 \qquad (11)$$

The modal filter is a series combination of a second-order band-pass filter and two second-order band-stop filters, as described in (12):

$$f_{\mathrm{m}\,i}(s)=\frac{s/\omega_{\mathrm{P}i}}{1+8\pi s/\omega_{\mathrm{P}i}^2+(s/\omega_{\mathrm{P}i})^2}\frac{1+(s/\omega_{\mathrm{B}i1})^2}{1+6\pi s/\omega_{\mathrm{B}i1}^2+(s/\omega_{\mathrm{B}i1})^2}\frac{1+(s/\omega_{\mathrm{B}i2})^2}{1+6\pi s/\omega_{\mathrm{B}i2}^2+(s/\omega_{\mathrm{B}i2})^2} \qquad (12)$$

where $\omega_{\mathrm{P1}}=\omega_{\mathrm{B21}}=\omega_{\mathrm{B31}}=\omega_1$ =82.38 rad/s, $\omega_{\mathrm{P2}}=\omega_{\mathrm{B11}}=\omega_{\mathrm{B32}}=\omega_2$ =143.07 rad/s, $\omega_{\mathrm{P3}}=\omega_{\mathrm{B12}}=\omega_{\mathrm{B22}}=\omega_3$ =177.33 rad/s.

Then the SVC-SSDC can be reformulated into (13)

$$\Delta B_{\mathrm{SSR}}=H(s)\Delta\omega_{\mathrm{HIP}} \qquad (13)$$

where $H(s)$ is the integrated transfer function of SVC-SSDC.

4.2.3. Linearized model of the controlled system

To formulate the control-design problem, the whole system is linearized around a certain operating point to obtain an open-loop linear model [22-24], as described in (14).

$$\Delta\omega_{\mathrm{HIP}} = G(s)\Delta B_{\mathrm{SSR}} \tag{14}$$

where $G(s)$ represents the open-loop transfer function from ΔB_{SSR} to $\Delta\omega_{\mathrm{HIP}}$.

The closed-loop system model is obtained by combining (13) with (14), as shown in Fig.13, where the control reference r is set to 0; d denotes noise and/or model uncertainties. With some manipulation, the transfer functions between the input d and the outputs y, ΔB_{SSR} can be derived as:

$$y(s) = S(s)d(s) \tag{15}$$

$$\Delta B_{\mathrm{SSR}}(s) = R(s)d(s) \tag{16}$$

where $S(s) = \left[1 + G(s)H(s)\right]^{-1}$ is the sensitivity function, and $R(s) = H(s)S(s)$.

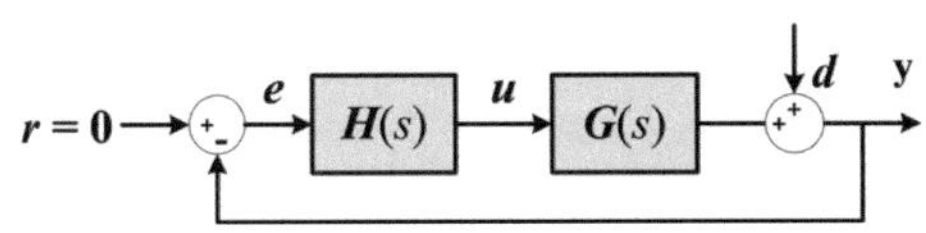

Figure 13. The close-loop system

For the target system, the open-loop and the closed-loop system function, i.e., $G(s), S(s)$ have a maximum order of 93 and 258. Obviously, their poles that correspond to torsional modes represent the stability or damping of these modes for the open-loop and the closed-loop systems, respectively. Hence, by eigen-analysis, the relationship between torsional damping with system variables and SVC-SSDC parameters can be traced out. With pre-determined filtering functions of SVC-SSDC, the selection of gains and phase-shifts or $k_i, T_i (i = 1,2,3)$ in Fig.12 is the key to solve the multimodal SSR problem.

4.3. Optimal design of the SVC-SSDC based on GA-SA

For the target system, sixty representative conditions are selected as the "evaluation conditions" (partly listed in Table 4), which cover the full range of generator output levels as well as the status of transmission lines. While these evaluation conditions by no means limit the situations under which the power system operates, together they form the edge of practical operating conditions and the most unfavorable situations relevant to SVC-SSDC. Therefore, the derived SVC-SSDC is robust enough to stabilize the system under all possible conditions.

The control-design task of SVC-SSDC can also be formulated into a nonlinear constrained optimization problem similar to that of (4)-(5). Specifically, it is expressed by (17)-(18).

$$\begin{aligned}&\max f\\&\text{subject to: } \left|R(j\omega)\right|_\infty \le R_{\mathrm{m}}, \left|k_i\right| \le k_{\mathrm{ub},i},\ 0 \le T_i \le T_{\mathrm{ub},i}\end{aligned} \tag{17}$$

$$\begin{aligned}&f = \sum_{i=1}^{3} w_i \eta_i + w_4 \min\{\eta_1, \eta_2, \eta_3\}, \sum_{i=1}^{4} w_i = 1, w_i > 0\\&\eta_i = \min_{j=1}^{j=60}\left\{\sigma_{ij}\right\},\ \sigma_{ij} = -\mathrm{Re}(\lambda_{ij}) / \left|\mathrm{Im}(\lambda_{ij})\right|,\\&\lambda_{ij} = \lambda_i\left\{S_j(k_1 \sim k_3, T_1 \sim T_3)\right\}; i = 1,2,3; j = 1,\ldots,60\end{aligned} \tag{18}$$

where k_i, T_i are control parameters; S_j is the sensitivity function; $\lambda_i\{\cdot\}$ means the calculation of the SSR mode of the included transfer function; the positive weight coefficients $w_1 = w_2 = w_3 = 0.2, w_4 = 0.4$; the subscripts *i, j* denote the number of SSR mode and the operating condition respectively.

In (17), $\left|R\right|_\infty$ is the H∞-norm of R, which represents the peak gain of R across all frequencies and can be calculated with the formula (19); R_{m} is the desired maximum value of $\left|R\right|_\infty$; and $k_{\mathrm{ub},i}$, $T_{\mathrm{ub},i}$ are the upper bounds of the gains and time constants.

$$\left|R\right|_\infty = \max_{\omega}\left|R(j\omega)\right| \tag{19}$$

Practically, the output of SSDC is restricted by the SVC capacity. Therefore, to limit the control output and to improve robustness to additive model uncertainty, the transfer function from disturbance to control output, i.e. $R(s)$ in (13), should not be too large. In the other side, SVC rating has great effect on the tuning of control gains and the effectiveness of the closed-loop system. Generally, larger size of SVC leads to higher gains and better torsional damping. However, equipment cost and operation loss increase accordingly. Thus, SVC size was determined for achieved the balance between cost and performance, 240MVA, about 10% of the rated power of all Jinjie machines.

The control-design problem (17)-(18) can be solved again by the proposed GA-SA method. In this case, GA-SA converges to satisfactory results after 40 - 60 iterations, consuming about 10-15 minutes on a modern computer. The optimized SVC-SSDC parameters are listed as follows: $k_1 = 25.46$, $k_2 = -26.03$, $k_3 = 25.02$, $T_1 = 0.0070\mathrm{s}$, $T_2 = 0.0053\mathrm{s}$, $T_3 = 0.0025\mathrm{s}$. The maximum fitness value generated is 0.00185.

4.4. Verification of the optimized SVC-SSDC

4.4.1. Small-signal eigen-analysis

A thorough eigenvalue analysis has been conducted, first in the absence and then in the presence of SVC-SSDC, to check the damping performance of the closed-loop system. Results of the specific operating conditions are presented in Table 5. Comparisons between damping values with and without SVC-SSDC suggest that all torsional modes see considerable improvement in modal damping by using the optimized SSDC and all unstable modes are well stabilized.

Oper. Con.		Mode 1 (s^{-1})		Mode 2 (s^{-1})		Mode 3 (s^{-1})	
Gen. output	lines	No SVC	With SVC	No SVC	With SVC	No SVC	With SVC
4×0%	2+3	-0.0370	-0.1524	-0.0060	-0.3903	**0.3202**	-0.7235
4×40%	2+3	-0.0378	-0.2174	-0.0262	-0.4796	**0.3099**	-0.6081
4×100%	2+3	-0.0367	-0.3256	-0.0468	-0.5863	**0.3062**	-0.4118
4×0%	2+2	-0.0323	-0.1566	**0.0001**	-0.4559	**0.5708**	-2.0244
4×40%	2+2	-0.0333	-0.2249	-0.0200	-0.5479	**0.5620**	-1.6510
4×100%	2+2	-0.0326	-0.3376	-0.0404	-0.6581	**0.5649**	-1.1766
4×0%	1+3	-0.0156	-0.1596	**0.0223**	-0.6877	**0.3795**	-0.5840
4×40%	1+3	-0.0175	-0.2437	-0.0023	-0.7942	**0.3620**	-0.8436
4×100%	1+3	-0.0182	-0.3792	-0.01760	-0.9205	**0.3480**	-1.2701
4×0%	1+2	-0.0132	-0.1645	**0.0305**	-0.7615	**0.2301**	-0.4166
4×40%	1+2	-0.0152	-0.2495	-0.0105	-0.8712	**0.2108**	-0.6349
4×100%	1+2	-0.0161	-0.3864	-0.0090	-1.0017	**0.1902**	-0.9860

Table 5. Real Parts of SSR modes without and with SVC-SSDC under the 4-machine operating conditions (Note: The "m+n" of the "lines" column means that there are "m" Jinjie-Xinzhou lines and "n" Jinzhou-Shibei lines in service.)

4.4.2. EMT study with real-time digital simulator

To further investigate the performance of the proposed control scheme, real SSDC controllers were developed using digital signal processors. The controllers were then connected to real-time digital simulator (RTDS), which thoroughly incorporated all system nonlinearities, including the power electronic SVC. Real-time electromagnetic simulations were conducted to check the response of the controlled system in the presence of various disturbances to validate the effectiveness of the developed controllers. Some typical results are presented here.

Preliminary study shows that, when only three Xin-Shi SCs are in service, the system is SSR-stable. However, once Jin-Xin SCs are switched on, the system becomes unstable. Fig. 14 depicts the responses of the system during consecutive switch-ons of two Jin-Xin SCs and SVC-SSDC. At the initial stage, Xin-Shi SCs are in service while Jin-Xin SCs and SVC-SSDC are out of service. When #1 Jin-Xin SC is switched on, the system is still stable, as indicated by the converging speed deviation in Fig.14(a). However, once #2 Jin-Xin SC is added in, SSR diverges rapidly and the generators experience violent torsional vibration, as indicated by the exponentially growing delta mechanical speed. This would probably lead to great damage to generator shafts. When SVC-SSDC is applied, SSR attenuates immediately. Fig.14(b) shows the current component of TCR corresponding to torsional mode 3, which is the most intractable among the three SSR modes.

To test the SVC-SSDC in the presence of large disturbances, a three-phase short-circuit is triggered first on the SSDC-free and then, on the SSDC-controlled system. The results are illustrated in Fig. 15.

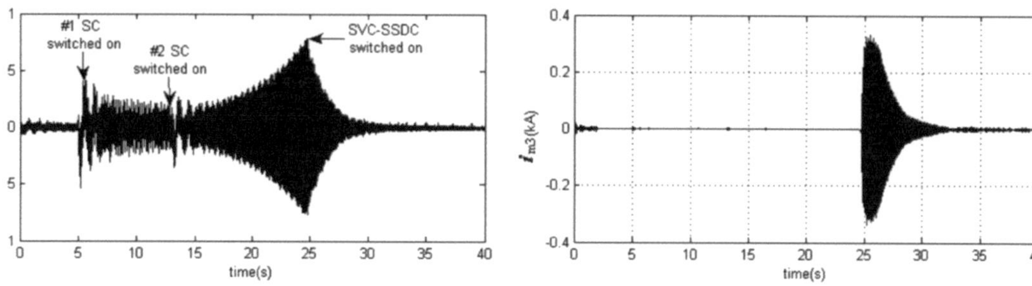

(a) Delta mechanical speed of the HIP turbine (b) The mode- 3component of TCR current

Figure 14. System response during the switching of the Jin-Xin SCs and SVC-SSDC

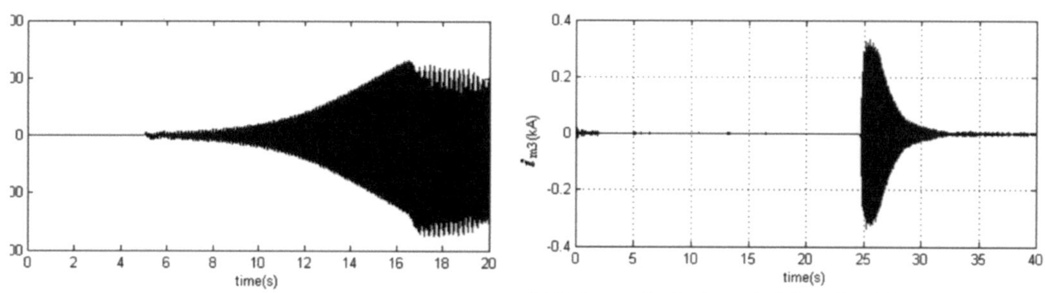

(a) Delta mechanical speed of the HIP turbine (without SVC-SSDC) (b) Delta mechanical speed of the HIP turbine (with SVC-SSDC)

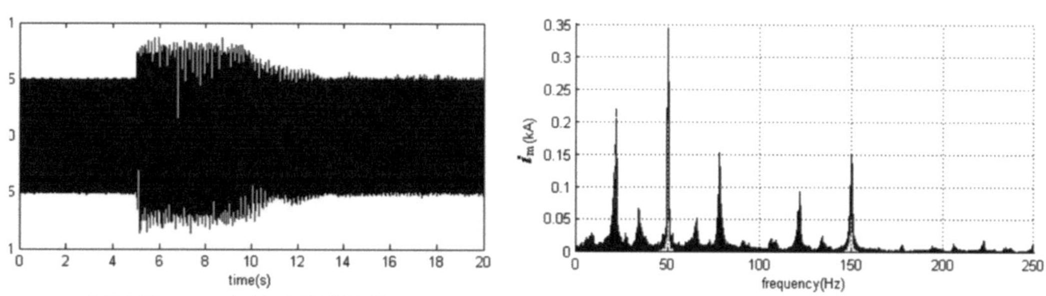

(c) TCR current (with SVC-SSDC) (d) Spectrum of TCR current (with SVC-SSDC)

Figure 15. System response following a three-phase fault

As shown in Fig.15 (a), in the absence of SSDC, the faulted system is unstable and soon collapses. However, when SVC-SSDC is put into service, the subsynchronous oscillation is successfully damped out (see Fig.15(b)). Fig.15(c) displays SVC's current output during the experiment. For a short period of time after the fault (about 5 seconds), the magnitude of the feedback is so large that the SSDC works nearly in a "bang-bang" mode due to the intense impact of the fault. As the vibration is weakened, the SSDC output becomes attenuated accordingly. From the spectrum of the TCR current shown in Fig.15(d), it can be seen that during the controlling process, the TCR current contains harmonics at the frequencies of $(2n \pm 1)\omega_0 \pm \omega_m$. This is in line with the previous theoretic analysis.

Besides the exemplary experiment, simulations under other operating conditions have been conducted in the presence of different large disturbances. From Figs.14-15 and other results not presented here, it is concluded that SVC-SSDC is effective in improving torsional damping and can satisfactorily mitigate subsynchronous vibrations caused by any type of disturbance.

5. Conclusions

In this chapter, a novel intelligent tuning technique based on the combined genetic algorithm and simulated annealing, or GA-SA, was proposed and explicated for the design of optimized controllers to depress SSR in practical multi-machine transmission systems compensated with fixed series capacitors. The advantage of the proposed method lies in that: by solving the constrained optimization problem, the multiple control parameters are simultaneously optimized and the obtained controller is robustified to damp multimodal SSR under a wide range of operating conditions. The proposed method is applied to the control-design problems of two practical power systems, i.e., SEDC of the Shangdu system and SVC-SSDC of the Jinjie system. The dynamics of the controlled system are investigated via eigenvalue analyses, electromagnetic simulation and/or field tests, the results of which fully demonstrate the effectiveness of the GA-SA tuned controllers.

Author details

Xiaorong Xie
State Key Lab. of Power System, Department of Electrical Engineering, Tsinghua University, Beijing, China

Acknowledgement

This work was supported in part by National Natural Science Foundation of China (Grant No. 51077080 and 51037002) and State Key Lab. of Power System (Grant No.SKLD11M02).

6. References

[1] Subsynchronous Resonance Working Group of the System Dynamic Performance Subcommittee. Reader's guide to subsynchronous resonance Power Systems. *IEEE Trans.Power System*. 1992; 7(1):150-157.

[2] J. W. Balance, S. Goldberg. Subsynchronous resonance in series compensated transmission lines. *IEEE Trans. Power Apparatus and Systems*. 1973, 92(5): 1649-1658.

[3] IEEE Committee Report. A bibliography for study of subsynchronous resonance between rotating machines and power systems. *IEEE Trans. Power Apparatus and Systems*. 1976, 95(1): 216-218.

[4] IEEE Committee Report. First supplement to a bibliography for study of subsynchronous resonance between rotating machines and power systems. *IEEE Trans. Power Apparatus and Systems*. 1979, 98(6): 1872-1875.

[5] IEEE Committee Report. Second supplement to a bibliography for study of subsynchronous resonance between rotating machines and power systems. *IEEE Trans. Power Apparatus and Systems*. 1985, 104(2): 321-327.

[6] IEEE Committee Report. Third supplement to a bibliography for study of subsynchronous resonance between rotating machines and power systems. paper no. 90 SM 328-5-PWRS presented at the PES Summer Meeting, Minneapolis, Minnesota, July 1990.

[7] IEEE Subsynchronous Resonance Working Group of the System Dynamic Performance Subcommittee Power System Engineering Committee. Countermeasures to subsynchronous resonance problems. *IEEE Trans. Power Apparatus and Systems*. 1980; 99(5): 1810-1818.

[8] IEEE Committee Report, Terms, definitions and symbols for subsynchronous oscillations. IEEE Trans. Power Apparatus and Systems. 1985, 104(3): 1326-1334.

[9] V. Pottakulath, E. P. Cheriyan, R. S. Kumar. Synthesis of power system model for SSR analysis. in *Proc. TENCON 2010 IEEE Region 10 Conf.*, 2010, pp. 545–550.

[10] P. M. Anderson, B. L. Agrawal, J. E. Van Ness, *Subsynchronous Resonance in Power Systems*. New York, NY: Wiley-IEEE Press, Feb., 1999.

[11] Liwei Wang, J. Jatskevich, H. W. Dommel, Re-examination of synchronous machine modeling techniques for electromagnetic transient simulations. *IEEE Trans. Power Systems*, 2007, 22(3): 1221–1230.

[12] U. Karaagac, J. Mahseredjian, O. Saad, S. Dennetiere. Synchronous machine modeling precision and efficiency in electromagnetic transients. *IEEE Trans. Power Delivery*, 2011, 26(2): 1072-1082.

[13] Donghui Zhang, Xiaorong Xie, Shiyu Liu, Shuqing Zhang. An intelligently optimized SEDC for multimodal SSR mitigation. *Electric Power Systems Research 2009*, 79(7): 1018-1024.

[14] Kit Po Wong, Yin Wa Wong. Combined genetic algorithm/simulated annealing/fuzzy set approach to short-term generation scheduling with take-or-pay fuel contract. *IEEE Transactions on Power System* 1996; 11(1): 128-136.

[15] Mantawy AH, Abdel-Magid YL, Selim SZ. Integrating genetic algorithms, tabu search, and simulated annealing for the unit commitment problem. *IEEE Transactions on Power Systems* 1999; 14(3): 829-836.

[16] Thompson M, Fidler JK. Application of the genetic algorithm and simulated annealing to LC filter tuning. *IEE Proceedings-Circuits Devices and System* 2001; 148(4): 177-182.

[17] Tang Renyuan, Yang Shiyou, Li Yan; Wen Geng, Mei Tiemin. Combined strategy of improved simulated annealing and genetic algorithm for inverse problem. IEEE Transactions on Magnetics 1996; 32(3): 1326-1329.

[18] Lee YK, Mohamed PS. A real-coded genetic algorithm involving a hybrid crossover method for power plant control system design. in *Proceedings of 2002 Congress on Evolutionary Computation*, pp.1069– 1074, Honolulu, Hawaii 2002.

[19] X. Xie, X. Guo, Y. Han. Mitigation of multimodal SSR using SEDC in the Shangdu series-compensated power system. *IEEE Trans. Power Systems*, 2011, 26(1): 384-391.

[20] Donghui Zhang, Xiaorong Xie, Shiyu Liu, Shuqing Zhang. An Intelligently Optimized SEDC for Multimodal SSR Mitigation. *Electric Power Systems Research*, 2009, vol. 7: 1018-1024.

[21] Hammad AE, El-Sadek M. Application of a thyristor-controlled var compensator for damping subsynchronous oscillations in power systems. *IEEE Trans. Power Apparatus and Systems*. 1984; 103(1): 198-206.

[22] Putman TH, Ramey DG. Theory of the modulated reactance solution for subsynchronous resonance. *IEEE Trans. Power Apparatus and Systems*. 1982; 101(6): 1527-1535.

[23] Hammad AE, El-Sadek M. Application of a thyristor-controlled var compensator for damping subsynchronous oscillations in power systems. *IEEE Trans. Power Apparatus and Systems*. 1984; 103(1): 198-206.

[24] Wang L, Hsu YY. Damping of subsynchronous resonance using excitation controllers and static var compensators: a comparative study. *IEEE Trans. Energy Conversion*. 1983; 3(1): 6-13.

Chapter 13

Fuzzy c-Means Clustering, Entropy Maximization, and Deterministic and Simulated Annealing

Makoto Yasuda

Additional information is available at the end of the chapter

http://dx.doi.org/10.5772/48659

1. Introduction

Many engineering problems can be formulated as optimization problems, and the deterministic annealing (DA) method [20] is known as an effective optimization method for such problems. DA is a deterministic variant of simulated annealing (SA) [1, 10]. The DA characterizes the minimization problem of cost functions as the minimization of Helmholtz free energy which depends on a (pseudo) temperature, and tracks the minimum of free energy while decreasing temperature and thus it can deterministically optimize the function at a given temperature [20]. Hence, the DA is more efficient than the SA, but does not guarantee a global optimal solution. The study on the DA in [20] addressed avoidance of the poor local minima of cost function of data clustering. Then it was extensively applied to various subjects such as combinational optimization problems [21], vector quantization [4], classifier design [13], pairwise data clustering [9] and so on.

On the other hand, clustering is a method which partitions a given set of data points into subgroups, and is one of major tools for data analysis. It is supposed that, in the real world, cluster boundaries are not so clear that fuzzy clustering is more suitable than crisp clustering. Bezdek[2] proposed the fuzzy c-means (FCM) which is now well known as the standard technique for fuzzy clustering.

Then, after the work of Li et al.[11] which formulated the regularization of the FCM with Shannon entropy, Miyamoto et al.[14] discussed the FCM within the framework of the Shannon entropy based clustering. From the historical point of view, however, it should be noted that Rose et al.[20] first studied the statistical mechanical analogy of the FCM with the maximum entropy method, which was basically probabilistic clustering.

To measure the "indefiniteness" of fuzzy set, DeLuca and Termini [6] defined fuzzy entropy after Shannon. Afterwards, some similar measures from the wider viewpoints of the indefiniteness were proposed [15, 16]. Fuzzy entropy has been used for knowledge retrieval from fuzzy database [3] and image processing [31], and proved to be useful.

Tsallis [24] achieved nonextensive extension of the Boltzmann-Gibbs statistics. Tsallis postulated a generalization form of entropy with a generalization parameter q, which, in a limit of $q \to 1$,reaches the Shannon entropy. Later on, Menard et.al.[12] derived a membership function by regularizing FCM with the Tsallis entropy.

In this chapter, by maximizing the various entropies within the framework of FCM, the membership functions which take the familiar forms of the statitical mechanical distribution functions are derived. The advantage to use the statistical mechanical membership functions is that the fuzzy c-means clustering can be interpreted and analyzed from a statistical mechanical point of view [27, 28]

After that, we focus on the Fermi-Dirac like membership function, because, as compared to the Maxwell-Boltzmann-like membership function, the Fermi-Dirac-like membership function has extra parameters α_ks (α_k corresponds to a chemical potential in statistical mechanics[19], and k denotes a data point), which make it possible to represent various cluster shapes like former clustering methods based on such as the Gaussian mixture[7], and the degree of fuzzy entropy[23]. α_ks strongly affect clustering results and they must be optimized under a normalization constraint of FCM. On the other hand, the DA method, though it is efficient, does not give appropriate values of α_ks by itself and the DA clustering sometimes fails if α_ks are improperly given. Accordingly, we introduce SA to optimize α_ks because, as pointed above, both of DA and SA contain the parameter corresponding to the system temperature and can be naturally combined as DASA.

Nevertheless, this approach causes a few problems. (1)How to estimate the initial values of α_ks under the normalization constraint ? (2)How to estimate the initial annealing temperature? (3)SA must optimize a real number α_k[5, 26]. (4)SA must optimize many α_ks[22].

Linear approximations of the Fermi-Dirac-like membership function is useful in guessing the initial α_ks and the initial annealing temperature of DA.

In order to perform SA in a many variables domain, α_ks to be optimized are selected according to a selection rule. In an early annealing stages, most α_ks are optimized. In a final annealing stage, however, only α_ks of data which locate sufficiently away from all cluster centers are optimized because their memberships might be fuzzy. Distances between the data and the cluster centers are measured by using linear approximations of the Fermi-Dirac-like membership function.

However, DASA suffers a few disadvantages. One of them is that it is not necessarily easy to interpolate membership functions obtained by DASA, since their values are quite different each other. The fractal interpolation method [17] is suitable for these rough functions [30].

Numerical experiments show that DASA clusters data which distribute in various shapes more properly and stably than single DA. Also, the effectiveness of the fractal interpolation is examined.

2. Fuzzy c-means

Let $X = \{\mathbf{x}_1, \ldots, \mathbf{x}_n\}(\mathbf{x}_k = (x_k^1, \ldots, x_k^p) \in R^p)$ be a data set in a p-dimensional real space, which should be divided into c clusters $C = \{C_1, \ldots, C_c\}$. Let $V = \{\mathbf{v}_1, \ldots, \mathbf{v}_c\}(\mathbf{v}_i = (v_i^1, \ldots, v_i^p))$ be the centers of clusters and $u_{ik} \in [0,1](i = 1, \ldots, c; k = 1, \ldots, n)$ be the membership function. Also let

$$J = \sum_{k=1}^{n} \sum_{i=1}^{c} u_{ik}(d_{ik})^m \quad (m > 1) \tag{1}$$

be the objective function of the FCM where $d_{ik} = \|\mathbf{x}_k - \mathbf{v}_i\|^2$. In the FCM, under the normalization constraint of

$$\sum_{i=1}^{c} u_{ik} = 1 \ \forall k, \tag{2}$$

the Lagrange function L_{FCM} is given by

$$L_{FCM} = J - \sum_{k=1}^{n} \eta_k \left(\sum_{i=1}^{c} u_{ik} - 1 \right), \tag{3}$$

where η_k is the Lagrange multiplier. Bezdek[2] showed that the FCM approaches crisp clustering as m decreases to $+1$.

3. Entropy maximization of FCM

3.1. Shannon entropy maximization

First, we introduce the Shannon entropy into the FCM clustering. The Shannon entropy is given by

$$S = -\sum_{k=1}^{n} \sum_{i=1}^{c} u_{ik} \log u_{ik}. \tag{4}$$

Under the normalization constraint and setting m to 1, the fuzzy entropy functional is given by

$$\delta S - \sum_{k=1}^{n} \alpha_k \delta \left(\sum_{i=1}^{c} u_{ik} - 1 \right) - \beta \sum_{k=1}^{n} \sum_{i=1}^{c} \delta(u_{ik} d_{ik}), \tag{5}$$

where α_k and β are the Lagrange multipliers, and α_k must be determined so as to satisfy Eq. (2). The stationary condition for Eq. (5) leads to the following membership function

$$u_{ik} = \frac{\mathrm{e}^{-\beta d_{ik}}}{\sum_{j=1}^{c} \mathrm{e}^{-\beta d_{jk}}}. \tag{6}$$

and the cluster centers

$$\boldsymbol{v}_i = \frac{\sum_{k=1}^{n} u_{ik} \boldsymbol{x}_k}{\sum_{k=1}^{n} u_{ik}}. \tag{7}$$

3.2. Fuzzy entropy maximization

We then introduce the fuzzy entropy into the FCM clustering.

The fuzzy entropy is given by

$$\hat{S} = -\sum_{k=1}^{n} \sum_{i=1}^{c} \{\hat{u}_{ik} \log \hat{u}_{ik} + (1 - \hat{u}_{ik}) \log(1 - \hat{u}_{ik})\}. \tag{8}$$

The fuzzy entropy functional is given by

$$\delta \hat{S} - \sum_{k=1}^{n} \alpha_k \delta \left(\sum_{i=1}^{c} \hat{u}_{ik} - 1 \right) - \beta \sum_{k=1}^{n} \sum_{i=1}^{c} \delta(\hat{u}_{ik} d_{ik}), \tag{9}$$

where α_k and β are the Lagrange multipliers[28]. The stationary condition for Eq. (9) leads to the following membership function

$$\hat{u}_{ik} = \frac{1}{e^{\alpha_k + \beta d_{ik}} + 1}, \tag{10}$$

and the cluster centers

$$\boldsymbol{v}_i = \frac{\sum_{k=1}^{n} \hat{u}_{ik} \boldsymbol{x}_k}{\sum_{k=1}^{n} \hat{u}_{ik}}. \tag{11}$$

In Eq. (10), β defines the extent of the distribution [27]. Equation (10) is formally normalized as

$$\hat{u}_{ik} = \frac{1}{e^{\alpha_k + \beta d_{ik}} + 1} \Big/ \sum_{j=1}^{c} \frac{1}{e^{\alpha_k + \beta d_{jk}} + 1}. \tag{12}$$

3.3. Tsallis entropy maximization

Let $\tilde{\boldsymbol{v}}_i$ and $\tilde{u}_{ik}$ be the centers of clusters and the membership functions, respectively.

The Tsallis entropy is defined as

$$\tilde{S} = -\frac{1}{q-1}\left(\sum_{k=1}^{n}\sum_{i=1}^{c} \tilde{u}_{i\,k}^{q} - 1\right), \tag{13}$$

where $q \in \boldsymbol{R}$ is any real number. The objective function is rewritten as

$$\tilde{U} = \sum_{k=1}^{n}\sum_{i=1}^{c} \tilde{u}_{ik}^{q} \tilde{d}_{ik}, \tag{14}$$

where $\tilde{d}_{ik} = \|\boldsymbol{x}_k - \tilde{\boldsymbol{v}}_i\|^2$.

Accordingly, the Tsallis entropy functional is given by

$$\delta\tilde{S} - \sum_{k=1}^{n} \alpha_k \delta \left(\sum_{i=1}^{c} \tilde{u}_{ik} - 1\right) - \beta \sum_{k=1}^{n}\sum_{i=1}^{c} \delta(\tilde{u}_{ik}^{q} \tilde{d}_{ik}). \tag{15}$$

The stationary condition for Eq. (15) yields to the following membership function

$$\tilde{u}_{ik} = \frac{\{1 - \beta(1-q)\tilde{d}_{ik}\}^{\frac{1}{1-q}}}{\tilde{Z}}, \tag{16}$$

where

$$\tilde{Z} = \sum_{j=1}^{c} \{1 - \beta(1-q)\tilde{d}_{jk}\}^{\frac{1}{1-q}}. \tag{17}$$

In this case, the cluster centers are given by

$$\tilde{\boldsymbol{v}}_i = \frac{\sum_{k=1}^{n} \tilde{u}_{ik}^{q} \boldsymbol{x}_k}{\sum_{k=1}^{n} \tilde{u}_{ik}^{q}}. \tag{18}$$

In the limit of $q \to 1$, the Tsallis entropy recovers the Shannon entropy [24] and $\tilde{u}_{ik}$ approaches u_{ik} in Eq.(6).

4. Entropy maximization and statistical physics

4.1. Shannon entropy based FCM statistics

In the Shannon entropy based FCM, the sum of the states (the partition function) for the grand canonical ensemble of fuzzy clustering can be written as

$$Z = \prod_{k=1}^{n} \sum_{i=1}^{c} \mathrm{e}^{-\beta d_{ik}}. \tag{19}$$

By substituting Eq. (19) for $F = -(1/\beta)(\log Z)$[19], the free energy becomes

$$F = -\frac{1}{\beta} \sum_{k=1}^{n} \log \left\{ \sum_{i=1}^{c} \mathrm{e}^{-\beta d_{ik}} \right\}. \tag{20}$$

Stable thermal equilibrium requires a minimization of the free energy. By formulating deterministic annealing as a minimization of the free energy, $\partial F / \partial \boldsymbol{v}_i = 0$ yields

$$\boldsymbol{v}_i = \frac{\sum_{k=1}^{n} u_{ik} \boldsymbol{x}_k}{\sum_{k=1}^{n} u_{ik}}. \tag{21}$$

This cluster center is the same as that in Eq. (7).

4.2. Fuzzy entropy based FCM statistics

In a group of independent particles, the total energy and the total number of particles are given by $E = \sum_l \epsilon_l n_l$ and $N = \sum_l n_l$, respectively, where ϵ_l represents the energy level and n_l represents the number of particles that occupy ϵ_l. We can write the sum of states, or the partition function, in the form:

$$Z_N = \sum_{\sum_l \epsilon_l n_l = E, \sum_l n_l = N} e^{-\beta \sum_l \epsilon_l n_l} \tag{22}$$

where β is the product of the inverse of temperature T and k_B (Boltzmann constant). However, it is difficult to take the sums in (22) counting up all possible divisions. Accordingly, we make the number of particles n_l a variable, and adjust the new parameter α(chemical potential) so as to make $\sum_l \epsilon_l n_l = E$ and $\sum_l n_l = N$ are satisfied. Hence, this becomes the grand canonical distribution, and the sum of states (the grand partition function) Ξ is given by[8, 19]

$$\Xi = \sum_{N=0}^{\infty} (e^{-\alpha})^N Z_N = \prod_l \sum_{n_l=0}^{\infty} (e^{-\alpha-\beta\epsilon_l})^{n_l}. \tag{23}$$

For particles governed by the Fermi-Dirac distribution, Ξ can be rewritten as

$$\Xi = \prod_l (1 + e^{-\alpha-\beta\epsilon_l}). \tag{24}$$

Also, n_l is averaged as

$$\langle n_l \rangle = \frac{1}{e^{\alpha+\beta\epsilon_l} + 1} \tag{25}$$

where α is defined by the condition that $N = \sum_l \langle n_l \rangle$ [19]. Helmholtz free energy F is, from the relationship $F = -k_B T \log Z_N$,

$$F = -k_B T \left(\log \Xi - \alpha \frac{\partial}{\partial \alpha} \log \Xi \right) = -\frac{1}{\beta} \left\{ \sum_l \log(1 + e^{-\alpha - \beta \epsilon_l}) + \alpha N \right\}. \tag{26}$$

Taking that

$$E = \sum_l \frac{\epsilon_l}{e^{\alpha + \beta \epsilon_l} + 1} \tag{27}$$

into account, the entropy $S = (E - F)/T$ has the form

$$S = -k_B \sum_l \{ \langle n_l \rangle \log \langle n_l \rangle + (1 - \langle n_l \rangle) \log (1 - \langle n_l \rangle) \}. \tag{28}$$

If states are degenerated to the degree of ν_l, the number of particles which occupy ϵ_l is

$$\langle N_l \rangle = \nu_l \langle n_l \rangle, \tag{29}$$

and we can rewrite the entropy S as

$$S = -k_B \sum_l \{ \frac{\langle N_l \rangle}{\nu_l} \log \frac{\langle N_l \rangle}{\nu_l} + \left(1 - \frac{\langle N_l \rangle}{\nu_l}\right) \log \left(1 - \frac{\langle N_l \rangle}{\nu_l}\right) \}, \tag{30}$$

which is similar to fuzzy entropy in (8). As a result, u_{ik} corresponds to a grain density $\langle n_l \rangle$ and the inverse of β in (10) represents the system or computational temperature T.

In the FCM clustering, note that any data can belong to any cluster, the grand partition function can be written as

$$\Xi = \prod_{k=1}^{n} \prod_{i=1}^{c} (1 + e^{-\alpha_k - \beta d_{ik}}), \tag{31}$$

which, from the relationship $F = -(1/\beta)(\log \Xi - \alpha_k \partial \log \Xi / \partial \alpha_k)$, gives the Helmholtz free energy

$$F = -\frac{1}{\beta} \sum_{k=1}^{n} \left\{ \sum_{i=1}^{c} \log(1 + e^{-\alpha_k - \beta d_{ik}}) + \alpha_k \right\}. \tag{32}$$

The inverse of β represents the system or computational temperature T.

4.3. Correspondence between Fermi-Dirac statistics and fuzzy clustering

In the previous subsection, we have formulated the fuzzy entropy regularized FCM as the DA clustering and showed that its mechanics was no other than the statistics of a particle system (the Fermi-Dirac statistics). The correspondences between fuzzy clustering (FC) and the Fermi-Dirac statistics (FD) are summarized in TABLE 1. The major difference between fuzzy clustering and statistical mechanics is the fact that data are distinguishable and can belong to multiple clusters, though particles which occupy a same energy state are not distinguishable. This causes a summation or a multiplication not only on i but on k as well in fuzzy clustering. Thus, fuzzy clustering and statistical mechanics described in this paper are not mathematically equivalent.

- **Constraints:** (a) Constraint that the sum of all particles N is fixed in FD is correspondent with the normalization constraint in FC. Energy level l is equivalent to the cluster number

	Fermi-Dirac Statistics	Fuzzy Clustering
Constraints	(a)$\sum_l n_l = N$ (b)$\sum_l \epsilon_l n_l = E$	(a)$\sum_{i=1}^{c} u_{ik} = 1$
Distribution Function	$\langle n_l \rangle = \frac{1}{e^{\alpha+\beta\epsilon_l}+1}$	$u_{ik} = \frac{1}{e^{\alpha_k+\beta d_{ik}}+1}$
Entropy	$S = -k_B \sum_l \left\{ \frac{\langle N_l \rangle}{\nu_l} \log \frac{\langle N_l \rangle}{\nu_l} + \left(1 - \frac{\langle N_l \rangle}{\nu_l}\right) \log \left(1 - \frac{\langle N_l \rangle}{\nu_l}\right) \right\}$	$S_{FE} = -\sum_{k=1}^{n}\sum_{i=1}^{c}\{u_{ik}\log u_{ik} + (1-u_{ik})\log(1-u_{ik})\}$
Temperature(T)	(given)	(given)
Partition Function(Ξ)	$\prod_l \left(1+e^{-\alpha-\beta\epsilon_l}\right)$	$\prod_{k=1}^{n}\prod_{i=1}^{c}\left(1+e^{-\alpha_k-\beta d_{ik}}\right)$
Free Energy(F)	$-\frac{1}{\beta}\left\{\sum_l \log(1+e^{-\alpha-\beta\epsilon_l}) + \alpha N\right\}$	$-\frac{1}{\beta}\sum_{k=1}^{n}\left\{\sum_{i=1}^{c}\log(1+e^{-\alpha_k-\beta d_{ik}}) + \alpha_k\right\}$
Energy(E)	$\sum_l \frac{\epsilon_l}{e^{\alpha+\beta\epsilon_l}+1}$	$\sum_{k=1}^{n}\sum_{i=1}^{c}\frac{d_{ik}}{e^{\alpha_k+\beta d_{ik}}+1}$

Table 1. Correspondence of Fermi-Dirac Statistics and Fuzzy Clustering.

i. In addition, the fact that data can belong to multiple clusters leads to the summation on k. (b) There is no constraint in FC which corresponds to the constraint that the total energy equals E in FD. We have to minimize $\sum_{k=1}^{n}\sum_{i=1}^{c} d_{ik}u_{ik}$ in FC.

- **Distribution Function:** In FD, $\langle n_l \rangle$ gives an average particle number which occupies energy level l, because particles can not be distinguished from each other. In FC, however, data are distinguishable, and for that reason, u_{ik} gives a probability of data belonging to multiple clusters.
- **Entropy:** $\langle N_l \rangle$ is supposed to correspond to a cluster capacity. The meanings of S and S_{FE} will be discussed in detail in the next subsection.
- **Temperature:** Temperature is given in both cases [1].
- **Partition Function:** The fact that data can belong to multiple clusters simultaneously causes the product over k for FC.
- **Free Energy:** Helmholtz free energy F is given by $-T(\log\Xi - \alpha_k \partial \log \Xi / \partial \alpha_k)$ in FC. Both S and S_{FE} equal $-\partial F/\partial T$ as expected from statistical physics.
- **Energy:** The relationship $E = F + TS$ or $E = F + TS_{FE}$ holds between E, F, T and S or S_{FE}.

4.4. Meanings of Fermi-Dirac distribution function and fuzzy entropy

In the entropy function (28) or (30) for the particle system, we can consider the first term to be the entropy of electrons and the second to be that of holes. In this case, the physical limitation that only one particle can occupy an energy level at a time results in the entropy that formulates the state in which an electron and a hole exist simultaneously and exchanging them makes no difference. Meanwhile, what correspond to electron and hole in fuzzy clustering are the probability of fuzzy event that a data belongs to a cluster and the probability of its complementary event, respectively.

Fig.2 shows a two-dimensional virtual cluster density distribution model. A lattice can have at most one data. Let M_l be the total number of lattices and m_l be the number of lattices which

[1] In the FCM, however, temperature is determined as a result of clustering.

have a data in it (marked by a black box). Then, the number of available divisions of data to lattices is denoted by

$$W = \prod_l \frac{M_l!}{m_l!(M_l - m_l)!} \tag{33}$$

which, from $S = k_B \log W$ (the Gibbs entropy), gives the form similar to (30)[8]. By extremizing S, we have the most probable distribution like (25). In this case, as there is no distinction between data, only the numbers of black and white lattices constitute the entropy. Fuzzy entropy in (8), on the other hand, gives the amount of information of weather a data belongs to a fuzzy set (or cluster) or not, averaged over independent data $\mathbf{x}_k$.

Changing a viewpoint, the stationary entropy values for the particle system seems to be a request for giving the stability against the perturbation with collisions between particles. In fuzzy clustering, data reconfiguration between clusters with the move of cluster centers or the change of cluster shapes is correspondent to this stability. Let us represent data density by $\langle\rangle$. If data transfer from clusters C_a and C_b to C_c and C_d as a magnitude of membership function, the transition probability from $\{\ldots, C_a, \ldots, C_b, \ldots\}$ to $\{\ldots, C_c, \ldots, C_d, \ldots\}$ will be proportional to $\langle C_a\rangle\langle C_b\rangle(1 - \langle C_c\rangle)(1 - \langle C_d\rangle)$ because a data enters a vacant lattice. Similarly, the transition probability from $\{\ldots, C_c, \ldots, C_d, \ldots\}$ to $\{\ldots, C_a, \ldots, C_b, \ldots\}$ will be proportional to $\langle C_c\rangle\langle C_d\rangle(1 - \langle C_a\rangle)(1 - \langle C_b\rangle)$. In the equilibrium state, the transitions exhibit balance (this is known as the principle of detailed balance[19]). This requires

$$\frac{\langle C_a\rangle\langle C_b\rangle}{(1 - \langle C_a\rangle)(1 - \langle C_b\rangle)} = \frac{\langle C_c\rangle\langle C_d\rangle}{(1 - \langle C_c\rangle)(1 - \langle C_d\rangle)}. \tag{34}$$

As a result, if energy d_i is conserved before and after the transition, $\langle C_i\rangle$ must have the form

$$\frac{\langle C_i\rangle}{1 - \langle C_i\rangle} = e^{-\alpha - \beta d_i} \tag{35}$$

or Fermi-Dirac distribution

$$\langle C_i\rangle = \frac{1}{e^{\alpha + \beta d_i} + 1}, \tag{36}$$

where α and β are constants.

Consequently, the entropy like fuzzy entropy is statistically caused by the system that allows complementary states. Fuzzy clustering handles a data itself, while statistical mechanics handles a large number of particles and examines the change of macroscopic physical quantity. Then it is concluded that fuzzy clustering exists in the extreme of Fermi-Dirac statistics, or the Fermi-Dirac statistics includes fuzzy clustering conceptually.

4.5. Tsallis entropy based FCM statistics

On the other hand, $\tilde{U}$ and $\tilde{S}$ satisfy

$$\tilde{S} - \beta\tilde{U} = \sum_{k=1}^{n} \frac{\tilde{Z}^{1-q} - 1}{1 - q}, \tag{37}$$

which leads to

$$\frac{\partial \tilde{S}}{\partial \tilde{U}} = \beta. \tag{38}$$

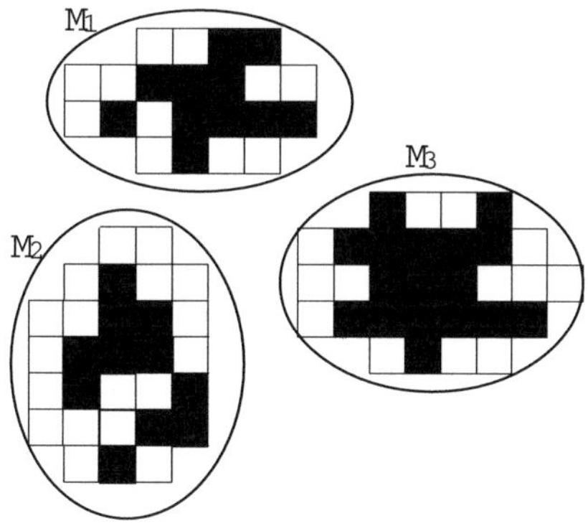

Figure 1. Simple lattice model of clusters. $M_1, M_2, \ldots$ represent clusters. Black and white box represent whether a data exists or not.

Equation (38) makes it possible to regard β^{-1} as an artificial system temperature T [19]. Then, the free energy can be defined as

$$\tilde{F} = \tilde{U} - T\tilde{S} = -\frac{1}{\beta}\sum_{k=1}^{n}\frac{\tilde{Z}^{1-q}-1}{1-q}. \tag{39}$$

$\tilde{U}$ can be derived from $\tilde{F}$ as

$$\tilde{U} = -\frac{\partial}{\partial\beta}\sum_{k=1}^{n}\frac{\tilde{Z}^{1-q}-1}{1-q}. \tag{40}$$

$\partial\tilde{F}/\partial\tilde{\boldsymbol{v}}_i = 0$ also gives

$$\tilde{\boldsymbol{v}}_i = \frac{\sum_{k=1}^{n}\tilde{u}_{ik}^{q}\boldsymbol{x}_k}{\sum_{k=1}^{n}\tilde{u}_{ik}^{q}}. \tag{41}$$

5. Deterministic annealing

The DA method is a deterministic variant of SA. DA characterizes the minimization problem of the cost function as the minimization of the Helmholtz free energy which depends on the temperature, and tracks its minimum while decreasing the temperature and thus it can deterministically optimize the cost function at each temperature.

According to the principle of minimal free energy in statistical mechanics, the minimum of the Helmholtz free energy determines the distribution at thermal equilibrium [19]. Thus, formulating the DA clustering as a minimization of (32) leads to $\partial F/\partial\mathbf{v}_i = 0$ at the current temperature, and gives (10) and (11) again. Desirable cluster centers are obtained by calculating (10) and (11) repeatedly.

In this chapter, we focus on application of DA to the Fermi-Dirac-like distribution function described in the Section 4.2.

5.1. Linear approximation of Fermi-Dirac distribution function

The Fermi-Dirac distribution function can be approximated by linear functions. That is, as shown in Fig.1, the Fermi-Dirac distribution function of the form:

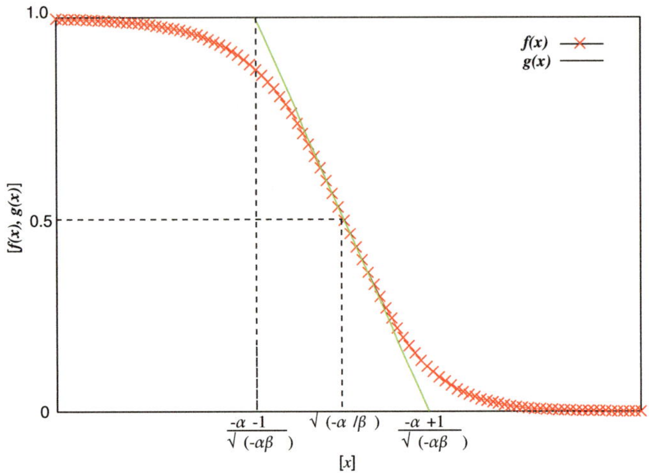

Figure 2. The Fermi-Dirac distribution function $f(x)$ and its linear approximation functions $g(x)$.

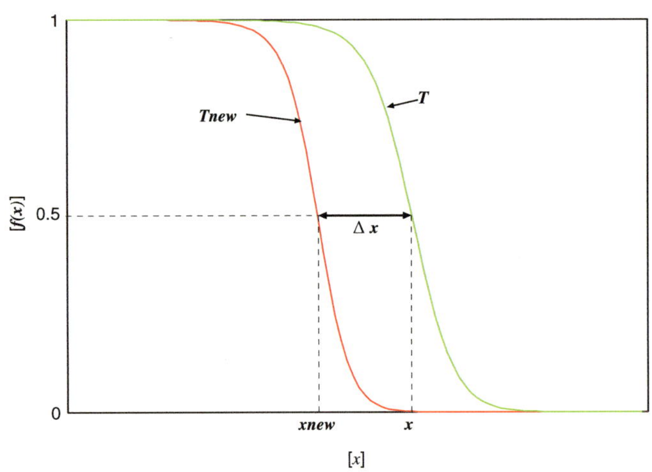

Figure 3. Decreasing of extent of the Fermi-Dirac distribution function from x to x_{new} with decreasing the temperature from T to T_{new}.

$$f(x) = \frac{1}{e^{\alpha+\beta x^2} + 1} \tag{42}$$

is approximated by the linear functions

$$g(x) = \begin{cases} 1.0 & \left(x \leq \frac{-\alpha - 1}{\kappa}\right) \\ -\frac{\kappa}{2}x - \frac{\alpha}{2} + \frac{1}{2} & \left(\frac{-\alpha - 1}{\kappa} \leq x \leq \frac{-\alpha + 1}{\kappa}\right), \\ 0.0 & \left(\frac{-\alpha + 1}{\kappa} \leq x\right) \end{cases} \tag{43}$$

where $\kappa = \sqrt{-\alpha\beta}$. $g(x)$ satisfies $g(\sqrt{-\alpha/\beta}) = 0.5$, and requires α to be negative.

In Fig.2, $\Delta x = x - x_{new}$ denotes a reduction of the extent of distribution with decreasing the temperature from T to T_{new} ($T > T_{new}$). The extent of distribution also narrows with increasing α. α_{new} ($\alpha < \alpha_{new}$) which satisfies $g(0.5)_\alpha - g(0.5)_{\alpha_{new}} = \Delta x$ is obtained as

$$\alpha_{new} = -\left\{\sqrt{-\alpha} + \sqrt{-\alpha\beta_{new}}\left(\frac{1}{\sqrt{\beta}} - \frac{1}{\sqrt{\beta_{new}}}\right)\right\}^2, \tag{44}$$

where $\beta = 1/T$ and $\beta_{new} = 1/T_{new}$. Thus, taking that T to the temperature at which previous DA was executed and T_{new} to the next temperature, a covariance of α_k's distribution is defined as

$$\Delta\alpha = \alpha_{new} - \alpha. \tag{45}$$

5.2. Initial estimation of α_k and annealing temperature

Before executing DA, it is very important to estimate the initial values of α_ks and the initial annealing temperature in advance.

From Fig.1, distances between a data point and cluster centers are averaged as

$$L_k = \frac{1}{c}\sum_{i=1}^{c} \|\mathbf{x}_k - \mathbf{v}_i\|, \tag{46}$$

and this gives

$$\alpha_k = -\beta(L_k)^2. \tag{47}$$

With given initial clusters distributing wide enough, (47) overestimates α_k, so that α_k needs to be adjusted by decreasing its value gradually.

Still more, Fig.1 gives the width of the Fermi-Dirac distribution function as wide as $2(-\alpha + 1)/(\sqrt{-\alpha\beta})$, which must be equal to or smaller than that of data distribution (=2R). This condition leads to

$$2\frac{-\alpha + 1}{\sqrt{-\alpha\beta}} = 2R. \tag{48}$$

As a result, the initial value of β or the initial annealing temperature is roughly determined as

$$\beta \simeq \frac{4}{R^2} \quad \left(T \simeq \frac{R^2}{4}\right). \tag{49}$$

5.3. Deterministic annealing algorithm

The DA algorithm for fuzzy clustering is given as follows:

1 *Initialize*: Set a rate at which a temperature is lowered T_{rate}, and a threshold of convergence test δ_0. Calculate an initial temperature $T_{high}(= 1/\beta_{low})$ by (49) and set a current temperature $T = T_{high}(\beta = \beta_{low})$. Place c clusters randomly and estimate initial α_ks by (47) and adjust them to satisfy the normalization constraint (2).
2 Calculate u_{ik} by (12).

3 Calculate $\mathbf{v}_i$ by (11).
4 Compare a difference between the current objective value $J_{m=1} = \sum_{k=1}^{n}\sum_{i=1}^{c} d_{ik}u_{ik}$ and that obtained at the previous iteration $\hat{J}$. If $\|J_{m=1} - \hat{J}\|/J_{m=1} < \delta_0 \cdot T/T_{high}$ is satisfied, then return. Otherwise decrease the temperature as $T = T * T_{rate}$, and go back to **2**.

6. Combinatorial algorithm of deterministic and simulated annealing

6.1. Simulated annealing

The cost function for SA is

$$E(\alpha_k) = J_{m=1} + S_{FE} + K\sum_{k=1}^{n}\left(\sum_{i=1}^{c} u_{ik} - 1\right)^2, \qquad (50)$$

where K is a constant.

In order to optimize each α_k by SA, its neighbor α_k^{new} (a displacement from the current α_k) is generated by assuming a normal distribution with a mean 0 and a covariance $\Delta\alpha_k$ defined in (45).

The SA's initial temperature $T_0(= 1/\beta_0)$ is determined so as to make an acceptance probability becomes

$$\exp\left[-\beta_0\{E(\alpha_k) - E(\alpha_k^{new})\}\right] = 0.5 \qquad (51)$$
$$(E(\alpha_k) - E(\alpha_k^{new}) \geq 0)$$

By selecting α_ks to be optimized from the outside of a transition region in which the membership function changes form 0 to 1, computational time of SA can be shortened. The boundary of the transition region can be easily obtained with the linear approximations of the Fermi-Dirac-like membership function. From Fig.1, data which have distances bigger than $\sqrt{-\alpha_k/\beta}$ from each cluster centers are selected.

6.2. Simulated annealing algorithm

The SA algorithm is stated as follows:

1 *Initialize*: Calculate an initial temperature $T_0(= 1/\beta_0)$ from (51). Set a current temperature T to T_0. Set an iteration count t to 1. Calculate a covariance $\Delta\alpha_k$ for each α_k by (45).
2 Select data to be optimized, if necessary.
3 Calculate neighbors of current α_ks.
4 Apply the Metropolis algorithm to the selected α_ks using (50) as the objective function.
5 If $max < t$ is satisfied, then return. Otherwise decrease the temperature as $T = T_0/\log(t+1)$, increment t, and go back to **2**.

6.3. Combinatorial algorithm of deterministic and simulated annealing

The DA and SA algorithms are combined as follows:

1 *Initialize*: Set a threshold of convergence test δ_1, and an iteration count l to 1. Set maximum iteration counts max_0, max_1, and max_2.
2 Execute the DA algorithm.
3 Set $max = max_0$, and execute the SA algorithm.
4 Compare a difference between the current objective value e and that obtained at the previous iteration $\hat{e}$. If $\|e - \hat{e}\|/e < \delta_1$ or $max_2 < l$ is satisfied, then go to **5**. Otherwise increment l, and go back to **2**.
5 Set $max = max_1$, and execute the SA algorithm finally, and then stop.

7. Experiments 1

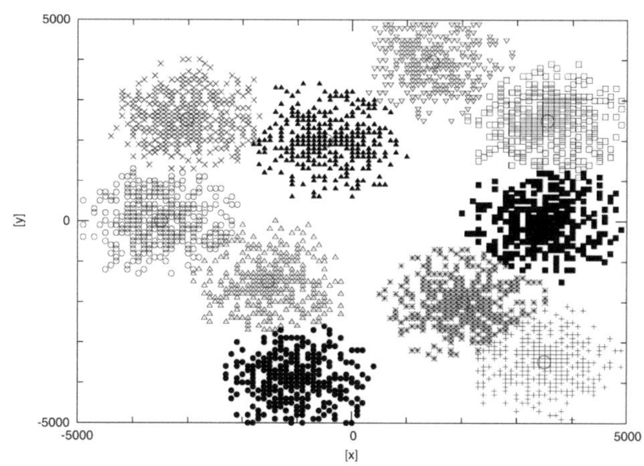

Figure 4. Experimental result 1. (Fuzzy clustering result using DASA. Big circles indicate centers of clusters.)

To demonstrate effectiveness of the proposed algorithm, numerical experiments were carried out. DASA's results were compared with those of DA (single DA).

We set $\delta_0 = 0.5$, $\delta_1 = 0.01$, $T_{rate} = 0.8$, $max_0 = 500$, $max_1 = 20000$, and $max_2 = 10$. We also set R in (48) to 350.0 for experimental data 1~3, and 250.0 for experimental data 4 [2].

In experiment 1, 11,479 data points were generated as ten equally sized normal distributions. Fig.4 shows a fuzzy clustering result by DASA. Single DA similarly clusters these data.

In experiment 2-1, three differently sized normal distributions consist of 2,249 data points in Fig.5-1 were used. Fig.5-1(0) shows initial clusters obtained by the initial estimation of α_ks and the annealing temperature. Fig.5-1(1)~(6a) shows a fuzzy clustering process of DASA. At the high temperature in Fig.5-1(1), as described in 4.3, the membership functions were widely distributed and clusters to which a data belongs were fuzzy. However, with decreasing of the temperature (from Fig.5-1(2) to Fig.5-1(5)), the distribution became less and less fuzzy. After executing DA and SA alternately, the clusters in Fig.5-1(6a) were obtained. Then, data to be optimized by SA were selected by the criterion stated in the section 4, and SA was executed. The final result of DASA in Fig.5-1(6b) shows that data were desirably clustered. On the contrary, because of randomness of the initial cluster positions and hardness of good estimation of the initial α_ks, single DA becomes unstable, and sometimes gives satisfactory

[2] These parameters have not been optimized particularly for experimental data.

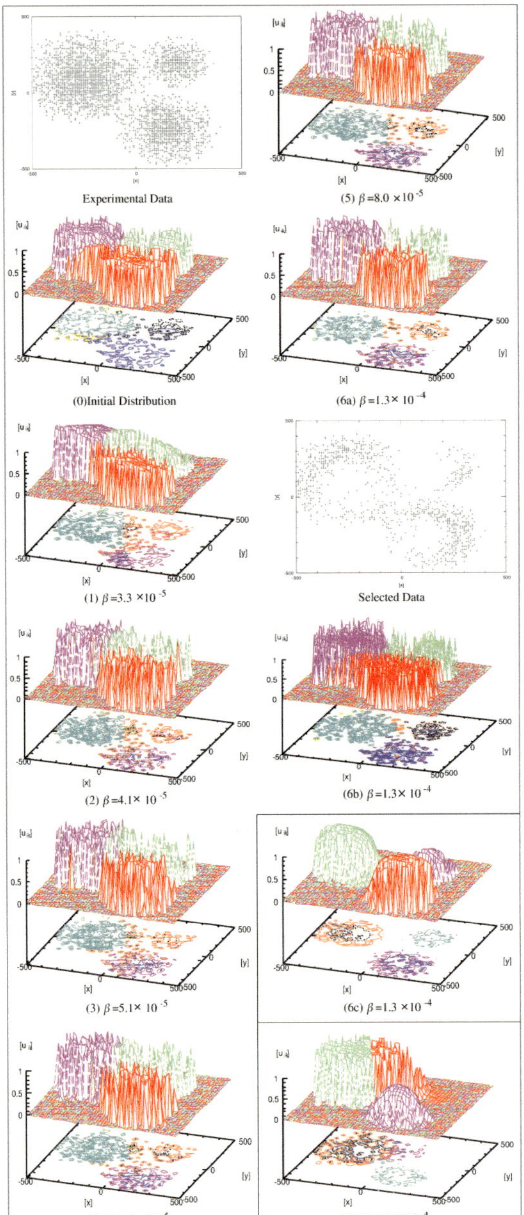

Figure 5-1. Experimental result 2-1. (Fuzzy clustering result by DASA and single DA. "Experimental Data" are given data distributions. "Selected Data" are data selected for final SA by the selection rule. (1)∼(6a) and (6b) are results using DASA. (6c) and (6d) are results using single DA (success and failure, respectively). Data plotted on the xy plane show the cross sections of u_{ik} at 0.2 and 0.8.)

results as shown in Fig.5-1(6c) and sometimes not as shown in Fig.5-1(6d). By comparing Fig.5-1(6b) to (6c), it is found that, due to the optimization of α_ks by SA, the resultant cluster shapes of DASA are far less smooth than those of single DA.

Changes of the costs of DASA ($J_{m=1} + S_{FE}$ for DA stage and (50) for SA stage (K was set to 1×10^{15} in (50)), respectively) are plotted as a function of iteration in Fig.5-2, and the both costs decreases with increasing iteration. In this experiment, the total iteration of SA stage was about $12,500$, while that of DA stage was only 7. Accordingly, the amount of simulation time DASA was mostly consumed in SA stage.

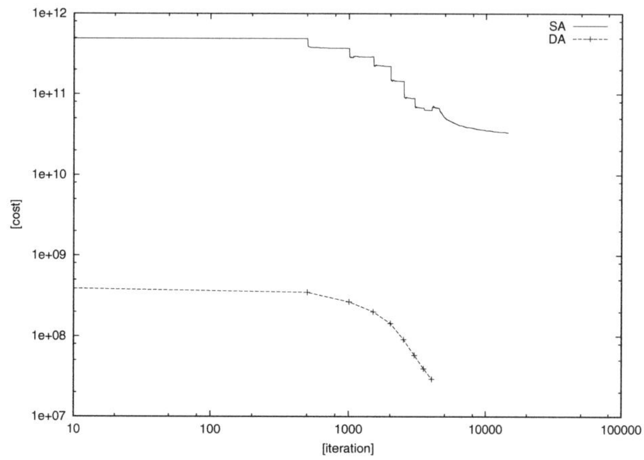

Figure 5-2. Experimental result 2-1. (Change of the cost of DASA as a function of iteration. $J_{m=1} + S_{FE}$ for DA stage and $J_{m=1} + S_{FE} + K \sum_{k=1}^{n} (\sum_{i=1}^{c} u_{ik} - 1)^2$ for SA stage, respectively.)

In experiment 2-2, in order to examine effectiveness of SA introduced in DASA, experiment 2 was re-conducted ten times as in Table 1, where *ratio* listed in the first row is a ratio of data optimized at SA stage. "UP" means to increase *ratio* as $1.0 - 1.0/t$ where t is a number of execution times of SA stage. Also, "DOWN" means to decrease *ratio* as $1.0/t$. Results are judged "Success" or "Failure" from a human viewpoint [3]. From Table 1, it is concluded that DASA always clusters the data properly if *ratio* is large enough ($0.6 < ratio$), whereas, as listed in the last column, single DA succeeds by 50%.

DASA						DA
ratio	0.3	0.6	1.0	UP	DOWN	
Success	6	9	10	6	7	5
Failure	4	1	0	4	3	5

Table 2. Experimental result 2-2. (Comparison of numbers of successes and failures of fuzzy clustering using DASA for $ratio = 0.3, 0.6, 1.0, 1.0 - 1.0/t$(UP), $1.0/t$(DOWN) and single DA. (t is a number of execution times of SA stage))

In experiments 3 and 4, two elliptic distributions consist of 2,024 data points, and two horseshoe-shaped distributions consist of 1,380 data points were used, respectively. Fig.5 and 6 show DASA's clustering results. It is found that DASA can cluster these data properly. In experiment 3, a percentage of success of DASA is 90%, though that of single DA is 50%. In experiment 4, a percentage of success of DASA is 80%, though that of single DA is 40%.

[3] No close case was observed in this experiment.

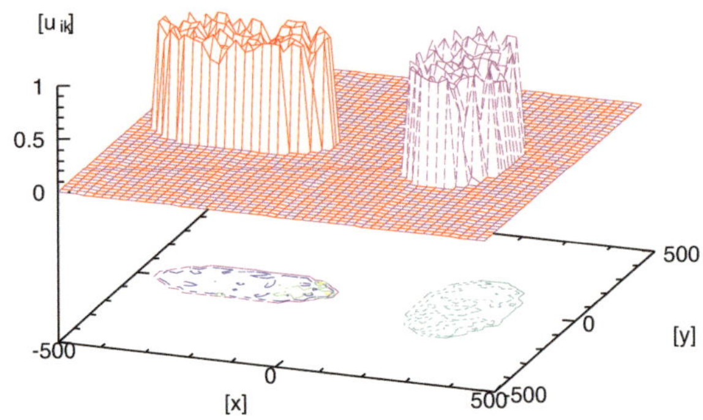

Figure 6. Experimental result 3. (Fuzzy clustering result of elliptic distributions using DASA. Data plotted on the xy plane show the cross sections of u_{ik} at 0.2 and 0.8.)

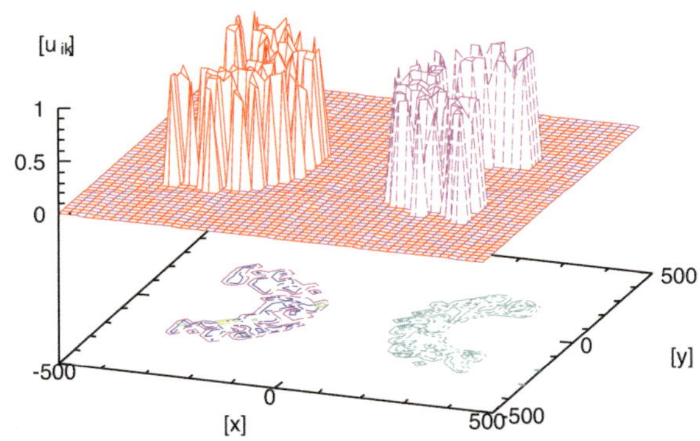

Figure 7. Experimental result 4. (Fuzzy clustering result of horseshoe-shaped distributions using DASA. Data plotted on the xy plane show the cross sections of u_{ik} at 0.2 and 0.8.)

These experimental results demonstrate the advantage of DASA over single DA. Nevertheless, DASA suffers two disadvantages. First, it takes so long to execute SA repeatedly that, instead of (10), it might be better to use its linear approximation functions as the membership function. Second, since α_ks differ each other, it is difficult to interpolate them.

8. Experiments 2

8.1. Interpolation of membership function

DASA suffers a few disadvantages. First, it is not necessarily easy to interpolate α_k or u_{ik}, since they differ each other. Second, it takes so long to execute SA repeatedly.

A simple solution for the first problem is to interpolate membership functions. Thus, the following step was added to the DASA algorithm.

6 When a new data is given, some neighboring membership functions are interpolated at its position.

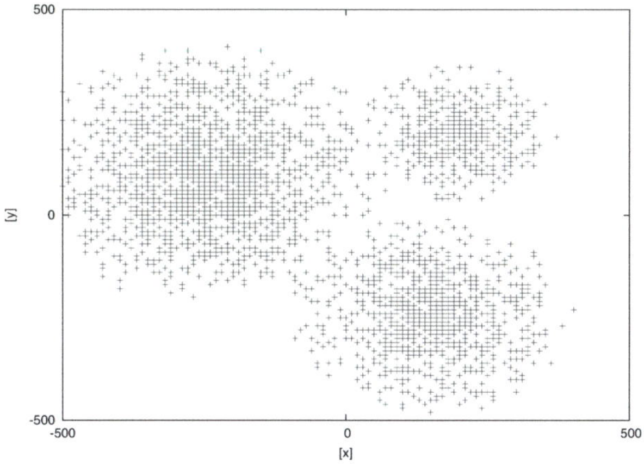

(a)Experimental data.

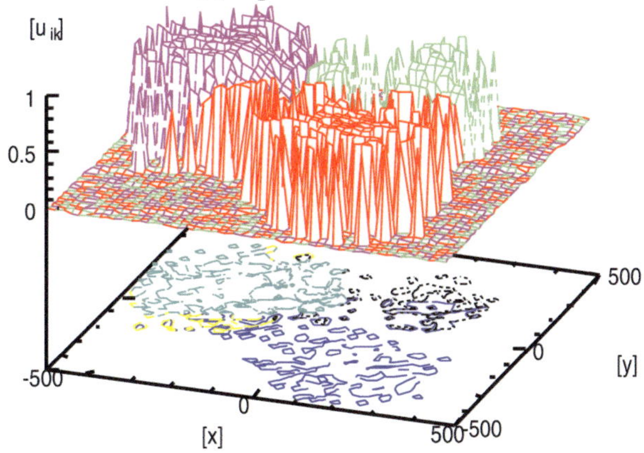

(a)Initial distribution.

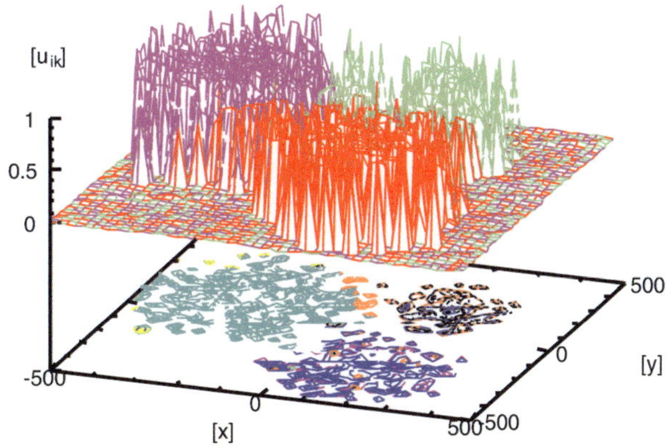

(b)Final distribution.

Figure 8. Experimantal data and membership functions obtained by DASA.(Data plotted on the xy plane show the cross sections of u_{ik} at 0.2 and 0.8)

To examine an effectiveness of interpolation, the proposed algorighm was applied to experimental data shown in Fig.8(a). For simplicity, the data were placed on rectangular grids on the xy plane.

First, some regional data were randomly selected from the data. Then, Initial and final memberhip functions obtained by DASA are shown in Figs.8(b) and (c) respectively.

After that, remaining data in the region were used as test data, and at each data point, they were interpolated by their four nearest neighboring membership values. Linear, bicubic and fractal interpolation methods were compared.

Prediction error of linear interpolation was 6.8%, and accuracy was not enough. Bicubic interpolation[18] also gave a poor result, because its depends on good estimated gradient values of neighboring points. Accordingly, in this case, fractal interpolation[17] is more suitable than smooth interpolation methods such as bicubic or spline interpolation, because the membership functions in Figs.8(c) are very rough.

The well-known InterpolatedFM (Fractal motion via interpolation and variable scaling) algorithm [17] was used in this experiment. Fractal dimension was estimated by the standard box-counting method [25]. Figs.9(a) and 3(b) represent both the membership functions and their interpolation values. Prediction error (averaged over 10 trials) of fractal interpolation was 2.2%, and a slightly better result was obtained.

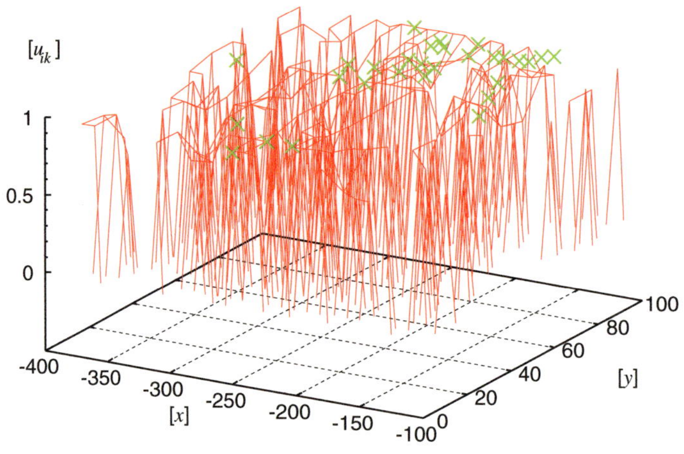

Figure 9. Plotted lines show the membership functions obtained by DASA . The functions are interpolated by the InterpolatedFM algorithm. Crosses show the interpolated data.

9. Conclusion

In this article, by combining the deterministic and simulated annealing methods, we proposed the new statistical mechanical fuzzy c-means clustering algorithm (DASA). Numerical experiments showed the effectiveness and the stability of DASA.

However, as stated at the end of **Experiments**, DASA has problems to be considered. In addition, a major problem of the fuzzy c-means methodologies is that they do not give a number of clusters by themselves. Thus, a method such as [28] which can determine a number of clusters automatically should be combined with DASA.

Our future works also include experiments and examinations of the properties of DASA, especially on an adjustment of its parameters, its annealing scheduling problem, and its applications for fuzzy modeling[29].

However, DASA has problems to be considered. One of them is that it is difficult to interpolate membership functions, since their values are quite different. Accordingly, the fractal interpolation method (InterpolationFM algorithm) is introduced to DASA and examined its effectiveness.

Our future works include experiments and examinations of the properties of DASA, a comparison of results of interpolation methods (linear, bicubic, spline, fractal and so on), an interpolation of higher dimensional data, an adjustment of DASA's parameters, and DASA's annealing scheduling problem.

Author details

Makoto Yasuda
Gifu National College of Technology, Japan

10. References

[1] E. Aarts and J. Korst, "Simulated Annealing and Boltzmann Machines", Chichester: John Wiley & Sons, 1989.
[2] J.C. Bezdek, "Pattern Recognition with Fuzzy Objective Function Algorithms", New York: Prenum Press, 1981.
[3] B.P.Buckles and F.E.Petry, "Information-theoretical characterization of fuzzy relational database", IEEE Trans. Systems, Man and Cybernetics, vol.13, no.1, pp.74-77, 1983.
[4] J. Buhmann and H. Kühnel, "Vector quantization with complexity costs", IEEE Trans. Information Theory, vol.39, no.4, pp.1133-1143, 1993.
[5] A.Corana, M.Marchesi, C.Martini, and S.Ridella, "Minimizing multimodal functions of continuous variables with the simulated annealing algorithm", ACM Trans. on Mathematical Software, vol.13, no.3, pp.262-280, 1987.
[6] A. DeLuca and S. Termini, "A definition of a nonprobabilistic entropy in the setting of fuzzy sets theory", *Information and Control*, vol.20, pp.301–312, 1972.
[7] A.P.Dempster, N.M.Laird, and D.B.Rubin, "Maximum likelihood from incomplete data via the EM algorithms", Journal of Royal Stat. Soc., Series B, vol.39, pp.1-38, 1977.
[8] W. Greiner, L. Neise, and H. Stöcker, "Thermodynamics and Statistical Mechanics", New York: Springer-Verlag, 1995.
[9] T. Hofmann and J. Buhmann, "Pairwise data clustering by deterministic annealing," IEEE Trans. Pattern Analysis and Machine Intelligence, vol.19, pp.1-14, 1997.
[10] S. Kirkpatrick, C.D. Gelatt, and M.P. Vecchi, "Optimization by simulated annealing", Science, vol.220, pp.671-680, 1983.
[11] R.-P. Li and M. Mukaidono, "A Maximum entropy approach to fuzzy clustering", Proc. of the 4th IEEE Int. Conf. Fuzzy Systems (FUZZ-IEEE/IFES'95), pp.2227-2232, 1995.
[12] M. Menard, V. Courboulay, and P. Dardignac, "Possibilistic and probabilistic fuzzy clustering: unification within the framework of the non-extensive thermostatistics", Pattern Recognition, vol.36, pp.1325–1342, 2003
[13] D. Miller, A.V. Rao, K. Rose, and A. Gersho, "A global optimization technique for statistical classifier design", IEEE Trans. Signal Processing, vol.44, pp.3108-3122, 1996.

[14] S. Miyamoto and M. Mukaidono, "Fuzzy c-means as a regularization and maximum entropy approach", Proc. of the 7th Int. Fuzzy Systems Association World Congress, vol.II, pp.86-92, 1997.
[15] N.R. Pal and J.C. Bezdek, "Measuring fuzzy uncertainty", IEEE Trans. Fuzzy Systems, vol.2, no.2, pp.107-118, 1994.
[16] N.R. Pal, "On quantification of different facets of uncertainty", Fuzzy Sets and Systems, vol.107, pp.81-91, 1999.
[17] H.-O. Peitgen, et.al., *The science of fractal images*, Springer-Verlag, 1988
[18] W.H. Press, S.A. Teukolsky, W.T. Vetteriling, and B.P. Flannery, *Numerical Recipes in C++*, Cambridge University Press, 2002.
[19] L. E. Reichl, *A Modern Course in Statistical Physics*, New York: John Wiley & Sons, 1998.
[20] K. Rose, E. Gurewitz, and B.C. Fox, "A deterministic annealing approach to clustering", Pattern Recognition Letters, vol.11, no.9, pp.589-594, 1990.
[21] K. Rose, E. Gurewitz, and G.C. Fox, "Constrained clustering as an optimization method", IEEE Trans. Pattern Analysis and Machine Intelligence, vol.15, no.8, pp.785-794, 1993.
[22] P.Siarry, "Enhanced simulated annealing for globally minimizing functions of many-continuous variables", ACM Trans. on Mathematical Software, vol.23, no.2, pp.209-228, 1997.
[23] D.Tran and M.Wagner, "Fuzzy entropy clustering", Proc. of the 9th IEEE Int. Conf. Fuzzy Systems (FUZZ-IEEE2000) , vol.1, pp.152-157, 2000.
[24] C. Tsallis, *Possible generalization of Boltzmann-Gibbs statistics*, Journal of Statistical Phys., vol.52, pp.479–487, 1988.
[25] R. Voss, *Random fractals: characterization and measurement*, Plenum Press, 1986.
[26] P.R.Wang, "Continuous optimization by a variant of simulated annealing", Computational Optimization and Applications, vol.6, pp.59-71, 1996.
[27] M. Yasuda, T. Furuhashi, and S. Okuma, *Statistical mechanical analysis of fuzzy clustering based on fuzzy entropy*, IEICE Trans. Information and Systems, Vol.ED90-D, No.6, pp.883-888, 2007.
[28] M. Yasuda and T. Furuhashi, *Fuzzy entropy based fuzzy c-means clustering with deterministic and simulated annealing methods*, IEICE Trans. Information ans Systems, Vol.ED92-D, No.6, pp.1232-1239, 2009.
[29] M. Yasuda and T. Furuhashi, *Statistical mechanical fuzzy c-means clustering with deterministic and simulated annealing methods*, Proc. of the Joint 3rd Int. Conf. on Soft Computing and Intelligent Systems, in CD-ROM, 2006.
[30] M. Yasuda, *Entropy based annealing approach to fuzzy c-means clustering and its interpolation*, Proc. of the 8th Int. Conf. on Fuzzy Sysmtes and Knowledge Discovery, pp.424-428, 2011.
[31] S.D. Zenzo and L. Cinque, "Image thresholding using fuzzy entropies", IEEE Trans. Systems, Man and Cybernetics-Part B, vol.28, no.1, pp.15-23, 1998.